Plant Breeding

Plant Breeding

2nd Edition of Introduction to Plant Breeding – revised and updated

Jack Brown
University of Idaho, USA

Peter D.S. Caligari
University of Talca, Chile

Hugo A. Campos
Monsanto, Chile

WILEY Blackwell

Library of Congress Cataloging-in-Publication Data

Brown, Jack, 1955–
  Plant breeding / Jack Brown, Peter D.S. Caligari, and Hugo A. Campos. – 2nd edition.
     pages cm
  "First edition published 2008 by Blackwell Publishing Ltd."
  Includes bibliographical references and index.
  ISBN 978-0-470-65829-1 (cloth) – ISBN 978-0-470-65830-7 (pbk.) 1. Plant breeding. I. Caligari, P. D. S. (Peter D. S.)
II. Campos, Hugo A., 1968– III. Title.
  SB123.B699 2014
  631.5′2 – dc23
                                                                                                    2014011407

Set in 9.5/11.5pt Times Ten by Laserwords Private Limited, Chennai, India.

Printed in Singapore by C.O.S. Printers Pte Ltd

1   2014

Contents

Preface

This book has its basis in an earlier text entitled *An Introduction to Plant Breeding* by Jack Brown and Peter Caligari, which was published in 2008. So why have we produced this current book barely six years after the publication of the previous one? The answer is multiple, but the main driving force is the increasing need of humankind for affordable food, fibres and renewable fuels, which in great part needs to be satisfied through the creation of improved cultivars. Current and future demands from agriculture and an increasing contribution of private breeding organizations, in addition to the more traditional efforts of long-standing public breeding, has significantly improved our knowledge base and the techniques we now have available to improve the efficiency of developing genetically superior cultivars. The challenges facing today's plant breeders have never been more overwhelming, yet the prospects to contribute significantly to global food security and farmers' quality of life have never been more exciting and fulfilling. Let us elaborate further.

1. As per United Nations' figures for the year 2013, world population is projected to exceed 8 billion people by 2025, and 9.6 billion by the year 2050. Although most of the projected population growth will take place in developing countries, where over 8 billion will live by 2050, the most significant increase in population will occur in the least developed countries of the world. To place this in simple numbers, every single day over 200,000 additional children needing to be fed are added to the world population.

2. The rate of accumulation of both fundamental and applied genetic and plant breeding knowledge is ramping up faster than ever before. Drawing from initial research in animal breeding, genomic selection approaches in crop species are being vigorously studied, and their adoption and use into crop breeding programmes has already began. In addition, insights into the molecular basis of genetic variation in plants is being quickly advanced because of the drastic reduction in the cost of DNA sequencing, heavy borrowing from medical translational research; and an increasing awareness of the critical importance of phenotypes to establish better genetic marker–trait associations. This holds the promise to significantly improve the efficiency of selecting for a wide range of qualitative and quantitative gene traits in cultivar development.

Furthermore, the traits developed through transgenic approaches are no longer constrained simply to insect or virus resistance or herbicide tolerance. New '*biotech traits*' addressing key global needs such as salt and drought tolerance, nitrogen and water use efficiency, or enhancing food quality are now being developed. Initial transgenic breeding efforts were directed towards improving some of the world's major crop species (i.e. maize, soybean, rice, cotton and canola) which are grown on the widest acreage. These techniques are now being applied to other staple crops, as demonstrated by the recent development of a virus-resistant common bean by public scientists based at EMBRAPA in Brazil.

3. Despite recent recognition of the need to increase public and private funding into cultivar development, over the last decades there has been a worrying decline in public funding for plant breeding-related research and to supporting international centres for germplasm development and crop improvement. In part, this has resulted in a serious reduction in the number of young people interested in devoting their professional careers to plant breeding, as well as the number of universities offering plant breeding courses or conducting relevant research in plant breeding for the next generation of plant breeders.

Although based on the earlier edition, this book has been completely reviewed, expanded and updated. We have also developed a companion website providing sections such as "Suggested answers" to the "Think Questions", "Further reading" and "Plant breeding teaching material" which can be used by lecturers and students alike.

We hope our book will appeal to undergraduate students of plant breeding, as well as providing valuable insight for graduate students, practicing plant breeders and those wishing to advance their understanding of plant breeding.

Our aim is to provide an integrated and updated view of the current scientific progress related to diverse plant breeding disciplines within the context of applied breeding programmes. By and large, plant breeders are integrators of multiple scientific disciplines, including genetics, biology, plant science, agronomy, food science and statistics, as well as having a keen appreciation and understanding of farmers' needs, both current and future. However, plant breeders also need to be aware of global trends affecting agriculture, such as the likely consequences of climate change for crop adaptation and performance.

We trust that this new book will encourage a new generation of students to pursue careers related to plant breeding. However, at the same time we hope it will assist a wider audience of agricultural students, agronomists, policy-makers and people with an interest in agriculture in gaining insight about the issues affecting plant breeding and its key role in improving the quality of life of people, and in securing sufficient food at the quality required at an affordable price.

In this latest book, the two original authors, Jack Brown and Peter Caligari, are joined by Hugo Campos, who adds further to the insights into the commercial aspects of plant breeding, along with his long experience in plant breeding.

All three of us would like to thank our families for their forbearance while we pursued the completion of this book, and a variety of mentors, colleagues and students who have all added to our knowledge and experience in a positive way.

About the Companion Website

This book is accompanied by a companion website:

www.wiley.com/go/Brown/Plantbreeding

The website includes:

- PDF version of the slides from Professor Caligari's Plant Breeding Course given at the Universidad de Talca, Chile
- Powerpoints of all figures from the book for downloading
- PDFs of all tables from the book for downloading
- Suggested answers to *Think Questions* from each chapter

1
Introduction

1.1 Requirements of plant breeders

The aim of plant breeding is to develop genetically superior cultivars that are adapted to specific environmental conditions and suitable for economic production in a commercial cropping system. These new, and more productive cultivars, are increasingly necessary to fulfil humankind's escalating needs for food, fibre and fuels.

The basic concept of varietal development is rather simple and involves three distinct operations:

- produce or identify genetically diverse germplasm from which segregating breeding populations are developed;
- carry out selection procedures on phenotypes or genotypes from within this germplasm to identify superior genotypes with specific improved characteristics;
- stabilize and multiply these superior genotypes and release cultivars for commercial production.

The general philosophy underlying any breeding scheme is to maximize the probability of creating, and identifying, superior genotypes which will make successful new cultivars; in other words, genotypes that will contain all the desirable characteristics/traits necessary for use in a given production system, or at least offer a beneficial trade-off between key advantageous characteristics compared with undesirable ones.

Plant breeders can be categorized into two types. One group of plant breeders is employed within private companies, while the other group works in the public sector (e.g. government-funded research institutes or universities). Private sector and public sector breeders often have different approaches to the breeding process. Many of the differences that exist between public and private breeding programmes are related to the time available for variety release, types of cultivar developed, and priorities for traits in the selection process.

Plant breeders within the public sector are likely to have a number of responsibilities related to academic activities or extension services, in addition to those solely directed towards producing new varieties. Public sector breeders also play an additional, often unappreciated yet critical role: the attraction, training and development of a younger generation of men and women interested in plant breeding. As plant breeding is a combination of science and art, the personal component of training plant breeders at the graduate level is generally recognized as more relevant and significant than in most other areas of science.

Private sector plant breeders tend to have a more clearly defined goal: developing new cultivars and doing it as quickly as possible. In addition, many private breeding organizations are, or are associated with, biotechnology and/or agrochemical companies. As a result, varietal development may be designed to produce cultivars suitable for

Plant Breeding, Second Edition. Jack Brown, Peter D.S. Caligari and Hugo A. Campos.
© 2014 John Wiley & Sons, Ltd. Published 2014 by John Wiley & Sons, Ltd.
Companion Website: www.wiley.com/go/brown/plantbreeding

integration within a specific production system. In many countries, including the US, the ratio of private to public breeders has increased over time, particularly in those highest acreage crops such as maize, soybeans and canola, to mention just a few, as well as in crops with a high profit, such as tomato, pepper and lettuce, where private companies can gain greatest financial returns from seed or chemical sales.

Despite the apparently simple description of the breeding process given above, in reality plant breeding involves a multidisciplinary and long-term approach. Regardless of whether a breeding scheme is publicly or privately managed, a successful plant breeder will require knowledge in many (if not all) of the following subjects:

Evolution It is necessary to have knowledge of the origins and past progress in adaptation of crop species if additional advances are to continue into the future. When dealing with a crop species, a plant breeder benefits from knowledge of the timescale of events that have modelled the given crop. For example, the time of domestication, geographical area of origin and prior improvements are all important and will help in setting feasible future objectives. In the case of crops where hybridization systems are commercially available, such as canola and corn, the evolution from OP (open pollinated) to hybrid cultivars represents a landmark driving profound changes in ensuing breeding programmes.

Botany The raw material of any breeding scheme is the available germplasm (lines, genotypes, accessions, etc.) from which agronomically relevant variation can be exploited. The biological relationship that exists within a species and with other species will be a determining factor indicating germplasm variability and availability.

Biology Knowledge of plant biology is essential to create genetic variation and formulate a suitable breeding and selection scheme. Of particular interest are modes of reproduction, types of cultivar and breeding systems.

Genetics The creation of new cultivars requires manipulation of genotypes and genes. The understanding of genetic processes is therefore essential for success in plant breeding. Genetics is an ever-developing subject, but knowledge and understanding that is particularly useful

will include single gene inheritance, population genetics, the expected frequencies of genotypes under selection, and the prediction of quantitative genetic parameters – all of which will underlie decisions on what strategy of selection will be most effective.

Pathology A major goal of plant breeding is to increase productivity and quality by selecting superior genotypes. A limiting factor in economic production is the impact of pests and diseases. Therefore developing cultivars that are resistant to detrimental pathogens has been a major contributor to most cost-effective production with reduced agrochemical inputs. Similarly, nematodes, insect pests and viruses can all have detrimental effects on yield and/or quality. Therefore plant breeders must also have knowledge of **phytopathology**, **nematology**, **entomology** and **virology**.

Weed science The response of a genotype to competition from weed populations will have an effect on the success of a new cultivar. Cultivars that have poor plant establishment, or lack subsequent competitive ability, are unlikely to be successful, particularly in systems where reduced, or no, herbicide applications are desirable, or their use is restricted. Similarly, in many cases genotypes respond differently, even to selective herbicides. Herbicide tolerance in new crops is looked upon favourably by many breeding groups, although cultivar tolerance to broad-spectrum herbicides can cause management difficulties in crop rotations. Herbicide tolerance brought about by traditional breeding as well as through recombinant DNA techniques has also driven changes in plant breeding approaches.

Food science Increasingly end-use quality is being identified as one of the major objectives of all crop-breeding schemes. As most crop species are grown for either human or animal consumption, knowledge of food nutrition and other related subjects is important.

Biometry Managing a plant-breeding scheme has aspects that are no different from organizing a series of large experiments over many locations and years. To maximize the probability of success it is necessary to use an appropriate experimental approach at all stages of the breeding scheme. Plant breeding is continually described as 'a numbers game'. In many cases this is true, and successful breeding will result in vast datasets on which selection decisions are to be made. These decisions often have to be made under significant time constraints, for instance between harvesting one crop and planting another. Therefore, plant breeders are required to be good **data managers**.

Agronomy It is the aim of crop breeders to predict how newly identified genotypes will perform over a wide range of environments. This will require research into agronomic features that may relate to stress tolerance, such as heat, drought, moisture, salinity and fertility. These experiments are essential in order that farmers (the primary customer) are provided with the optimal agronomic husbandry parameters, which will maximize the genetic potential of the new variety. Poor agronomic practice can detrimentally affect the expression of genotypes and the selection of appropriate phenotypes. Despite increasingly powerful statistical tools, poorly managed field trials represent a waste of effort and a risk of reducing the rate of genetic gain expected.

Molecular biology Advances in molecular biological techniques are having an increasing role in modern plant breeding. Molecular markers are increasingly used by plant breeders to help select (indirectly and directly) for characteristics that are difficult to evaluate in the laboratory, or are time-consuming or expensive to determine accurately on a small plot scale. Genetic engineering and tissue culture operations, such as the development of double haploid populations, are becoming standard in many plant-breeding schemes, and it is likely that further advances will be made in the future. Knowledge of all these techniques and continued awareness of ongoing research will be necessary so that new procedures can be integrated into the breeding scheme where appropriate. In those crops where transgenic variation is commercially available, such as herbicide or insect tolerance in corn, the availability of molecular markers has enabled the development of effective schemes collectively known as trait integration, where a small genomic region carrying a transgenic trait is very quickly introgressed into elite germplasm without detrimental effects on other traits which are also important for farmers.

Love for working in the field Despite continuing progress in biotechnology and the availability of statistical tools, there remains no substitute for applied field research and crop performance evaluation. Field crops, by definition, are produced commercially under field conditions, and plant breeders will rely heavily on field plot research for phenotypic evaluations. Therefore, plant breeders must be willing to spend significant time visiting field trials and nurseries, seeing how different phenotypes develop, learning how biotic and abiotic stresses build up in the field, and gaining a deep insight into how plants interact with and respond to the environment. The late Nobel Peace Prize winner Norman Borlaug said: *"Plants speak to human beings. However, they only whisper softly like a woman of great culture. They hence reveal their secrets only to those who are close to them. Those who spend their time in office rooms naturally think that plants are dumb."* This is as relevant today as it was many years ago.

Production The contributions that farmers and other growers have made to varietal development should never be underestimated. It should also be noted that growers are the first customers for plant breeding products. The probability of a new cultivar being successful will be maximized (or at least the probability of complete failure reduced) if growers and production systems are considered as major factors when designing breeding systems.

Management There is a need to manage people, time and money. It has already been stated that plant breeding is a multidisciplinary science, and this means being able to integrate and optimize people's effort to effectively use breeders' time. The length of most breeding programmes means that small proportional savings in time can be valuable, and it hardly needs emphasizing that breeding needs to be cost-effective and therefore the cost of the programme is always going to be important.

Communication Most varietal development programmes consist of inputs from more than one scientist, and so it is necessary that plant breeders are good communicators. Verbal and written communication of results and test reports will be a feature of all breeding schemes. Research publications and grant proposals are of major importance, particularly to public breeders, if credibility and funding is to be forthcoming. In the case of private breeders, they have to be able to establish good communication links with their business counterparts. In addition, there has been an increasing need for plant breeders to be able to communicate with other stakeholders, like administrators, politicians and society at large, in order to raise awareness about the importance of plant breeding and therefore attract the funding needed to sustain breeding efforts. Finally, at least some plant breeders must be good educators and have the ability to pass on essential information about the subject to future plant breeders.

Information technology The science underlying plant breeding is continually advancing, agronomic practices are continually being upgraded, the end-users' choices change, and the political context continually affects agriculture. This means that it is vital that breeders talk to these different groups of people, but also use whatever technology is available to keep up to date with developments as they occur – or better, before they occur! Information technology enables breeders to capture, manage and analyze ever-increasing datasets, and also to identify the genetic signal buried in experimental noise. As an example, bar code devices are playing an increasing role in the high throughput collection of plant phenotypes, and so reducing the rate of involuntary mistakes.

Psychic The success of most plant breeding schemes is not realized before many years of breeding and trialling have been completed. It may be 10 or more years between the initial crosses and varietal release. In the extreme case of, say, oil palm or fruit tree breeders, they might not live to see the results of the plant selections they have made. Plant breeders therefore must be *crystal ball gazers* and try to predict what the general public and farming community will need in the future, what progress biotechnology will bring about in the years to come, what diseases will persist in years ahead, and what quality characters will command the highest premiums. The experience that plant breeders accumulate enables them to augment the quality of decisions purely based on those scientific disciplines supporting any breeding programme.

In summary, therefore, successful plant breeders need to be familiar with a range of scientific disciplines and management areas. It is not, however, necessary to be an expert or indeed an authority in all of these. However, greater knowledge of the basic science underlying the techniques employed, and of the plant species concerned, in terms of the biology, genetics, history and pathology, will increase the chances of a breeder succeeding in developing the type of cultivars most suitable for future exploitation.

1.2 Evolution of crop species

Plant breeding consists of the iterative process of creation and manipulation of genetic variation within a crop species, and selection of desirable recombinants from within that variation. The process is therefore an intensification of a natural process, which has been ongoing since humans first appeared on earth. Plant breeders utilize and accelerate the evolutional process by directing it towards the increasing needs of humankind. As soon as humans began settled agriculture they effectively, albeit unconsciously, started plant breeding. In this section the main features of crop plant evolution will be covered briefly. The study of evolution is a vast and detailed subject in itself, and it will not be possible to cover more than an introduction to it in this book. Emphasis will be on the areas that are most important from a plant breeding standpoint.

Knowledge of the evolution of a plant species can be invaluable in breeding new cultivars. Studies of evolution can provide knowledge of the past changes in the genetic structure of the plant, an indication of what advances have already been achieved or might be made in the future, and help to identify relatives of the domesticated plant which could be used in interspecific or intergeneric hybridization to increase genetic diversity or introduce desirable characters not available within existing crops.

1.2.1 Why did hunter-gatherers become farmers?

It is difficult to arrive at a firm understanding as to why humans became a race of farmers. Early humans are believed to have been foragers and later hunters. Why then did they become crop producers? Farming is believed to have started shortly after the last ice age, about 10,000 years ago.

At that time there may have been a shortage of large animals for hunters to hunt, due to extinctions. Indeed, little is known about the order of agricultural developments. Did man domesticate animals and then domesticate crops to feed these beasts, or were crops first domesticated, and from this the early farmers found that they could benefit from specifically growing sufficient food to feed livestock? The earliest farmers may also have been fishermen who tended not to travel continually and were more settled in one region. In this latter case, perhaps the first farmers were women who took care of the farming operation while the males fished and hunted locally. It may simply have been that some ancient people became tired of nomadic travel in search of food, became bored with living in tents and opted for a quiet life on the farm! The answers are not known, although it can often be interesting to postulate why this change occurred. One misconception about the switch from hunting-gathering to farming is that farming was easier. It has been shown that gathering food requires considerably less energy than cultivating and growing crops. In addition, skeletal remains show that the initial farmers were smaller-framed and more prone to illness than their hunter-gatherer counterparts.

Regardless of the reason that caused mankind to cultivate crops, few would question that the beginning of farming aligned with the beginning of what most of us would consider civilization. Farming created communities, community structure and economies, group activities, enhanced trade and monetary systems, to name but a few. There is also little doubt that the total genetic change achieved by early farmers in moulding our modern crops has been far greater than that achieved by the scientific approaches that have been applied to plant breeding over the past century. Given that these early farmers were indeed cultivating crops, it is not surprising that they would propagate the most productive phenotypes, avoid the individuals with an off-taste or hard seeds, and choose not to harvest those plants that were spiny. Even today among peasant farmers there is a general trend to select the *best plants* for resowing the next year's crops. Early farmers may have used relatively sophisticated plant breeding techniques as there is evidence that some Native Americans have a long-established understanding of maintaining pure line cultivars of maize by growing seed crops that are isolated from their production fields.

1.2.2 What crops were involved? And when did they arise?

Today's world food production is dominated by small grain cereal crops, with world production of maize (*Zea Mays*), rice (*Oryza sativa*) and wheat (*Triticum* spp.) each being just above 600 million metric tonnes annually (Figure 1.1). Major root crops include potato (*Solanum tuberosum*), cassava (*Manihot esculenta*), and sweet potato (*Ipomoea batatas*). Oilseed crops include soybean oil (*Glycine max*), oil palm (*Elaeis guineensis*), coconut palms (*Cocos nucifera*) and canola (*Brassica napus*). World production of fruit and vegetables are similar, where tomato (*Lycopersicum esculentum*), cabbage (*Brassica oleracea*) and onion (*Allium* spp.) are leading vegetable crops, whilst orange

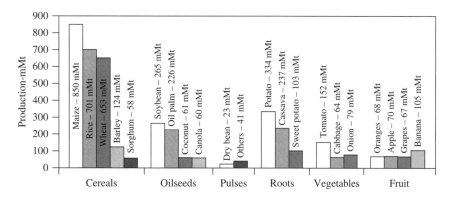

Figure 1.1   World production of major crops.

(*Citrus sinensis*), apple (*Malus* spp.), grape (*Vitis* spp.) and banana (*Musa acuminata* and *M. balbisiana*) predominate amongst the fruits. Many of these modern day crops were amongst the first propagated in agriculture.

Many studies have been made to determine the date when humans first cultivated particular crops. The accuracy of dating early remains of plant tissue has improved over the past half century with the use of radiocarbon methods. It should, however, be noted that well-preserved archaeological plant material has not proved easy to find. Many of the most significant findings have been from areas of arid environments (e.g. the eastern Mediterranean and Near East, New Mexico and Peru). These arid regions favour the preservation of plant tissue over time, and, not surprisingly, are the areas where most archaeological excavations have taken place. Conversely, there is a lower probability of finding well-preserved plant remains in regions with wetter and more humid climates. Therefore, archaeological information may provide an interesting, but surely incomplete, picture.

A summary of the approximate time of domestication and centre of origin of the world's major crop species, and a few recent crop additions, is presented in Table 1.1. It should be reiterated that many crop species have more than one region of origin, and that archaeological information is continually being updated. This table is therefore very much an oversimplification of a vast and complex picture.

Some of the earliest recorded information showing human domestication of plants comes from the region in the Near East known as the 'Fertile Crescent' (including the countries of Turkey, Syria, Israel, Iran and Iraq). Domestication of crops in this region surrounding the *Tigris River* began before 6000–7000 BC. Two of the world's leading cereal crops, einkorn and emmer wheat, as well as barley, have their centre of origin in this region. In addition, archaeological remains of onion, peas and lentils, dating back over 7,000 years have all been found within the Fertile Crescent. In the Americas, similar or slightly later dates of cultivation have been shown for beans and maize in central Mexico and Peru, and potato, cassava, and sweet potato in Peru, Chile and western South America. Sunflower (*Helianthus*) is the only major crop species with a centre of origin in North America, and indeed most other crops grown in the US and Canada

evolved from other continents. Rice, soybean, sugarcane (*Saccharum* spp.), and the major fruit species (orange, apple and banana) were all first domesticated in China and the Asian continent a few millennia BC. Examination of archaeological remains shows that the dates of crop domestication in Africa were later, yet sorghum, oil palm and coffee are major world crops that have their centre of origin in this continent. Similarly, cabbage and a few other vegetable crops have their centre of origin in Europe. Given more research, it may be found that many more of today's crops were domesticated at even earlier periods.

Several crops of importance have been domesticated relatively recently. Sugar beet was not grown commercially in Europe until the 18th century, while rubber, oil palm and coconut palms were not domesticated until the end of the 19th century. The forage grasses, clovers, and oilseed rape (*Brassica napus* L. or *B. rapa* L.) are also recently domesticated crops, although some researchers would argue that these crops have yet to make the transition necessary to be classified as truly domesticated. New crops are still being recognized today. The advent of bioenergy crops has identified the oilseed crop camelina (*Camelina* spp.), and the biomass crop switchgrass (*Panicum virgatum*) as potential new crops species which have yet to be grown in large-scale commercialization.

A high proportion of today's major crops come from a very small subsample of possible plant species (Figure 1.1). It has been estimated that all the crop species grown today come from 38 families and 91 genera. This restriction in our plant-based diet is of concern because it potentially limits the consumption of the key nutrients supplied by plants and increases our dependence on animal products. Therefore, although the source of our present-day crops may be more diverse than we have shown, they still only represent a fraction of the total families and genera which have been estimated to exist within the angiosperms as a whole. Also, it should be noted that the sources of origin of these crops are spread over Europe, the Near East, Asia, Africa and America.

At some time in the past, each of our present-day crop species must have originated in one or more specific regions of the world. Originally it was thought that there were only 12 major centres of origin including the Near East, Mediterranean,

Table 1.1 Estimated time of domestication and centre of origin of major crop species.

Crop	Length of time domesticated (years)	Possible region of origin
Cereals		
Maize, *Zea mays*	7,000	Mexico, Central America
Rice, *Oryza sativa*	4,500	Thailand, Southern China
Wheat, *Triticum* spp.	8,500	Syria, Jordan, Israel, Iraq
Barley, *Hordeum vulgare*	9,000	Syria, Jordan, Israel, Iraq
Sorghum, *Sorgum bicolor*	8,000	Equatorial Africa
Oilseeds		
Soybean, *Glycine max*	2,000	Northern China
Oil palm, *Elaeis guineensis*	9,000	Central Africa
Coconut palm, *Cocos nucifera*	100	Southern Asia
Rapeseed, *Brassica napus*	500	Mediterranean Europe
Sunflower, *Helianthus annuus*	3,000	Western United States
Pulses		
Beans, *Phaseolus* spp.	7,000	Central America, Mexico
Lentil, *Lens culinaris*	7,000	Syria, Jordan, Israel, Iraq
Peas, *Pisum sativum*	9,000	Syria, Jordan, Israel, Iraq
Root crops		
Potato, *Solanum tuberosum*	7,000	Peru
Cassava, *Manihot esculenta*	5,000	Brazil, Mexico
Sweet potato, *Ipomoea batatas*	6,000	South Central America
Sugar beet, *Beta vulgaris*	300	Mediterranean Europe
Vegetables		
Tomato, *Lycopersicum esculentum*	3,000	Western South America
Cabbage, *Brassica oleracea*	3,000	Mediterranean Europe
Onion, *Allium* spp.	4,500	Iran, Afghanistan, Pakistan
Fruit		
Orange, *Citrus sinensis*	9,000	South-east Asia
Apple, *Malus* spp.	3,000	Asia Minor, Central Asia
Grape, *Vitaceae* spp.	7,000	Eastern Asia
Banana, *Musa acuminata, M. balbisiana*	4,500	South-east Asia
Others		
Cotton, *Gossypium* spp.	4,500	Central America, Brazil
Coffee, *Coffea* spp.	500	West Ethiopia
Rubber, *Hevea brasiliensis*	200	Brazil, Bolivia, Paraguay
Alfalfa, *Medicago sativa*	4,000	Iran, Northern Pakistan

Afghanistan, the Pacific Rim, China, Peru, Chile, Brazil/Paraguay and the US. More recent research has altered this original view, and it is now apparent that:

- Crops evolved in all regions of the world where farming was practised.
- The centre of origin of any specific crop is not usually a clearly defined geographical region. Today's major crops are more likely to have evolved over large areas.
- Early farmers and nomadic travellers would have been responsible for widening the region where

early crops have been found and added confusion concerning the true centres of origin.
- Regions of greatest crop productivity are rarely related to the crop's centre of origin.

Overall, therefore, domesticated crops have originated from at least four of the six world continents (America, Europe, Africa and Asia). Australian aborigines remained hunter-gatherers and did not become farmers, and indeed farming in Australia is a relatively new activity started after western settlers arrived there. Not surprisingly, therefore, few of today's major agricultural crops originated in Australia; however, a recently domesticated crop (macadamia nuts) does have its origin in this continent.

1.3 Natural and human selection

All domesticated crops have been developed from wild, "weedy" ancestors. Early farmers modified wild species into modern-day crops through a process of genetic manipulation and selection. As a result these crop species have been sufficiently altered such that they can be considered to be domesticated. A definition of domestication has been given by the late Professor N.W. Simmonds as follows: "*a plant population has been domesticated when it has been substantially altered from the wild state and certainly when it has been so altered to be unable to survive in the wild*". The first part of this definition can certainly be readily accepted for almost all modern-day agricultural crops, although we still propagate many crops (e.g. date palm) where the crop species are modified only slightly from ancient ancestors. It is not always possible to relate domestication to a lack of potential to survive in non-cultivated situations, since many commercially grown plants survive as volunteer weeds, or "escapes", in either the same, or different, regions to those in which they are most commonly grown commercially. Nevertheless, their ability to compete against other plants, to withstand pests and/or diseases or to prosper in soils lacking mineral fertilizers or water is less than that observed in feral populations of the same plant species.

In the evolution of crop species we can often distinguish between natural and human selection. Natural selection tends to favour the predominance of the most adapted plant types, which manage to reproduce and disperse their progeny while tolerating the stress factors that prevail in a particular environment. Therefore natural selection favours plant phenotypes which have the greatest chance of survival, reproduction, and distribution of progeny. For example, wild cereal plants tend to have many small seeds at maturity and disperse their seed by shattering. These seeds are also likely to be attached to a strong awn to aid dispersal. Similarly, wild potato species produce many small tubers, have their tubers develop at the end of very long stolons (so that daughter plants do not occupy ground too close to the parent), and many have tubers with high levels of toxin, which discourage animals from eating them.

Human selection is the result of conscious decisions by a farmer or plant breeder to keep the progeny of a particular parent and discard others. Human selection is not usually directed to better survival in the wild (and indeed is often detrimental to survival outside cultivation). As an example, breeders have developed cereal cultivars which have fewer but larger seeds that do not shatter their seeds at maturity, and that have a non-persistent awn. Similarly, potato breeders have selected plants with fewer but larger tubers, shorter stolons, and with reduced levels of toxins in the tuber. Human selection has also produced crops that are more uniform in the expression of many of their characteristics. For example, they have selected seeds that all mature at the same time, with uniform germination, and fruits with uniform fruit size, colour and shape. In more recent times plant breeders' selection has tended to result in shorter plants, greater harvest index (the ratio of harvested product dry matter to the total dry matter produced to sustain such harvested product), and increased ease of harvest. In addition, plant breeders have been able to either reduce or extend growth cycles, or to develop photoperiod insensitive cultivars. A large number of our crop species that used to require harvest by hand can now be harvested by machine, mainly as a consequence of their small stature and uniform ripening.

There is of course a range of characteristics that would have been positively selected both by natural evolution and early plant breeders. These might include aspects of yield potential, tolerance to stress factors, and resistance to pests and diseases.

1.4 Contribution of modern plant breeders

Around the turn of the 20th century the foundation of modern plant breeding was laid. Darwin's ideas on the differential survival of better adapted types were combined with those of Mendel on the genetic basis for the inheritance of plant characteristics. These two theories, combined with the research of scientists such as Weissman on the continuity of germplasm, the analyses of Johannsen resulting in the idea of genotype/phenotype relationships, and the rediscovery of Mendel's laws by Bateson, provided the scientific foundation of modern plant breeding.

There is little doubt that mankind has had a tremendous influence in moulding the morphology, plant types, end uses, and productivity of most crop species. Early farmers have taken wild, weedy plants and developed them into domesticated crops. The contribution of modern plant breeding efforts is not always clearly defined, nor can the achievements easily be measured.

Over the past century the world's human population has risen dramatically (Figure 1.2). World human population first exceeded one billion in 1804. It took a further 118 years of population expansion to double the world population. The human generation born after World War II (1945 to 1955) are often referred to as 'baby boomers'. Interestingly, this is the first generation to witness the world's population double, from 3 billion to 6 billion individuals. It has further been estimated that within the next 20 years another 2 billion people will inhabit this earth, and that the Earth's population is predicted to reach over 9 billion people by 2050. At the time of writing, the seventh billion person has already arrived on planet Earth.

Population explosion, combined with mass urbanization, and proportionally fewer farmers, led to fears from world population specialists of worldwide hunger and famine. However, since the start of the 'baby boom' era, the yield of almost all of our major agricultural crops has increased as dramatically as the human population. Cereal and oilseed crop production have doubled since 1955 (Figure 1.3). Similar increases in vegetable production of 235%, fruit production of 28%, pulses by 51% and root crops by 50% have taken place in a 56 year time span. When world agricultural production is adjusted according to population increase (Figure 1.4), cereal production per capita has increased by 19%, and fruit, vegetable, and oilseed production per capita has increased by 57%, 146% and 200%, respectively, while production of root crops and pulses has reduced per head of capita in the world. However, these significant increases in food production cannot distract us from the overwhelming challenges lying ahead: if populations increase at the predicted rate, then it has been estimated that during the next 50 years agriculture will have to produce an equivalent amount of food to that produced during the last 10,000 years combined.

These yield increases have been brought about by a combination of improved soil fertility (mainly due to additions of inorganic nitrogen fertilizers), improved chemical control of diseases and pests, better weed control through improved agronomic practices and herbicides, and better crop agronomic practices (e.g. correct plant densities), as well as by growing genetically improved cultivars.

So, how much of the improved yield can be attributed to the plant breeder (i.e. genetic change) and how much to better farming practices (i.e. environmental change)?

Yield increases of more than 100% have been found between single cross maize hybrid cultivars

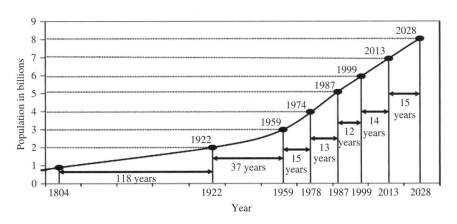

Figure 1.2 World population increase.

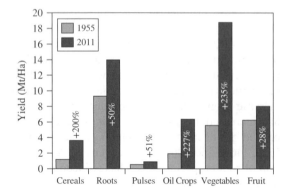

Figure 1.3   Total world crop production, 1955 and 2011.

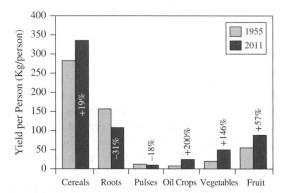

Figure 1.4   Total world crop production per capita, 1955 and 2011.

over the traditional open pollinated varieties. Many researchers have attributed this increase to the heterotic advantage of single crosses over homozygous inbred lines, and therefore conclude that the contribution of plant breeding must be very high. However, a complication arises when comparing single cross hybrids, where selection has been aimed at maximum hybrid productivity, against inbred lines which have been chosen for their combining ability rather than their own performance *per se*.

It might be suggested that the question could only be answered properly by growing a range of old and new varieties under identical agricultural conditions. Since most modern cultivars are dependent upon high levels of soil fertility and the application of herbicides, insecticides and fungicides, these

would have to be used in the comparison trial. However, older cultivars were not grown under these conditions. Certainly, older cereal varieties tend to be taller than newer ones and are therefore more prone to lodging (flattening by wind or rain) when grown under conditions of high soil fertility. These considerations also show that cultivars are bred to best utilize the conditions under which they are to be cultivated. Nevertheless, several attempts to compare old and new cultivars have been undertaken in an attempt to determine the contribution of modern plant breeding to recent yield increases.

In one comparison carried out in the United Kingdom, winter wheat cultivars ranging in introduction date from 1908 to 1980 were simultaneously evaluated in field trials. In a similar experiment, spring barley cultivars ranging in introduction dates from 1880 to 1980 were compared. From the wheat cultivars available in the mid-1940's the grain yield from this study was about 5.7 t ha^{-1} but from the most recently introduced cultivars from 1980, yields were about 50% higher. There was a similar improvement in barley yield over the same period of about 30%. Therefore, considering these studies, we can reasonably conclude that breeding has contributed about half to the more-than-doubled cereal yield between 1946 and 1980.

In contrast, a study carried out in potatoes, with cultivars with dates of introduction from 1900 to 1982 (Figure 1.5), found that modern plant breeding

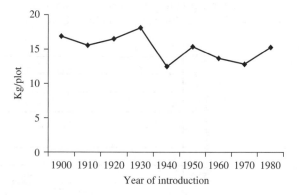

Figure 1.5   Saleable yield of tubers from potato cultivars grown over a three-year period. Yield is related to the year that each cultivar was introduced into agriculture.

had been responsible for a very small contribution to the more-than-doubled potato yield in the UK. This study in potato may, in part, reflect why 'Russet Burbank', introduced before 1900, still dominates potato production in the US, while the cultivar 'Bintji', introduced in 1910, remains a leading potato cultivar in the Netherlands.

In conclusion, modern-day crops have shown significant yield increases over the past century. It would be mistaken to suggest that the major contributor to this increase has simply been a direct result of plant breeding. Increases have rather resulted from a combination of plant breeding and improvements in crop husbandry. For example, the increased use of inorganic nitrogen fertilizer has greatly increased wheat (and other cereal) yield. However, this was allied with the introduction of semi-dwarf and dwarf wheat cultivars that allowed high nitrogen fertilizer application without detrimental crop lodging. Without the addition of nitrogen fertilizers, would the dwarf wheat cultivars have been beneficial? Probably not. However, would high nitrogen fertilizer application have been possible without the introduction of dwarf wheat cultivars? It is difficult to know. The overall increase achieved to date has resulted from both genetic and non-genetic changes in agriculture. In addition, it has been shown that significant, consistent genetic progress has been achieved not only under well managed conditions, but also under stressful conditions such as drought tolerance.

In the future the same is likely to be true: that the next leap in crop productivity will result from a marked change in agronomic practice, plus the introduction of plant types that can best utilize this husbandry change. What changes will these be? It is impossible to know with any certainty. Recent moves to reduced/no-tillage systems may be one option that could be considered, and that would require specific cultivars to maximize performance under these situations.

Similarly, advances in biotechnology may result in the development of crops with markedly different performances and adaptations to those available today. Introduction of these crop types may necessitate a major (or minor) change in crop husbandry to utilize the potential of these genetically modified crops.

Think questions

(1) Different crop species originated in different regions of the world. List the centres of origin of the following ten crop species: onion (*Allium* spp), alfalfa (*Medicago sativa*), rice (*Oryza sativa*), potato (*Solanum tuberosum*), soybean (*Glycine max*), sorghum (*Sorghum bicolor*), cotton (*Gossypium* spp.), sunflower (*Helianthus* spp.), wheat (*Triticum* spp.), and apple (*Malus* spp.).

(2) A combination of natural selection, and selection directed by plant breeders (early and modern), has influenced the crops we now grow. List five characteristics that mankind has selected which would not have been selected by a process of natural evolution.

(3) "The yield of many crops species has risen dramatically over the last 50 years. This has been the direct result of plant breeding during this period and hence the trend is likely to continue over the next 50 years." Briefly discuss this statement.

(4) "The place of origin of crops, their history and evolution are events from the past and therefore have no relevance to modern plant breeding." True or False? Discuss your answer.

(5) Many believe that civilization (of humans) started with the beginnings of agriculture. Basically there are two forms of agriculture: (1) rearing animals for meat, milk, etc., and (2) raising crops for human or animal feed. No one knows which form of agriculture evolved first (or maybe both types started together). Explain why (in your opinion) one form came before the other or both forms evolved at the same time.

(6) A combination of natural selection, and selection directed by plant breeders (early and modern) have influenced the crops we now grow. Have modern plant breeders improved the genetic fitness of our agricultural crop species, or have they simply selected plant types that are more suited to modern agricultural systems?

(7) A tremendous amount of plant diversity exists within Australia. Why, therefore, are there few Australian aboriginal farmers, and almost none

of today's world crop species had their centre of origin in the Australian continent?

(8) You are drifting your way through life when there is a crash and a tremendous flash of light, and from nowhere an alien spaceship lands. The alien, speaking perfect English, says *"I bring to you a gift: the Universe's most productive and perfect crop species"*. Thereafter the alien gives you a bag of 200 seeds, returns to the spacecraft and whoosh, it is gone. Describe what you would do with the seeds and what information you would collect that would allow you to develop new cultivars from these seeds.

2
Modes of Reproduction and Types of Cultivar

2.1 Introduction

The most appropriate type of cultivar that can be developed to best fit the needs of a production situation will be determined, in part, by the breeding system and mode of reproduction of the species involved.

A cultivar (or variety) is defined as a group of one or more genotypes with a combination of characters giving it **distinctness**, **uniformity** and **stability** (DUS).

Distinctness When a cultivar is 'released' for production it has to be proven that it is indeed new and that it is distinct from other already available cultivars. Distinctness is often defined on the basis of morphological characters that are known not to be greatly influenced by the environment. But other features such as physiology, disease or virus reaction, insect resistance and chemical quality may be used as well as, increasingly, molecular characterization in some countries (i.e. DNA markers). In the "Guide to UK National Listings", distinctness is described as follows: "*The variety, whatever the origin, artificial or natural, of the initial variation from which it has resulted shall be clearly distinguishable, by one or more important characteristics, from any other plant variety*". This requirement is in part to ensure that new cultivars have not previously been registered by other breeding organizations. Information used to determine distinctness also can be used later to identify and protect proprietary ownership of that cultivar so that other organizations cannot register the same cultivar or trade without permission (within a set period) in that cultivar.

Uniformity Uniformity is related to the level and type of variation that is exhibited (usually phenotypic) between different plants within the cultivar. Any such variation should be predictable and capable of being described by the breeder. The variation should also be commercially acceptable and occur with no greater a frequency than that defined for that type of cultivar (as we will see below). The amount of variation that is permitted to exist in released cultivars varies according to the country of release. For example, in the US, provided the degree and type of variation is clearly stated when the cultivar is released, the breeders can decide exactly how much heterogeneity exists for any character. In European countries, regulations regarding uniformity are more clearly defined and requirements for these to be adhered to. In the UK the guidelines read: "*The plant variety shall be such that the plants of which it is composed are, apart from a very few aberrations, and account being taken of the distinctive features of the reproductive system of the plants, similar or genetically identical as regards the characteristics, taken as a whole, which are considered by the Ministers for the purpose of determining whether the variety is uniform or not*".

Plant Breeding, Second Edition. Jack Brown, Peter D.S. Caligari and Hugo A. Campos.
© 2014 John Wiley & Sons, Ltd. Published 2014 by John Wiley & Sons, Ltd.
Companion Website: www.wiley.com/go/brown/plantbreeding

Stability   Stability of a cultivar means that it must remain true to its description when it is reproduced or propagated. Again, the requirements for this differ between countries – in Europe it is generally by statute, while in the US and Canada it is usually considered to be the responsibility of the breeder to ensure stability. Again the UK guidelines give a description of what they mean by stability, and it is: "*The plant variety shall continue to exhibit its essential characteristics after successive reproductions or, where the breeder has defined a particular cycle of reproduction, at the end of each cycle of reproduction*".

The requirement of distinctness is needed to protect proprietary cultivars and ensure that different organizations are not trying to claim the same cultivar and identify such cultivars as to their proprietary ownership. The requirements of uniformity and stability are there to protect the growers and to ensure that they are being sold something that will grow and exhibit the characteristics described by the breeder.

The further requirement of any new cultivar is perhaps obvious, but nevertheless is a statutory requirement in many countries and referred to as **value** for **cultivation** and **use** (VCU). VCU can be determined by two primary methods and there will always be debate regarding which system is better. In the UK, which organizes statutory trials, VCU is described as follows:

> "*The quality of the plant variety shall, in comparison with the qualities of other plant varieties in a national list, constitute either generally or as far as production in a specific area is concerned, a clear improvement either as regards crop farming or the use made of harvested crops or of products produced from these crops. The qualities of the plant variety shall for this purpose be taken as a whole, and inferiority in respect of certain characteristics may be offset by other favourable characteristics.*"

In a few countries (including the US) any plant breeder can sell seed from a cultivar developed and registered, irrespective of how well adapted it is to a given region or how productive the cultivar is likely to be. The choice of which cultivar to grow is left entirely to farmers and producers. It is common that farmers will allot a small proportion of the farm to plant a new cultivar, and if acceptable, will increase acreage with time. Obviously, unadapted cultivars or those with inferior end-use quality are unlikely to gain in acreage in this way. Similarly, companies (seed or breeding) and organizations rely on their reputation to sell their products. As is obvious, reputations can easily be tarnished by releasing and selling inferior products.

However, it is more common that countries have statutorily organized trialling schemes to determine the VCU of cultivars that are to be released. This testing is usually conducted over two or three years, in a range of environments in which the cultivars are likely to be grown. If breeding lines perform better than cultivars already available in that country, then government authorities will place that cultivar on the National List. Only cultivars that are included on the National List are eligible for propagation in that country. In some countries, newly listed cultivars are also entered into further statutory trials for one or two additional years. Based on performance in these extra years' trials, cultivars may be added to a *Recommended Varieties List*. This effectively means that the government authority, or testing agency, is recommending that it would be a suitable new cultivar for farmers to consider. The theoretical advantage of statutory VCU testing is that it only allows 'the very best cultivars' to be grown and prevents unadapted cultivars from being sold to farmers. The major drawbacks of the regulatory trialling schemes are:

- Mistakes are inevitably made (although it is difficult to estimate at what rate) in that potentially desirable adapted new cultivars simply do not do well in the test conditions, for whatever reason. In this situation the cultivar is removed from further screening and all the time and effort expended by the breeder on that genotype's development would have been wasted.

- Authorities (or their agents) who organize these trials are often limited by resources and cannot always evaluate the number of test entries that may be submitted as thoroughly as might be desirable. In these situations there is often a 'lottery system' introduced where: companies are allowed to enter *a certain number* of test entries; not all entries are grown in all trial sites; and a non-comprehensive set of control cultivars is included.

- Statutory trials suffer the same deficiencies as all small plot evaluation tests: they do not always reproduce or mimic the conditions or situations that occur on a realistic scale of production.

- They delay the period from a cultivar being developed to when it is released for commercial production. In some crop species (e.g. potato) this is not a great problem as the rate of increase of seed tubers in potato is low and it would normally take several years to produce sufficient tuber seeds to be planted on a commercial scale (a rate of increase of approximately 10:1 by traditional propagation, although now tissue culture does allow a considerable increase in multiplication rates). However, in other seed crops, for example canola (also known as oilseed rape in certain European countries), the rate of seed increase (particularly if off-season increase is possible) can be considerable, around 1000:1, and a three- to four-year delay in release can be costly to breeding companies.

The criteria for judging both DUS and VCU will be strongly determined by the type of species, particularly its mode of reproduction and multiplication for production.

2.2 Modes of reproduction

It is essential to have an understanding of the mode(s) of reproduction prior to the onset of a plant breeding programme. The type of reproduction of the species (at least in commerce) will determine the way that breeding and selection processes can be maximized to best effect. There are two general types of plant reproduction, **sexual** and **asexual**.

2.2.1 Sexual reproduction

Sexual reproduction involves fusion of male and female gametes that are derived either from two different parents or from a single parent. Sexual reproduction is, of course, reliant on the process of **meiosis**. This involves megaspores within the ovule of the pistil and the male microspores within the stamen. In a **typical diploid species**, meiosis involves reduction division of the $2n$ female cell to form four haploid megaspores, by the process of *megasporogenesis*. This process in male cells, to form microspores, is called *microsporogenesis*. Fertilization of the haploid female cell by a haploid

male pollen cell results in the formation of a diploid $2n$ embryo. The endosperm tissue of the seed can result from the union of two haploid nuclei from the female with another from the pollen, and hence ends up as being $3n$.

Asexual reproduction is the multiplication by use of plant parts (vegetative propagation, by tubers, by cuttings, in other words by cloning) or by the production of seeds (apomixis) that do not involve the union of male and female gametes. In general, organisms grow by cell division in a process called **mitosis**. The process of mitosis will result in two cells that are identical in genetic make-up and of the same composition as the parental cell.

Seeds are effectively classified according to the source of pollen that is responsible for the fertilization. In the case of self-pollination, the seeds are a result of fertilization of female egg cells by pollen from the same plant. Cross-pollination occurs when female egg cells are fertilized by pollen from a different plant, usually one that is genetically different. As a result plant species are usually classified into **self-pollinating** and **cross-pollinating** species. This is of course a gross generalization. There are species which are effectively 100% self-pollinating, those that are 100% cross-pollinating, but there exists a whole range of species that cross-pollinate or self-pollinate to varying degrees. From the top 122 crop plants grown worldwide, 32 are mainly self-pollinating species, 70 are predominantly cross-pollinating, and the remaining 20 are cross-pollinating but do show a degree of tolerance to successive rounds of inbreeding.

The method of pollination will be an important factor in determining the type of cultivar that can, or will, be most adapted to cultivation. For example, most species that can be readily used in F_1 hybrid production are generally cross-pollinating but need to be tolerant of inbreeding by selfing. This is because the hybrids are effectively the cross-pollinated progeny between two inbred parents.

Self-pollinating species:

- are tolerant of inbreeding and consequently deleterious recessive alleles are not common;
- tend to have flower structures and behaviours that promote selfing;
- have individual lines of descent that tend to approach homozygosity;

- show little heterotic advantage when out-crossed; and
- individually, tend to have a narrower range of adaptation.

Cross-pollinating species tend to be intolerant to inbreeding, principally because they carry many deleterious recessive alleles (these exist in the populations since they can be tolerated in heterozygous form). Generally, cross-pollinating species:

- have a crossing mechanism that promotes out-crossing;
- show greater heterotic effect;
- are more widely adaptable to many different environments;
- have individual plants that are highly heterozygous at many loci.

Particularly important are the outcrossing mechanisms. Cross-pollinating species often have distance barriers, time barriers or other mechanisms that limit, reduce or prevent self-pollination. Plants may be monoecious, where separate male and females flowers are located on different parts of the plant (e.g. maize) or indeed dioecious, where male and female flowers occur on different plants. Cross-pollination is also favoured in many cases where male pollen is shed at a time when the female stigma on the same plant is not receptive.

Other, more clearly defined sets of mechanisms are those termed as *self-incompatibility*. Self-incompatibility occurs when a plant with fully functional male and female parts will not produce mature seed by self-pollination. There is a set of mechanisms that have naturally evolved to prevent self-pollination and hence to increase cross-pollination within plant species, thus promoting heterozygosity. Adaptation to environmental conditions is greater if wider ranges of genotypes are produced in a progeny (i.e. the progeny shows greater genetic variation). Thus the chances of survival of at least some of the progeny will be enhanced, and conversely the chance of extinction will be reduced.

There are a number of mechanisms than can determine self-incompatibility in higher plants:

- Pollen may fail to germinate on the stigma of the same plant flower.

- Pollen tubes fail to develop down the style and hence do not reach the ovary.
- Pollen tube growth is not directed towards the ovule and hence pollen tubes fail to enter the ovary.
- A male gamete that enters the embryo sac may fail to unite with the egg cell.
- Fertilized embryos result from self-pollination but these do not produce mature seeds.

In several species (e.g. *Brassica* spp.) self-incompatibility can be overcome by bud-pollination, where pollen is applied to receptive stigmas of plants before the flowers open, as the self-incompatibility mechanism is not functional at this growth stage. Self-incompatibility is rarely complete, and usually a small proportion of selfed seed can be produced under certain circumstances. For example, it has been found that environmental stress factors (particularly caused by applying salt solution to developing flowers) tends to increase the proportion of self-seed produced.

2.2.2 Asexual reproduction

Asexual reproduction in plants produces offspring that are genetically identical to the mother plant, and plants that are produced this way are called **clones**. Asexual reproduction can occur by two mechanisms: *reproduction through plant parts* that are not true botanical seeds and *reproduction through apomixis*.

Reproduction through plant parts

A number of different plant parts can be responsible for asexual reproduction. For example, the following are some of the possible organs that are reproductive propagules of plants:

- A **bulb** is a modified shoot consisting of a very much shortened stem enclosed by fleshy leaves (e.g. a tulip or an onion).
- A **corm** is a swollen stem base bearing buds in the axils of scale-like remains of leaves from the previous year's growth (e.g. gladiolus).
- A **cutting** is an artificially detached part of a plant used as a means of vegetative propagation.

- A **rhizome** is an underground stem with buds in the axils of reduced leaves (e.g. mint or couch grass).
- A **stolon** is a horizontally growing stem that roots at nodes (e.g. strawberry runners).
- A **tuber** is a swollen stem that grows beneath the soil surface bearing buds (e.g. potato).

Reproduction by apomixis

Asexual production of plant seeds can occur in obligate and facultative apomicts. In obligate apomicts the seed that is formed is asexually produced, while in facultative apomicts most seeds are asexually produced, although sexual reproduction can occur.

Apomixis can arise by a number of mechanisms that differ according to which plant cells are responsible for producing an embryo (i.e. androgenesis from the sperm nucleus of a pollen grain; apospory, from somatic ovary cells; diplospory, from $2n$ megaspore mother cells; parthenogenesis from an egg cell without fertilization; and semigamy from sperm and egg cells independently without fusion). Apomixis can occur spontaneously, although in many cases pollination must occur (pseudogamy) if viable apomictic seeds are to be formed. Although the role of pseudogamy is not understood in most cases, pollination appears to stimulate embryo or endosperm development.

2.3 Types of cultivar

It may seem obvious that modes of reproduction determine the type of cultivar that is produced for exploitation. Cultivar types include pure-lines, hybrids, clones, open-pollinated populations, composite-crosses, synthetics and multilines. Obviously it would be difficult, if not impossible to develop a pure-line cultivar of a crop species like potato (*Solanum tuberosum*) as it is mainly reproduced vegetatively, and has many deleterious (or lethal) recessive alleles. Similarly, pea (*Pisum sativum*) is almost an obligate self-pollinator and so it would be difficult to develop hybrid pea, if nothing else seed production is likely to be expensive. A brief description of the different types of cultivar is presented below.

2.3.1 Pure-line cultivars

Pure-line cultivars are homozygous, or near--homozygous, lines. Pure-line cultivars can be produced most readily in naturally self-pollinating species (e.g. wheat, barley, pea and soybeans). But they can also be produced from species that we tend to consider as cross-pollinating ones (e.g. pure-line maize, gynoecious cucumber and onion). There is no universally agreed definition of what constitutes a pure-line cultivar, but it is generally accepted that it is normally one in which the line is homozygous for the vast majority of its loci (usually 90% or more).

The most common method used to develop pure-line cultivars from inbreeding species is to artificially hybridize two chosen (usually) homozygous parental lines, allow the heterozygous first filial generation (F_1) to self to obtain F_2 seed, and continue to allow selfing in future generations, up to a point where the line is considered to be 'commercially true breeding', maybe the F_6 or F_7. At the same time, it has been common to carry out recurrent phenotypic selection on the segregating population over each generation. As described in Chapter 9, the use of double haploid lines is gaining popularity in a number of crop species as a fast method to gain homozygosity.

2.3.2 Open-pollinated cultivars

Open-pollinated cultivars are heterogeneous populations comprised of different plants which are genetically non-identical. The component plants tend to have a high degree of heterozygosity. Open-pollinated cultivars are almost exclusively from cross-pollinating (out-breeding) species. These are populations that have been selected to a standard that allows for variation in many traits, but which shows 'sufficient' stability of expression in the characters of interest. Stability of these traits can be used to pass the DUS requirements necessary for cultivar release. Examples of open-pollinated cultivars would include onions, rye, herbage grass, non-hybrid sweetcorn, sugar beet and oil palm.

In developing open-pollinated cultivars, the initial hybridization (the point at which the genetic diversity and variation is created) is usually between two open-pollinated populations. In this case segregation is apparent at the F_1 generation. Desirable populations are identified and improved by increasing the frequency of desirable phenotypes within them.

2.3.3 Hybrid cultivars

Hybrid cultivars (single cross, three-way cross and double cross hybrids) have different levels of homogeneity but, importantly, are highly heterozygous. An F_1 hybrid (single cross) cannot be reproduced from seed collected from them because the progeny would then effectively be an F_2 and would segregate, thus resulting in a different and non-uniform population. However, the other types of hybrid (three-way cross and double cross) are produced and grown, and although they give segregating populations of plants, they do so in a predictable way and at a predictable level.

Hybrid breeding is perhaps the most complex of the breeding methods. The process of cultivar development involves at least two stages. The first stage is to select desirable inbred lines from chosen out-pollinated populations. These inbred selections are then used in test crosses to allow their comparison and assessment in relation to their general or specific combining ability. Superior parents are selected, and these are then hybridized to produce seed of the hybrid cultivar. The parent lines are then maintained and used to continually reproduce the F_1 hybrids. Despite its complexity, hybrid breeding has been the method of choice in many out-breeding species because, through the exploitation of heterosis, hybrid cultivars often yield more than other types of cultivars.

2.3.4 Clonal cultivars

Clonal cultivars are genetically uniform but tend to be highly heterozygous. Uniformity of plant types is maintained through vegetative rather than sexual reproduction. Cultivars are vegetatively propagated by asexual reproduction (cloning) including cuttings, tubers, bulbs, rhizomes and grafts (e.g. potatoes, bananas, peaches, apples, cassava, sugarcane, strawberries, blueberries and chrysanthemums). A cultivar can also be classified as a clone if it is propagated through obligate apomixis (e.g. buffelgrass).

Clonal varietal development begins by either sexual hybridization of two parents (often clones) or the selfing of one of them to generate genetic variability through controlled crosses but using the normal process of sexual reproduction. Most of the parental lines will be highly heterozygous and segregation will be observed even in the first generation. Desirable recombinants are selected from between the various clones. Breeding lines are maintained and multiplied through vegetative reproduction, and hence the genetic constitution of each selection remains '*fixed*'.

2.3.5 Synthetic cultivars

Intercrossing a set number of defined parent lines generates a synthetic variety, and so it can be maintained by reconstruction of the population from the parents. In the simplest case a 'first generation, two-parent synthetic' is very similar to an F_1 hybrid. Synthetic lines can be derived using parents that are clonally propagated or inbred lines, although the latter instance is not common. From the initial open pollination, synthetic cultivars are labelled as a series of generations/categories (Syn.1, Syn.2, ... , Syn.n) according to the number of open-pollinated generations that have been grown since the synthetic line was first generated. So, the first generation of a synthetic variety is classified as Syn.1; if this population is then open pollinated, the next generation is classified as Syn.2.

The use of synthetic cultivars has been most successful in cases where crop species show partial self-incompatibility (e.g. alfalfa). Examples of other crops where synthetic varieties have been released include canola (*B. rapa* cultivar types), rye, pearl millet, broom grass and orchard grass.

2.3.6 Multiline cultivars

Multiline cultivars are mixtures or blends of a number of different cultivars or breeding lines. By definition it is accepted that each genotype in the mixture will be represented by at least 5% of the total seed lot. Many multilines are the result of developing near-isogenic lines and mixing these to form a population of lines. These cultivars are usually self-pollinating species. A multiline is therefore not the same as a synthetic, where the aim is to maintain heterozygosity by inter-crossing between the parent lines. Multilines became popular because of the wish to increase disease resistance but reduce the selective pressure on the pathogen to evolve/mutate in order to overcome the biological resistance that had been bred into the lines. For example, near-isogenic lines of barley, which differ in terms of possessing qualitative disease resistance genes, could be mixed to make a multiline. The main thought is to affect the epidemiology of the pathogen such that it would be less likely to evolve virulence to all resistance genes in the mixture.

2.3.7 Composite-cross cultivars

Composite-cross cultivars are populations derived by intercrossing two or more cultivars or breeding lines. These cultivar types have all tended to be inbreeding species (e.g. barley or lima beans). After the initial hybridizations have been carried out, the composite-cross population is multiplied in a chosen environment such that the most adapted segregants will predominate and those less adapted to these conditions will occur at lower frequencies. A composite-cross population cultivar is therefore continually changing and can be considered (in a very loose sense) similar to the old land races. These cultivars are difficult to commercialize as they cannot be maintained with the same composition as the originally released cultivar.

2.4 Annuals and perennials

Plant species are categorized into annuals and perennials. World crop plants are fairly evenly distributed between annuals (approximately 70 species) and perennials (approximately 50 species). All major self-pollinating crop species are annuals, while the greatest majority of cross-pollinating crops are perennials. Perennials pose greater difficulty in breeding than most annuals. For example, most perennials do not become reproductive within the first years of growth from seed, and also most perennials are clonally propagated, which can cause additional difficulties in maintaining disease-free parental lines and breeding material. On the other hand, winter annuals also present difficulties as they require vernalization or a chilling treatment before moving from vegetative to reproductive growth, and so can increase the time necessary for developing new cultivars.

2.5 Reproductive sterility

Female and male sterility has been identified in many crop species, and both genetic and cytoplasmic sterility has been identified in various species. Such sterility can, of course, cause problems to breeders and limit the choice of parental cross combinations that are possible. But it also can be exploited when, for example, developing hybrid cultivars.

Think questions

(1) Complete the following table by assigning a **YES** or **NO** to each of the 16 cells.

	Pure-line cultivar	Open-pollinated cultivar	F_1 Hybrid cultivar	Clonal cultivar
Are these cultivars composed of only one single genotype?				
Is heterosis a major yield factor in resulting cultivars?				
Are resulting cultivars propagated by means of botanical seeds?				
Can the seed, or plant parts, of a cultivar be used for its own propagation?				

(2) Inbreeding and out-breeding species tend to have different characteristics. Explain factors that would determine whether a given species should be classified as inbreeding or out-breeding.

(3) A number of different cultivar types are available in agriculture. Outline the major features of the following cultivar types:
 (a) hybrid cultivars;
 (b) pure-line cultivars;
 (c) clonal cultivars;
 (d) multiline cultivars;
 (e) open-pollinated cultivars; and
 (f) synthetic cultivars.

(4) List two inbreeding crop species and two out-breeding crop species that have been exploited as:
   (a) pure-line cultivars;
   (b) open-pollinated cultivars;
   (c) hybrid cultivars;
   (d) clonal cultivars;
   (e) multiline cultivars; and
   (f) synthetic cultivars.
(5) Describe the major features of the following types of apomixis:
   (a) diplospory;
   (b) semigamy;
   (c) advantageous embroyony;
   (d) pseudogamy;
   (e) parthenogenesis;
   (f) apospory.
(6) List five different plant parts that can be used for asexual reproduction.

3
Breeding Objectives

3.1 Introduction

The first exercize that must precede any of the breeding operations (and indeed a task that should be continually updated) is preparing a breeding plan or setting **breeding objectives**. Every breeding programme must have well defined objectives that are both economically and biologically feasible. It can be argued that instead of '*breeding objectives*', which implies that such objectives are useful only for breeding, a more appropriate name would be '*cultivar improvement objectives*', because this reflects the benefit of changing a trait by any means.

In practice, many new cultivars fail to be commercialized successfully after they are introduced to large-scale agricultural production. In some cases these failures are associated with the programme having the wrong economic objectives. Similarly, many excellent new genotypes fail to become successful cultivars because of some unforeseen defect which was not considered important or was overlooked in the breeding scheme.

Objectives, then, are the first of the plant breeder's decisions. The breeder will have to decide on such considerations as:

- What political, social and economic factors are likely to be of greatest importance in future years?
- What criteria will be used to determine the yielding ability required of a new cultivar?

- What end-use quality characters are likely to be of greatest importance when the newly released cultivars are at a commercial stage?
- What diseases or pests are likely to be of greatest importance in future years?
- What type of agricultural system will the cultivar be developed for?

All these will need to be considered and extrapolated ahead to a time that is likely to be 8 to 14 years from the onset of the breeding process. It should also be noted that politics, economics, yield, consumer preferences, quality and plant resistance are not independent factors, and that interactions between all these factors are likely to have an effect on the breeding strategy. It is only after answering these questions that breeders will be able to ask:

- What type of cultivar should be developed?
- How many parents to include in the crossing scheme? Which parents to include? How many crosses to examine? To examine two-way or three-way parent cross combinations and why?
- How should progenies progress through the breeding scheme (pedigree system, bulk system, etc.)?
- What characters are to be selected for, or against, in the breeding scheme, and at what stage should selection for these characters take place?
- How to release the variety and promote its use in agriculture and its benefits to end-customers?

Plant Breeding, Second Edition. Jack Brown, Peter D.S. Caligari and Hugo A. Campos.
© 2014 John Wiley & Sons, Ltd. Published 2014 by John Wiley & Sons, Ltd.
Companion Website: www.wiley.com/go/brown/plantbreeding

3.2 People, politics and economic criteria

The final users of almost all agricultural and horticultural crops are consumers (humans or other animals) who are increasingly removed from agricultural production systems. In 1863, the United States Department of Agriculture (USDA) (www.usda.gov) was created, and at that time 58% of the US population were actually farmers. Indeed, only a few decades ago the vast majority of people in the Western world were directly involved in agricultural and food production. However, in 2010, less than 1.5% of the US population was directly involved with agriculture. This past century, therefore, has resulted in a dramatic shift away from working on the land to living and working in cities. Agricultural output has, and continues to, increase almost annually despite fewer and fewer people working directly in agriculture. The major consumers of agricultural products are therefore city dwellers who are remote from agricultural production but obviously have a large influence on the types of food that are purchased. In addition, these non-agriculturists have a tremendous influence on the way that agricultural products are grown and processed, and plant breeders would be foolish not to consider end-users' likes and dislikes when designing future crops. Although it is the case that, in developing countries, the proportion of the population involved in agriculture is generally higher than in developed ones, nevertheless in 2012 it was announced that for the first time in history more than 50% of the world's population lived in urban areas.

Despite an overall shortfall in world food supply, many developed countries have an overabundance of agricultural food products available, and consumers have become used to spending a lower proportion of their total earnings on food than ever before. In addition, consumers have become more interested in the way that food products are grown and processed. Many consumers are interested in eating '*healthy*' food, often grown without agrochemicals and without subsequent chemical colourings or preservatives. Obviously breeding for cultivars that are resistant to diseases, which can be grown without application of pesticides, fits the needs of these consumers. The desire of customers for what they call '*more natural*' food has,

in recent years, had a large impact on agriculture and intercountry trade in agricultural commodities, particularly with the advent of large-scale commercialization of genetically modified organisms (GMOs).

The development of recombinant DNA techniques that allow the transfer and expression of a gene from one species, or organism, into another represented a breakthrough in modern plant breeding, and offers enormous potential advances in crop development as it increases the genetic variation available to plant breeders. However, the first GMO crops to be commercialized conveyed advantages more appreciated by the farmer growing them than by end-users. In general this technology has had a mixed reception by consumers. The reasons for this are numerous, but are related to a general mistrust among consumers regarding the health and safety of these *novel* products. In addition there are concerns amongst many groups that there might be environmental risks associated with GMO crops, and that transgenes might *escape* into the ecosystem as or by intercrossing with related weeds or wild species. Other issues obviously are involved in this barrier of acceptance, including free trade agreements and monopolies of transformation technology by a few companies worldwide. As scientists, we must also accept that we do not know all the possible facts or outcomes. None the less, during the 2011 growing season, over 10% (i.e. 160 million hectares) of the world's 1,389 million hectares of total crop-land planted were cultivated with transgenic cultivars, and the rate of adoption of this technology by large and small farmers shows no sign of slowing down thus far.

How many plant breeders, say, 10 to 15 years ago had the insight to foresee such public opposition in acceptance of GMO products, that many consider as showing "*a significant advantage over proceeding cultivars*"? However, as noted, almost all of the first GMO crops had genetic advantages that were "hidden" to the final consumer. Nevertheless, on closer inspection, research recently carried out at Iowa State University in the US suggests that world prices of corn, soybeans and canola would probably be, respectively, 6%, 10% and 4% higher on average than 2007 baseline levels if GMO cultivars of these crops were not available to farmers.

Therefore, even though the most direct beneficiaries of the first wave of GMO crops have been farmers and biotechnology-based private breeding companies, there are already spillovers directly benefiting the purchasing power of end-consumers. The main reason why the first GMO releases had grower advantages rather than benefits to the end-user is the sheer genetic complexity of modifying the metabolic pathways involved in quality traits important for final consumers. Nevertheless, the first GMO cultivars with improved oil fatty acid profiles, enriched with omega-3 fatty acids (associated with a reduced cardiovascular risk) which show a clear advantage for end-users, are now commercially available. In addition, transgenic maize hybrids tolerant to drought are already at the commercialization stage in the US, which will provide benefits to farmers by stabilizing the production of grain under water-limiting growing conditions, but could also contribute to a more efficient use of an ever-more valuable resource, water.

Including recombinant DNA technologies as tools in cultivar development is one issue that many breeders and breeding programmes are at present giving serious consideration. Consumers, farmers or housewives, also have a tremendous impact on all aspects of everyday life, as they are usually involved in democratically electing politicians that govern the nations of the world.

Many people would consider it impossible to try to predict what is in the mind of a politician. Politicians and political forces will, however, continue to be a large determining factor in shaping agriculture. For example, there may be a very cheap and 'safe' chemical available that controls a certain disease in your crop of interest. With this in mind, the breeding objectives may not include selection or screening for biological resistance to this disease. Several years into the breeding scheme, it may be demonstrated that the use of this chemical is harmful to the environment and government policy responds by withdrawing the use of the chemical, and suddenly the need for resistance to this disease in breeding lines is vital. As an example, it is likely that over the years the majority of organophosphate-based insecticides will not be registered or re-registered for application to many of the crops that today depend on them for successful production. It has further been suggested that over 80% of all agrochemicals used in the US might not be registered or re-registered

for use in agriculture at some point in the future. Another example of the far-reaching influence of political decisions on breeding outcomes is the current debate of food versus fuel. Through the development of economic incentives like subsidies, politicians can raise (or reduce) the demand for a given type of crop, and thus affect the breeding approach required to fulfil these requirements.

Many soil fumigants are highly toxic, volatile chemicals that, amongst other things, have adverse depletion effects on our atmosphere. As a result many governments worldwide have banned the use of soil fumigants such as methyl bromide. One possible alternative to using synthetic fumigation is Brassicaceae cover crops (plough-down crops) or high glucosinolate-containing seed meal soil amendments, an approach called *biofumigation*. Glucosinolates *per se* are not toxic, but when mixed with water in the presence of myrosinase they degrade into a number of toxic compounds including isothiocyanates and ionic thiocyanate. Past research has shown that different Brassicaceae species produce different quantities and types of glucosinolate which have greater or lesser pesticidal activity on different pests. Far-sighted plant breeders have recognized the potential of biopesticides and have found that interspecific hybridization between different Brassicaceae species offers an opportunity to develop '*designer glucosinolate*' plants with specific pesticidal effects. It is fascinating to note that some hydrolytic byproducts of specific glucosinolates have also been shown to play a significant cancer preventative role, and very recently broccoli cultivars that contain higher levels of those glucosinolates have been made available to farmers, as a means to improve the diet of end-users and to encourage the adoption of healthier dietary habits.

Government agencies (e.g. the Food and Drug Administration in the US (www.fda.gov), or the European Food Safety Authority in Europe (www.efsa.europa.eu)) can greatly influence consumer acceptance and choice. One recent example of this relates to *trans* fats in cooking oils. Research has shown that hydrogenated vegetable oils which contain *trans* fats have adverse effects on human health when included in diets. Indeed, the Food and Drug Administration now requires *trans* fat content to be listed on all food products. Traditional vegetable oils like oilseed rape (canola) and soy, although relatively low in saturated fats, are usually hydrogenated to avoid off-flavours in high-temperature frying and to increase the shelf life of the oil products. Rancidity and off-flavours in vegetable oil are caused by high concentrations of polyunsaturated fats. This has greatly raised the consumer awareness of *trans* fats, and as a result there is now high demand for vegetable oils which have a low polyunsaturated fat content. Perceptive canola and soy breeders had anticipated *trans* fat labelling and had low polyunsaturated (high oleic acid) cultivars of canola and soy available to meet market needs. These low polyunsaturated fat cultivars produce oils that show higher thermal stability, lower levels of oxidation products, and increased shelf life with minimal hydrogenation. Because health-conscious consumers and food companies want to avoid *trans* fats in foods, farmers willing to grow these soybean cultivars under contracts can often receive a premium price for their crop compared to other cultivars.

Political pressure can also have an influence on the types of crops that are grown. Within several countries in the world, and also groups of countries (e.g. the European Union), the farming community are offered subsidies to grow certain crops. As a result, over-production can occur, which can affect the world price of the crop, and hence influence the economics of farming outside the subsidized regions. If crop subsidies are reduced or stopped, this can also have a similarly large but opposite effect on the economics and hence directly affect acreage of the crop in these other regions. Crop price is always driven by demand, greater demand resulting in a higher price. However, this can give rise to increased acreage, which may then mean over-production, which in turn usually leads to reduced crop prices.

The US, like many other Western countries, has become increasingly dependent upon imported oil to satisfy energy demand. It is possible to substitute oils from fossil fuels with renewable agricultural products (Figure 3.1). Therefore, bio-ethanol and biodiesel fuel, lubricating oil, hydraulic oil and transmission oil can all be derived from plants. At present the agricultural substitutes are still higher in cost than traditional fossil-derived equivalents. However, in the future this may change either as a result of a change in fossil oil costs or in further breakthroughs in either increasing crop productivity

Figure 3.1 Volkswagen Beetle ('Bio-Bug') powered by biodiesel produced from mustard oil.

or the processes needed to obtain these substitutes from agriculture. Governments in several countries have mandated that liquid fuel should contain a certain proportion of biodiesel or bio-ethanol, and public transport vehicles in inner cities are being encouraged to use biofuels as these have fewer emission problems. Similarly, in many countries (particularly Northern Europe) governments are legislating that certain operations (e.g. chainsaw lubricants) should use only biodegradable oils. Also taxation decisions made by relevant countries can affect the relative cost and hence use of fossil- and plant-derived fuels and oils.

The recent spike in food prices due to climatic disasters such as severe drought spells in Russia and Australia, and diminishing grain stocks, triggered rioting and civil unrest in many countries. This is a sober reminder about the importance of a supply of cheap food and the close link existing between food stability and political stability. Additionally it highlights the overreaching impact plant breeding imparts, which extends well beyond the purely agricultural realm. This further evidences the impact that plant breeding has on food production, the quality of livelihoods and even on political stability in many parts of the world.

Economic criteria are important because the breeder must ensure that the characteristics of cultivars that are to be developed are the ones that will satisfy not only the farmers, but also the end-users, and that can be produced in an agricultural system at an economic level. The supply of a product and the consumer's demand for that product are inter-related. If there is over-production of a crop then there is a tendency for the purchase price to be lower. Conversely, of course, when a product is in high demand and there is limited production, then the product is likely to command high premiums. In general, however, there is a tendency for an equilibrium: that farmers will only produce a volume that they think they can sell according to the needs of the end-user.

Although plant breeding is influenced by economic activity, unfortunately financial concerns are rarely considered in setting breeding objectives or setting a breeding or selection strategy. It is often assumed that increased yielding ability, better quality and greater disease or pest resistance are going to be associated with improved economics of the crop. Unfortunately this topic has been examined by very few researchers and is an area where greater examination would increase the probability of success in plant breeding and therefore its contribution to society and mankind.

Private breeding companies have developed economic breeding objectives that have had greatest influence on breeding objectives or strategy. This is perhaps not surprising as these groups require developing better cultivars and selling seeds or collecting royalties on these genotypes in order to sustain the ever-increasing research and development expenses required to survive in the industry. This, for example, has been one of the main drivers to develop so-called hybrid cultivars, which generally allow farmers to achieve higher levels of yield and greater uniformity than with other types of cultivars, but require fresh seed to be purchased every growing season.

Public breeding programmes have in the past been better placed to carry out breeding with objectives that might be considered to have a greater social than economic impact, for example breeding varieties that might suit small-holder growers, concentrating on species where there was a limited seed market, or crops with small overall acreages. They also usually were associated with universities, colleges or government departments which had, as a background, research at a more fundamental level or which could be carried out over a longer timescale without immediate financial return. This gave them greater freedom to carry out research into aspects that were not necessarily 'mainstream'

to the variety production process and investigate such aspects as wild germplasm characteristics, plant breeding strategies, novel crops or novel uses for crops. This has, however, become much less the case in recent times, with more pressure being exerted in public systems for near-market research to be privatized and for public research to be self-funding. So, for example, they are often encouraged to develop breeding material which can be licensed to private breeding companies. Therefore the differences between the two types of breeding programme have been severely diminished.

3.3 Grower profitability

In general yield is the most important character of interest in any plant breeding programme. Therefore, increasing crop yield will always be a sensible strategy. There are rare cases, however, when farmers are willing to trade-off yield with quality, if the end product premium can offset the reduced productivity. One such case is the production of grapes for premium wine-making, where farmers deliberately avoid maximizing production in order to reach the quality standards and higher prices under which their grapes are purchased by wineries. In general terms, though, there would be only limited use for a new cultivar unless it has the potential to at least yield at a comparable level with existing varieties, unless, as noted above, the harvestable product fits a particular niche market and hence can attract a higher 'per unit price'. Plant breeders tend therefore to select for increased profitability. Farmers' profits are related to input costs and gross returns on the crop. Plant breeders can increase grower's profitability by:

- increasing the yield per planted area, assuming input costs remain constant;
- expanding the region of crop production;
- reducing input costs while maintaining high yield per unit area (input costs will include herbicides, insecticides and fungicides);
- increasing the inherent quality component of the end product so that growers receive a higher unit price when the harvestable product is sold.

All crops have restricted ranges of environments to which they are adapted. Bananas and sugarcane are unlikely to be grown as commercial crops in the Pacific Northwest region of the US or in northern

Europe. However, one attribute to increasing yield may be related to increasing the range of environments in which a crop can be grown. For example, the development of earlier maturing *Brassica napus* lines has extended the canola (oilseed rape) acreage in Canada to include regions further west than was previously thought possible. A similar extension of adaptation must have been involved with movement of wheat and maize to northern temperate regions over the past decades. For example, potato production in many world regions is difficult, as healthy seed tubers cannot be produced or made available when required for planting. Developing potato cultivars that are propagated from true botanical potato seed (TPS) would overcome many difficulties that occur in these regions. Few potato diseases are transmitted through TPS. In addition, small quantities of TPS would be required for planting compared with traditional seed tubers, which are bulky and usually require refrigerated storage. Further examples of the expansion of cropping enabled by plant breeding are the expansion of soybean production in North Dakota from 2.5 million acres in 2002 to 4 million acres in 2010, the development of corn hybrids able to perform well in the summer crop known as '*safrinha*' in Brazil and Paraguay, and the development of short-cycle wheat varieties in Argentina enabling a second summer crop of soybeans or maize. The latter example not only brings about economic benefits to farmers; in addition it creates alternative crop rotations which enhance soil organic matter content and a more efficient use of fertilizers.

These examples are, however, perhaps exceptions, and it should be noted that the majority of breeding programmes are concerned with increasing yield potential within an already well-established growing region.

Crop profitability is based on net profit rather than gross product. By reducing input costs or labour requirements and at the same time maintaining high yield per unit area, breeders can increase crop profits. Input costs in crop production include herbicides, insecticides and fungicides. Therefore developing cultivars that are more competitive with weeds, resistant to damage by insect pest, or have disease resistance, reduces inputs and alleviates the need to purchase and apply chemicals. This rationale has supported the development of herbicide-tolerant and pest-resistant crops through

plant breeding in many crop species. In the case of these herbicide-tolerant crops, which have been developed not only through biotechnology but also through traditional means, they have in addition supported the use of herbicides with a lesser environmental footprint, such as glyphosate.

Other inputs will include nutrients (mainly nitrogen) and water (which can have a high price in irrigated farming regions). It follows, of course, that developing cultivars that require less nitrogen or are more tolerant to drought and other stress factors will result in greater profitability to growers and will also contribute to stabilizing food supply.

3.3.1 Increasing harvestable yield

Yield, in the eyes of breeders, is considered to have two main components: *biomass*, the ability to produce and maintain an adequate quantity of vegetative material, and *partition*, the capacity to divert biomass to the desired harvestable product (seeds, fruits, or tubers etc.). Therefore, the partitioning of assimilate is very important in obtaining maximum yielding ability. Partition, in general, takes the form of enhancement of yield of desired parts of the plant product at the expense of unwanted plant parts (sometimes referred to as increasing the harvest index, defined as the ratio between harvestable dry matter and total dry matter). This can take three main forms:

- Vegetative growth is reduced to a minimum compared to reproductive growth. In many crop species plant breeders have selected plants that have short stature, or indeed are dwarf mutants. This was mainly driven by the need to reduce lodging (plants falling over before harvest) under increasing levels of applied nitrogen. However, short plant stature also allows more convenient harvest and sometimes (e.g. in fruit trees) allows for mechanical harvest, hence avoiding more expensive harvest by hand-picking. This strategy has been adopted in breeding objectives of many crops such as wheat, barley, oats, sunflower, several legumes, along with fruit trees like apple, orange, peach and cherry. In the case of rice and wheat, the exploitation by breeders of naturally occurring dwarfing alleles enabled the Green Revolution.
- Reproductive performance is suppressed in favour of a vegetative product. This has been

applied to a number of vegetatively repro- duced crops where sexual reproduction has been selected against in plant breeding programmes in favour of vegetative growth (e.g. potato, sug- arcane, sugar beet and various vegetables). The rationale behind such efforts has been to divert the energy naturally used to develop and main- tain reproductive organs towards those organs needed and appreciated by customers.

- Vegetative production to different vegetative parts can be used to increase yield of root and vegetable crops like potato, rutabaga (swede) and carrot. In this case, the breeder's task is to maximize the partition towards one type of vegetative yield (e.g. tubers in the case of potato) while maintaining the minimum biomass of unused plant parts (e.g. the haulm of potato).

3.3.2 Selection for yield increase

It is perhaps ironic that harvestable yield is arguably the most important factor in all plant breeding schemes, and yet it is possibly the most difficult to select for. Increasing yield is complex and involves multiple modifications to the plant's morphology, physiology and biochemistry. Yield is, not sur- prisingly, quantitatively inherited (i.e. under the control of many genes, each with small phenotypic effect) and highly modifiable by a wide range of environmental factors. Evaluating accurately the genotypic response to differing environments and genotype–environment interactions are the major limiting factors to maximizing selection response in plant breeding. Despite advances in molecular marker selection (mainly quantitative trait loci, and more recently haplotypes and the development of molecular markers-based breeding values used in genomic selection efforts), increased yield is achieved by evaluating the phenotype of breeding lines under a wide range of rather atypical environ- ments, sometimes called the target population of environments.

Yield potential will be one of few characters that is evaluated (or at least considered) at virtually all stages of plant breeding programmes. Plant breeding schemes begin with many (often many thousands) genotypes on which selection is carried out over years and seasons until the 'best' cultivar is identified, stabilized and increased. Usually the size of plots used for field evaluation trials increases with increasing rounds of selection. On completion of the selection process, surviving breeding lines must have produced phenotypically high yield in small unreplicated plots (often a single plant), and a vari- ety of increasing plot sizes associated with advanc- ing generations in the selection scheme. Towards the most advanced stages, a few breeding lines will be grown in farm-scale tests. Irrespective of the crop species involved, most plant breeders have been successful by selecting for yield *per se* of the plant part of importance, rather than by selecting for modified assimilate partition within a new crop cultivar, albeit that selection for the first resulted in a difference in the second. A number of different plant breeders have selected for yield components (i.e. attributes that contribute to total yield such as number of ears per plant, number of seeds per ear and seed weight in the case of cereals, or number of tubers per plant and tuber weight in potato) rather than for yield itself. In most of these cases, however, there has been little achieved in respect of increas- ing overall yield. The reasons behind this failure are complex, but one factor is related to the negative relationships between many yield components. Therefore positive selection for one component is counterbalanced by a negative response in another. One of the few examples of successful selection based on partitioning has been the efforts of the International Maize and Wheat Improvement Cen- ter (commonly called by its Spanish acronym CIM- MYT for *Centro Internacional de Mejoramiento de Maíz y Trigo*) to improve drought tolerance in tropical corn, where the trait selection imposed is directly involved in increasing the partitioning of dry matter into kernels, resulting in corn breeding lines able to produce more kernels, and therefore more yield, under a severe stress at flowering.

It is the actual yield obtained by the farmer that is clearly an important criterion, and therefore factors such as 'harvestability' come to the fore. Mechanical harvesters now carry out many harvest operations. If a given genotype is not suited to mechanical harvest, its usefulness can be greatly limited. Therefore characters such as plant lodging, precocious sprouting, seed shattering or fruit drop are all factors that will reduce the harvestable yield.

In many instances the uniformity of morphologi- cal characters (seed/tuber/fruit – size, shape, colour, etc.) have a great effect on 'useable yield'. Obvi- ously, the end-user has a preference for a product which has a certain size, shape or colour, and any deviation (either genetic or environmental) from this appearance will reduce useable yield. Similarly, if a crop product is prone to develop defects when processed, or in storage, this will affect useable yield as the defects will not meet the required legislative standards, or customer expectations, and will need to be culled out. A secondary factor regarding defective products is related to the cost that is incurred in having the defective fruits/tubers/seeds removed.

Uniformity of yield is more difficult to evaluate than yield itself. Often it is not possible to evaluate product uniformity with any accuracy in small plot trails, and therefore many potentially highly uniform breeding lines may be wrongfully discarded in the early stages of selection. At the same time, it is important to consider that in cases like tomato and other horticultural crops, uniformity is of paramount importance to the vast majority of consumers purchasing them at retail points such as in supermarket outlets. Uniformity can also be important to processors or corporate customers such as restaurant chains.

Research by crop physiologists has provided a great deal of information regarding plant growth models for yield, and we have developed the ability to predict actual yield from a wide range of different physiological measurements. In the latter half of the last century, many plant breeders believed that input from crop physiologists and physiological biotype models of our crop plants would assist plant breeders to identify superior cultivars. Crop physiologists believed that photosynthetic or net assimilation rates could be used as selection tools to increase plant productivity and hence increase yield of crop plants. Some successes do exist, and an example is afforded by lupins. Physiologically-based research and modelling led to the proposal it would be beneficial to aid the development of the crop into northern Europe to breed for a particular crop architecture, using genotypes with a determinate growth habit. Suitable mutants were found, and indeed proved to be a marked improvement on the traditional lupin types in the new target environments. Despite the success in lupin, however, the impacts of physiological biotype models in plant breeding are rare. More progress has been observed with the use

of crop models to support environmental classification approaches. For instance, work carried out in Australia at the University of Queensland has shown that the crop model APSIM is able to model the interaction of crops like wheat with its environment, and to establish environmental classifications able to reduce the relative effect of Genotype × Environment interactions and therefore to increase expected genetic gains and suggest breeding approaches.

3.4   Increasing end-use quality

Regardless of the yielding potential of a newly developed cultivar, success of a new cultivar will also be determined by the end-use quality of the saleable product. Demand for sale of year-round fruits and vegetables has resulted in food products being shipped greater and greater distances to arrive fresh almost on a daily basis. In addition, greater emphasis is now, and will continue, to be placed on storage of perishable agricultural products to make them available at times of shortage of local supplies or, as noted above, simply to make them available on a year-round basis.

There are two main types of end-use quality:

- **Organoleptic** – consumer acceptance or preference for taste, size, texture and colour. Although many people differ in their preference, there is often general agreement within taste panels as to preference towards certain levels of expression of these attributes, thus 'liking' some genotypes over others (even disregarding 'off-tastes'). Similarly, the visual appearance of a product (particularly with fruit, vegetables and pulses) can have a large influence on which ones are preferred over others (Figure 3.2). An increasingly important factor for today's agricultural products is their ability to be stored for long periods without loss of quality. Hence many vegetable or fruit crops have a certain '*harvest window*' where the majority of the crop is harvested. The end-user, however, demands that the product retains its quality characteristics over whatever storage period is necessary before it reaches the consumer.
- **Chemical** – where quality is determined by chemical analyses of the harvestable product. This is perhaps easily understandable in, say, oil crops, where the quality of the oil can be determined

Figure 3.2 Visual appearance of saleable products is an important characteristic for breeders of fruits and vegetables.

with great accuracy by determination of the oil fatty acid profile, or in the pharmaceutical (or *farm-aceutical*) industry where the quality of drugs is chemically determined. However, this is also true for fibre plants like cotton, as well as fodder crops where protein content and digestibility are determined by analytical methods. Increasingly, the 'functionality' (in other words, what positive effects it can have on health) of any human food is also of interest, and consumers are turning towards taking this into account in choice of purchases.

Several crop species have been utilized for a range of different uses according to variation in their physical or chemical characteristics. Take, for example, a potato crop. The end-uses of potatoes are either as raw tubers to be cooked or through industrial processors, via retail purchasers. The needs and requirements of a potato will be different depending upon the use that the product will be put to. For example, potatoes can be boiled, mashed, baked, chipped, canned, dried or fried. Each cooking method (or use) will demand certain quality characteristics. Boiled potatoes need to remain relatively firm and not disintegrate on boiling. This trait is related to the 'solids' content of the

tubers, the lower proportions of solids being associated with less disintegration. Conversely, potato chip (crisp) processors do not wish to purchase potatoes with low solids as these have a higher water content, which has to be turned into steam (and hence waste) in the frying process. 'Chippers' also require potatoes with low reducing sugar content, which ensures that the chips (crisps) produced will have a pale golden colour.

Many crop species therefore have several end-uses, and specific quality characteristics will be required in cultivars bred for these uses. Cultivars of bread wheat are required to have hard seed and high seed protein, while those of biscuit wheat should have soft seed and low protein content. Canola (edible rapeseed oil) needs a fatty acid composition low in erucic acid (22:1 fatty acid) and high in oleic acid (18:1), while industrial rapeseed cultivars need to have oil that is high in erucic acid content. The determination of most quality traits is genetic, although the growing conditions of soil type, climate, irrigation management and nitrogen application can all have large influences on the level at which these characters are expressed and hence on the final crop quality. Similarly, mechanical damage (particularly in vegetables and fruits) and crop disease can both greatly reduce the overall quality of the product irrespective of what the end use will be.

Determining *'desirable quality'* characteristics can be difficult and requires close integration of the breeding team with end-users and processors. In some countries (e.g. the US) government authorities have laid down rules for quality standards. In these cases it is often easier to set standards for the acceptable level of quality required from breeding lines. Caution, however, needs to be exercized since it is unlikely that these standards will remain constant over time; indeed, they may change dramatically even before the new cultivar is even released.

3.4.1 Testing for end-use quality

If new cultivars are released that have special quality characters, there may be justification and economic merit in introducing this as a *'specialty'* product, even if the overall yielding ability is not high. This would be justified if economic returns were sufficiently enhanced such as to overcome the deficiencies in total yield. It should also be noted that competitors and other breeders will, of course, be quick to notice the market opportunity that has been opened and will focus on rapidly superseding such introductions, perhaps overcoming any of the obvious defects present in the original cultivar (e.g. pink grapefruit).

It is usually difficult (and most often impossible) to simulate an exactly similar processing operation as carried out commercially on a very large number of breeding lines and at the speed required in a plant breeding programme. For example, in order to obtain the true quality potential of a new potato line with regards to French fry production (taking into account quality of end product, oil uptake, ease of processing, etc.), it would be necessary to produce several hundred tons of tubers and make French fries from them in a commercial processing plant. Similarly, in order to determine malting potential of barley for whisky (including all operations through to consumer acceptance) would require large quantities of grain and considerable time. Obviously both these would be impossible or too expensive in all but the very last stages of a breeding scheme. The basic features of any effective quality assessment in a plant breeding programme

is that they should be **quick**, **cheap** and **use very little material**. These three criteria are important because:

- A plant breeding programme will involve screening many thousands of breeding lines each year or growing cycle.
- A plant breeder is often working against time. Many quality traits are assessed post-harvest and it is often important to make selection decisions quickly, before large-scale quality is determined. Moreover, many breeding programmes use off-season nurseries (i.e. in the southern hemisphere during the northern hemisphere's winter) to shorten the time needed to release new cultivars, which further constrains the time available to run quality tests on breeding lines before advancement decisions are made.
- In most stages of a plant breeding scheme there are only limited amounts of material available for testing. In addition, this is often the reproductive parts of the plants (seeds or tubers) that are used for testing, and so a further complication is that many of the quality tests available are destructive of the very parts of the plant that are required to be grown to provide the next generation for selection.

It is clear, therefore, that it is important to determine at what stage in the breeding scheme it is best to begin various quality screens. Obviously, quality evaluation should be included as early in the selection process as possible to avoid discarding some breeding lines carrying high-quality characteristics. However, often this decision must be based on the cost of the test, volume of material needed, accuracy of the test, and the importance of the quality trait for the success of any new cultivar.

Taste panels (groups of experienced (or sometimes inexperienced) people who assess the food quality of new products) are often used. It is, however, impossible to compare more than a modest number of types or breeding lines with a taste panel. These tests must also include some standard control lines, for comparative purposes, which further reduces the number of new lines that can be tested.

Most other quality assessments are, at best, estimates of what will happen in the *'real world'*. They tend to be mini-reconstructions of parts of a larger scale commercial process or operation.

When carrying out these assessments, great care should be taken to ensure that the test follows as accurately as possible the actual process as it is carried out in industry. It is therefore essential that good links are set up with industry partners and that the breeding programme tries to integrate ideas from the processing industry into the breeding strategy as much as is feasible. This will also allow experiments to establish the levels of relationship (the correlations) between the 'lab tests' and the behaviour of the lines in commercial practice.

In other instances it is easier to record a related character than to record the trait itself. For example, in canola breeding it is desirable to have low glucosinolate content in seed meal. Glucosinolate breakdown products are highly toxic and can cause dietary problems when seed meal is fed to livestock. Determining glucosinolate content is an expensive, two-day process requiring rather sophisticated equipment such as gas chromatography. A much quicker and less expensive alternative is available. One of the breakdown products of glucosinolates is glucose. It is possible to obtain a good estimate of glucosinolate content simply by crushing a few seeds, adding water and estimating glucose concentration, using glucose-sensitive paper. Similarly, malt barley breeders evaluate and select breeding lines for seed nitrogen content and soluble carbohydrates, which are highly related to the malting quality traits of malt extract and oligosaccharides, but these latter two characteristics are difficult to assess with small quantities of seed. Finally, the quality objectives of forage/fodder breeding programmes are biological in character and would ideally be met by testing the growth of animals fed on the breeding lines, but these large-scale feed studies are virtually never carried out, for reasons of time and cost. It should always be remembered, however, that these quality determinations are, at best, predictions, and in many cases, only a crude estimate of the character that is actually to be selected for.

Plant breeders seek to predict quality, however complex, by relatively simple and cheap measurements or organoleptic tests. Often small-scale testing units based on the larger operation are used, but in many cases quality assessment is determined by the correlation or relationship between an easily measured character and the more difficult to assess trait. However, before a new cultivar is released into agriculture it is desirable that new genotypes be actually tested on a commercial scale process. So wheat should be milled in a commercial mill, barley should be malted and beer made, potatoes should be fried and sold in fast food stores, onions stored in commercial storage and fodder fed to livestock before the product is released.

It is only after several rounds of testing at the commercial level that a secondary factor can be accurately estimated – that is, the uniformity of quality. Uniformity of quality is as important as the actual quality character itself. A cultivar that produces excellent quality in one environment or year but unacceptable quality in others will have little merit in commercial production. Unfortunately, uniformity in quality (although one of the most important characters of a new cultivar) is difficult to assess within the restrictions of a feasible sized breeding scheme.

Overall, quality is what creates the demand for a product and what allows differentiation in the market, thus reducing the commoditization that often leads to low prices. It is the end-user who will mostly determine whether that crop will be grown in future years. It is a very naïve breeder who ignores the fact that consumer preference is continually changing and that the quality standards of today may be superseded by a new set of standards in the future. It is therefore imperative to organize a breeding scheme to be flexible and to try to cover as many potential aspects of yield, quality and other factors which may be important in the next two decades.

3.5 Increasing pest and disease resistance

A major limiting factor affecting both harvested yield and end-use quality of agricultural and horticultural crops is infection or infestation by plant pests and diseases. Breeding cultivars that are genetically resistant to pests and diseases is still a primary objective of plant breeding.

The development of resistant cultivars involves consideration of the genetic variability of the pest or disease as well as the variability in resistance (or sometimes tolerance) that exists within the crop species (or related species from which resistance can often be obtained). The durability of resistance of developed cultivars can be affected by the emergence of new races of the disease/pest that are able to overcome the resistance mechanism in the host plants. It has been argued that the environmental changes brought about by global warming will also affect the dynamics of plant interactions with pests and diseases, perhaps rendering crops more or less susceptible to these biotic stresses. Thus the longevity of disease resistance that can be achieved in a new cultivar is often as important as the extent, or degree, of resistance that the new cultivar actually exhibits. In many cases the source of alleles conferring disease resistance have been from genetic resources such as landraces, old cultivars or materials kept by small farmers, which highlights the importance that the conservation of genetic resources represents for plant breeders.

The major forms of disease and pests include: fungi (air- and soil-borne), bacteria, viruses, nematodes and insects. However, this is not an exhaustive list. Other damage can occur (e.g. bird damage and mammal foraging), although in many cases it is difficult to imagine how biological plant resistance can greatly reduce such damage (cashew trees being knocked over by hippopotami is difficult to breed against!).

It can be assumed that pests and diseases will cause damage, and almost all important diseases have been given attention by plant breeders. Crops differ greatly in the number of diseases that attack them and similarly in the exact damage that infection can cause. Small grain cereals are particularly susceptible to air-borne fungal epidemics; most Solanaceous crops are especially affected by viruses; while cotton is particularly affected by insect attack.

It is difficult to assign importance to any class of plant diseases. In some cases there are interactions between diverse species preying on plants. For instance, the damage created by an insect chewing on a growing corn kernel can pave the way for opportunistic fungi to colonize such damaged tissues, further increasing the potential economic losses incurred. In economic terms the soil- and air-borne fungi may be more important than all other diseases, so much so that many breeding textbooks consider breeding for disease resistance to actually be simply breeding for resistance to fungal disease. Because of the harsh tropical environments, namely both high temperatures and humidity, breeding tropical crops is, to a larger extent than breeding for temperate crops, fundamentally breeding to better withstand diseases and pests. This is, of course, an over-generalization, and there is no doubt that other disease types also have potentially significant impacts on breeding objectives and goals, depending on the crop being bred. Indeed, it is recognized that virus diseases and many soil infestations are problematic because there are few treatments (especially agrochemicals) that can be used to treat crops once plants become infected.

Any breeder trying to develop new cultivars with specific disease resistance must have knowledge of the particular disease or pest and its effect on the crop. One of the obvious and most important effects of almost all crop pests and diseases is reduction in yield. This is caused in four main ways:

- Destruction of leaf tissue and hence reducing plant photosynthetic capacity or efficiency (e.g. many rusts, mildew and blight).
- Stunting plants by metabolic disturbance, nutrient drain or root damage (e.g. many viruses, aphids or nematodes).
- Reducing plant stands by killing whole plants and leaving gaps in the crop which cannot be compensated for by increased productivity of neighbours (e.g. vascular wilts, soil-borne fungi and boring insects).
- Killing parts of plants (e.g. boring or feeding insects).

Killing plant tissue or causing reduced plant vigour can reduce yields *per se*, although reduced yield can result from other factors like increased weed infestation through reduced crop competition.

Other impacts of plant pests and diseases relate to damage to the end-use product of the crop. These infestations are often initiated in the field but often become more apparent after harvest (e.g. cereal smuts, various rots and insect boring of fruits and tubers), but can also arise during postharvest life. Many pests/diseases also reduce the quality of harvested crops (e.g. insect damage in fruit or fungal blemishes in fruit or tubers). Some can even lead to the accumulation of potentially carcinogenic compounds like mycotoxins in grains.

The first task, which must often be carried out prior to screening for natural resistance to diseases, is to determine:

- Which diseases can affect the crop?
- What is the effect of these diseases, for example on yield or quality?

- From amongst the several diseases that can attack a given crop, which are the most relevant and damaging, and which are likely to become so?

Others who have been working with the crop in particular areas can often answer the first question. For example, if a relatively well established crop is to be bred (e.g. wheat in the Pacific Northwest of the US), there will already be a large body of data that have been collected regarding particular diseases and an indication of the frequency of disease attack.

The exact yield, quality or economic effect that different pests or diseases have on a crop can be used to partition the degree of effort that is exerted in breeding for resistance. Obviously, if a particular disease does not exist within the region there may be little point in devoting a large effort towards screening for natural resistance, unless this situation is predicted (e.g. due to climate change) to alter in coming years. Similarly, if a certain pathogen does not recognize your crop as a host, any attempt to increase resistance would be a waste of time and effort. In reality, the availability of a cheap and effective control measure will also decrease the priority a breeder assigns to tackling the resistance or tolerance to that particular disease or pest, unless there is an environmental priority involved.

The most common means to determine the effect of a disease is to grow a series of genotypes under conditions where disease is artificially managed. In most cases the simplest way that this is done is to grow plots where disease is chemically controlled next to others where disease is allowed to occur naturally (or indeed artificially infected to ensure high disease pressure). In the case of wheat and rust tolerance, differential genotypes carrying loci conferring resistance to specific rust races can be used to gain insight about the complex of rust races existing in a given growing region, and therefore help to predict which resistance loci are needed to withstand that rust population in that particular region.

For example, the effect on yield caused by infestation by cabbage seedpod weevil (Figure 3.3) on four Brassicacea species (*Brassica napus*, *B. juncea*, *B. rapa* and *Sinapis alba*) was examined in field trials in 1992 and 1993. Forty genotypes were grown in a pseudo-split-plot design where each entry was grown under three treatments: full weevil control with several insecticides, partial control with one insecticide, and no chemical control of the pest (Figure 3.4). The results differed between the four species investigated. However, without chemical control three of the four species showed yield reductions (some to a large degree). It is also obvious that *Sinapis alba* has more insect resistance (or tolerance) than the other species, and indeed offered breeders a source of resistance through intergeneric hybridization (see later). Additional

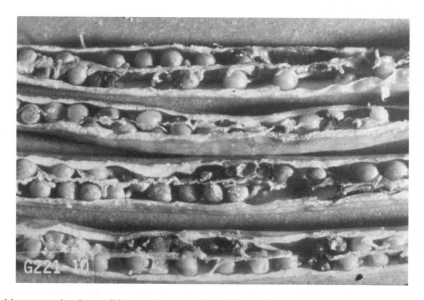

Figure 3.3 Cabbage seedpod weevil larvae damage of canola seeds.

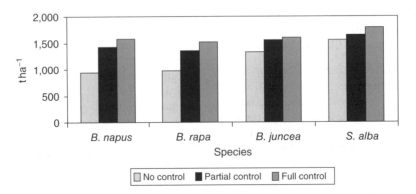

Figure 3.4   Seed yield (t ha^{-1}) from four *Brassicaceae* species as affected by late-season insect infestation when grown with full insect control, partial insect control and no insect control.

data regarding cost of chemical application, and so on, can then be used in coordination with these data to estimate the actual economic effect of this pest on *Brassica* crop production.

A major difficulty in carrying out effective disease impact trials is to remove variation in as many other factors that may interact with those under study as possible. For example, if the effect of a particular air-borne fungus is to be studied, then attempts must be made to ensure that other yield-reducing factors (e.g. other air-borne fungi or other pests and diseases) are kept under control in the trial.

3.6   Types of plant resistance

It has been claimed that for each gene in the host that controls resistance there is a gene in the pathogen that determines whether the pathogen will be **avirulent** (unable to overcome the resistance and hence unable to infect or injure the host) or **virulent** (able to infect or injure the host). This gene-for-gene hypothesis, first postulated by H.H. Flor over half a century ago, has been likened to a set of locks and keys. A simple example of this lock and key situation is shown in Table 3.1. In the first example, the host plant genotype has no resistance genes, so any pest genotype will be able to infect the host plant irrespective of the presence or absence of virulence alleles. In the second example, the host plant has one (or two) dominant alleles for resistance to pest A strain (i.e. *A_bb*), and as the pest genotype has no virulence genes, the plant is resistant to the disease. In the third example, the

Table 3.1   Possible phenotypic plant responses (i.e. resistant to disease or susceptible to disease) in various combinations of dominant alleles conferring single gene plant resistance (capital letters A or B represent resistance genes), and recessive alleles conferring susceptibility (a or b) and the 'matching' alleles in the pest where a′ or b′ confer virulence and A′ and B′ give avirulence.

Plant genotype	Pest genotype	Plant response
aabb	Any virulence gene	Susceptible
A_bb	No virulence gene	Resistant
A_bb	a′a′B′B′	Susceptible
A_B_	a′a′B′B′	Resistant
A_B_	A′a′b′b′	Resistant
A_B_	a′a′b′b′	Susceptible

host plant again has one (or two) dominant resistance alleles against the A strain of pest, but now the pest genotype is homozygous recessive (i.e. two copies) of the virulence gene (*a′a′*) and therefore unlocks (or can overcome) the resistance gene (*A*) in the host plant, and the plant is susceptible to the disease. In the last three examples the host plant has one or more copies of the dominant resistance alleles to both pest A and B strains, (i.e. *A_B_*). When the pest genotype is homozygous for the recessive virulent alleles *a′a′* but has no *b′* virulence alleles (i.e. *a′a′B′B′*) the plant is resistant to the disease, as the pest opened (or overcame) the *A* resistance gene, but could not open the *B* resistance gene. Similarly, when the pest genotype is homozygous for the recessive virulent alleles *b′b′* but has no *a′* virulence alleles (i.e. *A′A′b′b′*) the plant is also resistant to the disease, in this case because the pest opened (or overcame) the *B* resistance gene, but could not open the *A* resistance gene. In the last example, the pest genotype is homozygous recessive for both the $a'a'$ and $b'b'$ virulence alleles (i.e. $a'a'b'b'$) and can open (or overcome) both the *A* and *B* resistance genes in the host, and hence the host plant is susceptible to infection by the disease.

However, the situation is far from being this simple. Resistance to pests or diseases can be the result of either qualitative (single gene) or quantitative (multiple gene) determination. Resistance that is controlled by a single gene will result in distinct classes of resistance (usually resistant or susceptible) and are referred to as **specific** or **vertical resistance**. Resistance that is controlled by many genes will show a continually variable degree of resistance and is referred to as **non-specific, field, general** or **horizontal resistance**. Throughout this text the terms used will be vertical or horizontal resistance.

Vertical resistance is associated with the ability of single genes to control specific races of a disease or pest. The individual alleles of a major gene can be readily identified and transferred from one genotype to another. In many cases the source of the single gene resistance is derived from a wild or related species, and backcrossing is the most common method to introduce the allele into an elite commercial background. Segregation of single genes can be predicted with a good degree of reliability, and the selection of resistant genotypes can be relatively simply achieved by infection tests with specific pathogen races.

The primary disadvantage of vertical disease resistance is that new races of the pathogen are quite likely to arise that will be able to completely overcome the resistance. These new races may, in fact, have existed at a low level within the population of the pathogen before the resistance was even incorporated into the new cultivar or that cultivar was grown in agriculture. Thus, of course, the 'resistance' can be overcome relatively quickly. In addition, introduction of vertical resistance will increase the selective advantage of any mutant that arises in the pathogen population which can overcome the resistance. And as only a single mutation is required, the pathogen population may be many millions in size, so such a mutant will arise! New races of pathogen have overcome vertical resistance to air-borne diseases particularly quickly. Other cases, for example the single gene (H_1) in potato which gives vertical resistance to potato cyst nematode (*Globodera rostochiensis*), have proved very durable, probably as a result of the much lower degree of mobility of the earth-dwelling nematode pest.

One technique used by plant breeders is to *pyramid* single gene resistance where there are a number of qualitative genes available. This technique was attempted in potato for late blight (*Phytophtora infestans*) using a series of single resistance genes (R genes) derived from the wild potato species *Solanum demissum*. To date, nine R genes have been identified, and up to six of these combined into a single potato clone. However, the late blight pathogen was able to overcome the pyramiding of R genes quickly and the technique was not successful. Pyramiding single gene resistance to diseases and pests, with the use of molecular markers, which avoids the need to have suitable virulent pathotypes to screen for multiple resistance genes, has recently kindled interest.

Horizontal resistance is determined by many alleles acting collectively, with each allele only having a small contribution to overall resistance. Because of the multiplicity of genes involved, horizontal resistance tends to be far more durable than vertical resistance. The advantage of horizontal resistance is in its ability to control a wide spectrum of races, and new races of the pathogen have difficulty overcoming the alleles at all loci controlling the resistance. The main disadvantage of horizontal resistance is that it is often difficult to transfer from parent to offspring. The probability of transferring all the resistant alleles from a resistant parent to a susceptible one is generally very low. Breeding for horizontal resistance therefore tends to be a cyclic operation with the aim of increasing the frequency of desirable resistant genes.

3.7 Mechanisms for disease resistance

Two main disease resistance mechanisms exist. These are:

- resistance due to lack or reduction of infection;
- resistance due to lack or reduction of subsequent growth or spread after initial infection.

By far the most numerous examples of inhibition of infection in crop plants are related to hypersensitivity. Infection of the host plant causes a rapid localized reaction at the infection site. Host plant cells surrounding the infection point die, and hence the pathogen is effectively isolated from the live plant tissue and cannot spread further into the host plant. Hypersensitivity is usually associated with a necrotic flecking at the infection site, and the host plant is totally immune to the pathogen as a result. Plant resistance through hypersensitivity is controlled by single genes and hence can usually be easily incorporated into breeding lines. In cases of high levels of disease infection, cell death in the host plant can cause a significant reduction in the plant's photosynthetic area and, in extreme cases, plant death through lethal necrosis.

Other examples of disease infection inhibitors are less numerous and are usually associated with physical or morphological barriers. For example, the resistance to cabbage seedpod weevil found in yellow mustard (*Sinapis alba*) has been attributed to the very hairy surface of its pods and other parts, a feature not appreciated by the weevils, which are deterred from laying their eggs. Similarly, leaf wax mutants of cabbage can deter insect feeding, and tightly wrapped corn husks can prevent insect pests from feeding on the developing seed.

Growth inhibition after infection is caused by the host plant restricting the development of a pathogen after initial infection. The pathogen is not able to reproduce in a resistant host plant as rapidly after infecting compared with a susceptible host. For example, Russian wheat aphids feeding on susceptible wheat plants inject toxins into leaves, causing the leaves to fold. Adult Russian wheat aphids lay eggs in the folded leaves and the developing larvae gain protection from within the folds. Resistance genes have been identified that do not deter the adult Russian wheat aphids from feeding or injecting toxins into the leaves. However, the toxins do not cause the leaves of resistant wheat to fold, and hence there is greater mortality of developing Russian wheat aphid larvae, and reduced populations of the pest. Resistance to lack of spread of disease after infection can result from **antibiosis**, where the resistance reduces survival, growth, development or reproduction of the pathogens or insects feeding on the plant, or by **antixenosis**, where the resistant host plant has reduced preference or acceptance to the pest, usually insects. Resistance due to growth inhibition can be controlled by either qualitative or quantitative genes.

A complication in screening and determining plant disease resistance is related to **tolerance**. Tolerance is the ability of a genotype to be infected by a disease and yet not have a marked reduction in productivity as a result. Therefore, despite plants showing disease symptoms (e.g. fungal lesions or insect damage), the plant compensates for the infection or damage. Tolerance to disease has been related to plant vigour, which may be associated with other physiological stresses. For example, genotypic tolerance in potato to infection by potato cyst nematode, late blight, early blight (Alternaria) and wilt (Verticillium) are all highly correlated to drought stress or salinity tolerance.

One other factor needs to be considered in relation to disease resistance, and that is **escape**. This is where a genotype (although not having any resistance genes) is not affected by a disease because the infective agent of the disease is not present during the growth period of the genotype. Disease escape is most often related to maturity or other growth parameters of the plant and phenology of the pest. For example, potato cultivars that initiate tubers and mature early are unlikely to be affected by potato late blight (*Phytophthora infestans*) as the plants are mature before the disease normally reaches epidemic levels. The scope to use escape as a disease tolerance approach is rather limited because the life cycle of a crop is often constrained by the environment it is produced under, and also because of the strong association existing between early maturity and diminished yields, as the period of time to build up dry matter into the harvestable organ is reduced.

3.8   Testing plant resistance

In order to select for plant resistance to disease and pests, it is necessary to have a well-established testing scheme, one that truly mimics the disease or pest effects as they exist in an agricultural crop. If a plant's resistance to a pest or pathogen cannot be reliably measured, it will not be possible to screen germplasm for differential resistance levels; nor will it be possible to select resistant lines from amongst segregating populations.

Methods used for assessing disease and pest resistance in plant breeding are extremely varied but may be conveniently grouped into three categories.

- Plants can be artificially infected in a greenhouse or laboratory. This can be especially effective in screening for vertical resistance (single gene resistance) where an all-or-nothing reaction is expected. For example, a simple resistance or susceptible rating to pathogens of the air-borne fungi is commonly recorded on seedlings in the greenhouse or on detached leaves in the laboratory. Resistance screening to viruses is often carried out under greenhouse conditions where plants are 'hand-infected' with virus and the plant response noted. Similarly, resistance to potato cyst nematode (both qualitative and quantitative resistance) can be effectively screened in a greenhouse by growing test plants in soil with high cyst counts. These methods in general demand rather precise pathological control but are usually quick, accurate and give clear results. If greenhouse or laboratory testing is to be used, it is essential that the test results relate to actual resistance under field conditions. For example, seedling resistance to powdery mildew in barley is not always correlated with adult plant resistance observed under field conditions. The simplest method to authenticate small-scale testing is to carry out direct comparisons with field trials as an initial part of establishing the testing regime.
- *In vitro* testing and screening has been used, where the diseases that infect plant tissue are a result of toxins produced by the pathogens. These toxins can be extracted and *in vitro* plantlets, or plant tissue cultures, can be subjected to the toxin. As with testing in a greenhouse (above) it must be clearly shown that the *in vitro* plant tissue reaction to disease toxin is indeed related to whole plant resistance. It is never sufficient to observe phenotypic variation of *in vitro* material and to assume that this variation is useful *in vivo*. *In vitro* evaluation can be particularly effective if there are other reasons to propagate genotypes *in vitro*, for example, if embryo rescue is necessary (as in some interspecific crosses), if whole plants are being regenerated from single cells (protoplast fusion or transformation), or if initial

plant propagation is carried out *in vitro* to avoid disease (e.g. with potato).
- Field testing, using natural infection, artificial infection, or commonly a combination of natural and artificial infection, is widely carried out. The object is usually to ensure that the initial infection rates are not limiting so that those estimates of genotypic resistance levels are more reliable (rather than simply a result of being escapes due to lack of infection). In the field there can be no precise control of the pathogen in terms of race composition or level of infection. Artificial infection may be achieved by inoculating seed or planting infecting material before planting the test plants, by inter-planting 'spreader rows' of plants that are highly susceptible to the disease, by spraying or otherwise infecting the crop with the disease, or by artificial infection of the soil with nematode or fungi. Uneven inoculation or infection is almost always a problem, especially with some of the soil pathogens. It is essential, therefore, that these trials use adequate experimental designs and in many cases suitably high levels of replication.

In conclusion, breeding for disease or pest resistance is no different (in many ways) to breeding and selecting for other traits. The steps that must be taken in a breeding scheme include:

- Develop a means to evaluate germplasm and breeding lines. In many cases success will be directly related to how effective the screening methods are at detecting differences in resistance levels. If it is not possible to differentiate consistently and accurately the level of disease or pest present, it will not be possible to identify sources of plant resistance or to screen for resistance within segregating progeny.
- Evaluate germplasm and breeding lines to identify sources of plant resistance. In the first instance, evaluation of the most adapted lines should be carried out. If no resistance is identified, then more primitive or wild genotypes need to be screened.
- Examine, if possible, the mechanism of resistance to determine the mode of resistance being exploited (e.g. avoiding infection, limiting spread, non-preference and antibiosis).
- Determine the mode of inheritance (i.e. qualitative or quantitative) of any resistance detected.

The mode of inheritance can have a large influence on the method of introducing the resistance into the commercially acceptable gene pool (e.g. a single gene would perhaps be best handled by a back-crossing method).

- Introgress source of resistance into a new cultivar. Despite the importance of disease and pest resistance in plant breeding, it should always be remembered that a new cultivar will be unlikely to succeed simply because of disease resistance – new cultivars will also need to have acceptable levels of expression for the other important traits.

3.9   Conclusions

A house builder would not build a house without an architect first providing a plan, nor would an automobile producer build an automobile without having some form of test model. So also a plant breeder will not produce a successful cultivar development programme without suitable breeding objectives. This will involve:

- Examining the whole production and use system from farmer, through processor, to final product user. Determine the demands or preferences of each group in the production chain and take these into account in the selection scheme. Remember that there may be conflicts in the preferred requirements from different parts of this chain.
- Examine what is currently known about the crop. How is this end product produced at present? What types of cultivars presently predominate? What are the advantageous characters of these cultivars, and what are their defects? Is there any alternative type of cultivar that might offer more advantages to farmers and/or end-users?
- Examine how the operation of correcting present deficiencies can most effectively be addressed while always remembering that it is necessary to maintain at least the same level of acceptability for most other traits.

In order to set appropriate breeding objectives, the breeder needs to consider incorporating: yield potential, disease and pest resistance, end-use quality, and even the influence of potential political factors/decisions. Having taken these into account, it will be possible to design a successful plant breeding programme.

Think questions

(1) Explain the difference between vertical and horizontal disease resistance and outline the advantages or disadvantages of each form of resistance as it relates to cultivar development.

(2) Outline the major features involved in selecting for end-use quality, and indicate any particular problems that breeding for improved quality might cause.

(3) When breeding new cultivars it is often necessary to try to predict events that may occur in the future. Briefly outline four factors that may influence cultivar breeding and hence need to be considered in setting the breeding objectives of a cultivar development programme.

(4) In plant breeding, two main disease resistance mechanisms exist, ***inhibition of infection*** and ***inhibition of growth after infection***. Explain each of these mechanisms.

(5) Choose any agricultural/horticultural crop (e.g. wheat, barley, potato, apple, hops, etc.) and outline a set of breeding objectives to be used to develop new cultivars. Indicate potential markets for the new cultivars.

(6) Explain, using examples as necessary, the meaning of the terms ***plant tolerance*** and ***plant escape*** in relation to pest and disease resistance and plant breeding.

(7) Four loci affecting powdery mildew disease have been identified, at which single, dominant, disease-resistance alleles (coded, A, B, C and D) each offer complete immunity to a specific race of powdery mildew (with alleles a, b, c and d each showing susceptibility). Virulence genes have also been identified in the pest (A', a', B', b', C', c', and D', d'). Given the information below, indicate whether each genotype would be resistant or susceptible to the disease.

Plant genotype	Mildew genotype	Plant response
AAbbCCdd	A'a'B'B'c'C'd'd'	
AaBbCcDd	A'a'B'b'C'c'D'd'	
aaBBccDD	A'A'b'b'C'c'd'd'	
Aabbccdd	A'A'B'B'C'C'D'D'	

(8) You have been offered a job as Senior Plant Breeder with the '*Dryeye Onion Company*' in southern Idaho, United States. Your first task in this new position is to set breeding objectives for onion cultivar development over the next 12 years. Outline the main points to be considered when setting your breeding objectives; indicate what questions you would like answered to enhance the breeding objectives you will set. What might be the main sources of information that you would use to arrive at your response?

4
Breeding Schemes

4.1 Introduction

All successful breeding programmes have been designed around a breeding scheme. The breeding scheme determines the passage of breeding lines through the selection process, and through to the increase of planting material for cultivar release. The process of selection will be carried out over a number of years, and under differing environmental conditions. The early selection stages of breeding programmes will involve screening many thousands of different genotypes. The early screening is therefore relatively crude, and in many instances this involves only visual selection. After each round of selection, the '*better*', more adapted, or more disease-resistant genotypes will be retained for further evaluation while the least adapted lines will be discarded. This process will be repeated over a number of years, and at each stage the number of individual genotypes or populations is reduced and evaluation is conducted with greater precision in estimating the worth of each entry.

The breeding scheme used will be highly dependent upon the crop species and the type of cultivar (pure-line/self-pollinated, open-pollinated, hybrid, clonally propagated, synthetic, etc.) that is being developed. The general philosophy for developing a clonal cultivar like potato is therefore different from a pure-line cereal cultivar, say barley. In the former, breeding selections are genetically fixed through vegetative propagation, but there will be a low rate of multiplication of planting material. In the latter, there will be more rapid increase of planting material, although the segregating nature of the early-generation breeding lines will complicate the selection process.

The most effective breeding schemes will utilize the positive attributes of a crop species while minimizing difficulties that might arise through the selection process. In the following section the general breeding schemes for pure-line, open-pollinated, hybrid and clonally propagated cultivars will be explained, along with mention of the schemes used for developing multilines and synthetics.

4.2 Development of pure-line cultivars

Crops that are generally produced as pure-line cultivars include barley, chickpea, flax, lentil, millet, peas, soybean, tobacco, tomato and wheat.

One and a half centuries ago, most inbred crop species were grown in agriculture as 'landraces'. Landraces were locally grown populations which were, in fact, a collection of many different genotypes grown in mixture and which were, of course, both genetically and phenotypically variable. Pure-line (synonymous with inbred) cultivars were developed first from these landraces by farmers who selected specific (presumably more productive or least infected with disease or pests) plants from the mixed populations and maintained these in isolation, thus encouraging selfed

Plant Breeding, Second Edition. Jack Brown, Peter D.S. Caligari and Hugo A. Campos.
© 2014 John Wiley & Sons, Ltd. Published 2014 by John Wiley & Sons, Ltd.
Companion Website: www.wiley.com/go/brown/plantbreeding

progenies, and eventually developed homozygous, or near-homozygous, lines. It is reasonable to assume that these homozygous lines were indeed more productive than the original landraces because by the end of the 19th century, landraces had almost completely disappeared in countries with advanced agricultural systems.

These early 'pure-line breeders' used the naturally existing genetic variation within the landraces they were propagating and the natural tendency of some species to self-pollinate (e.g. wheat) at a high frequency. However, this strategy has a limited potential in terms of generating new variation, and so modern plant breeders have to continuously generate genetic variation and hence the three-phase breeding schemes were established to **create genetic variation**, **identify desirable recombinant lines within progenies** and **stabilize and increase the desired genotype**. It is interesting to note, however, that recently a number of plant breeders have returned to old landraces of wheat and barley to examine their wealth of genetic diversity, as well as to testing combinations of lines in '*modern*' landrace combinations (i.e. multilines). Unfortunately most of the landraces that existed even 100 years ago are no longer available and potentially valuable germplasm and adapted combinations have been lost.

By far the most commonly used method of generating genetic variation within inbreeding species is via sexual reproduction using artificial hybridization. There are, of course, other ways to produce genetic variation. For example, variation can be produced by induced mutation, somatic variation, somatic hybridization and recombinant DNA techniques (all discussed in later chapters).

After sexual crossing between genotypes and then selfing them to generate suitable genetic variation, plant breeders will then traditionally screen the segregating population for desirable '*segregants*' while continuing to self-pollinate successive generations, to produce homozygous lines. Thus, accomplishing the last two steps of the breeding scheme (selection and stabilization) can be achieved more or less simultaneously.

4.2.1 Homozygosity

One of the difficulties in selecting desirable recombinant lines in pure-line breeding is related to segregating populations and the masking of specific character expression as a result of the dominant/recessive nature of the segregating alleles in the heterozygotes. Another consideration is the relationship between genetic homozygosity and '*commercial inbred lines*'. The definition of complete homozygosity is that all the alleles at all loci are identical by descent, that is, there is no heterozygosity at **any** locus. However, for practical commercial exploitation, the level of homozygosity does not need to be complete. Clearly the lines must basically breed '*true to type*' but this is by no means absolute. The degree of homozygosity is determined by the level of inbreeding and directly related, for example, to the number of selfing generations that have been performed. Consider the simple case of just one locus with two possible alleles *A-a*:

Parents		AA × aa			Frequency of heterozygotes
F_1			Aa		100%
F_2 Frequency	AA ¼		Aa ½	aa ¼	50%
F_3 Frequency	AA ¼	AA ⅛	Aa ¼	aa ⅛	aa ¼	25%

Consider now a more complex situation where six loci are involved, as set out below. Of these six loci, two, loci *A* and *F*, have the same allele in both parents (and are both homozygous) and so the F_1 is also homozygous at these two loci. At the other four loci the parents are homozygous but have different alleles, and so the F_1 is heterozygous at these loci and subsequently these segregate in the F_2.

Parents	*AAbbCCDDeeff × AABBccddEEff*
F_1	*AABbCcDdEeff*
F_2	*AABBCcDdeeff, AABbCcDDEeff,* *AAbbCCEeDdff, AABBccDdEeff,* etc.

This can be generalized in mathematical terms as follows. Consider an F_1 that is heterozygous at *n* loci; heterozygosity (*h*) at any single loci after *g* generations ($g = 0$ at F_1) of selfing will be:

$$h = (1/2)^g$$

Therefore the probability (p) of homozygosity at all n loci will be:

$$p = (1 - h)^n$$

Hence after g generations:

$$p = [1 - (1/2)^g]^n$$

This can also be written as:

$$p = [(2^g - 1)/2^g]^n$$

The level of homogeneity (i.e. uniformity in appearance/phenotype as opposed to genetic homozygosity) required in an inbred cultivar will depend to a varying extent on the personal choice of the breeder, seed regulatory agencies, farmers and end-users. Almost all pure-line breeding schemes involve selection of individual plants at one or more stages in the breeding scheme. The stages of single plant selection will have a large impact regarding the degree of heterogeneity in the end cultivar. If single plant selection is carried out at an early generation, say F_2, there may be greater heterogeneity within the resulting cultivar compared with a situation where single plant selection was delayed until a later generation, say F_8, when individual plants would be more homozygous. Breeders must ensure that a level of uniformity and stability exists throughout multiplication and into commercialization. Farmers and processors (such as millers) will have preferences for cultivars that are homozygous, and hence homogeneous, for particular characters. These characters may be related to uniform maturity, plant height or other traits related to ease of harvest. Many believe that farmers are not concerned with uniformity of characters that do not interfere with end-use performance (e.g. flower colour segregation). However, farmers take a natural pride in their farms and, therefore, like to grow '*nice-looking*' crops, and these are ones which are uniform for almost all visible characters. In any case they certainly like uniformity in terms of characters such as harvest date! For most end-users there will be an obvious preference for cultivars with high uniformity of desirable quality characters. For example, there may be a premium for uniform germination in malting barley or more uniform characters relating to breadmaking in wheat.

In contrast, some breeders of pure-line cultivars like to maintain a relatively high degree of heterogeneity in their developed cultivars. They believe that this heterogeneity can help to '*buffer*' the cultivar against changes in environment and hence make the cultivar more *stable* over different environments. Often statutory authorities determine the degree of variability that is allowed in a cultivar. For example, all pure-line cultivars released in the European Union countries, along with Canada and Australia, must comply to set standards for **distinctness, uniformity** and **stability** (DUS) in Statutory National Variety Trials. In these cases it is common to have almost total homogeneity and homozygosity in released inbred cultivars.

For most breeders, time is at a premium. Therefore, some methods are commonly used to reduce the time taken to achieve homozygosity, and these include **single seed descent** and the use of **off-season sites** (this excludes the production of homozygous lines through doubled haploids, which is relevant here but will be discussed separately in a later chapter).

Single seed descent

Single seed descent involves repeatedly growing a number of individuals initiating from a segregating population, usually under high-density, low-fertility conditions, to accelerate seed-to-seed time. At maturity, a single seed from the natural self of every plant is taken and replanted. This operation is repeated a number of times to obtain homozygous plants. Single seed descent is most suited for rapid generation increase in a greenhouse where a number of growth cycles may be possible each year. Single seed descent in canola, wheat and barley can be further accelerated by growing plants under stress conditions of high density, high light, restricted root growth, and low nutrient levels, which result in stunted plants with only one or two seeds each, but in a shortened growing period compared with growth under normal conditions (up to three or four generations in a year are possible in barley or spring canola).

It is very important, however, when using single seed descent, that unintentional selection is not being carried out for adverse characters. For example, in a single seed descent scheme in winter wheat (where plants will require a vernalization period prior to initiating a reproductive phase), vernalization requirements may be overcome artificially in a cold room. If this is done, then care should be taken so that all seedlings do indeed receive sufficient cold treatment to overcome the vernalization requirement – otherwise the system will select the plant types with lower vernalization requirements. In addition, some genetic characteristics are not fully expressed when plants are grown under high competition stress conditions used for single seed descent. For example, the *erectoides* dwarfing gene (*ert*) in barley plants is not expressed under single seed descent in the glasshouse, and therefore genotypes cannot be selected for dwarfism under these conditions. In any case it is strongly advised that **no** selection be practised during this phase.

Off-season sites

Off-season growing sites can also reduce the time for achieving a desired level of homozygosity. This is possible by having more than one growing season per year. Dual locations at similar latitudes in the northern and southern hemispheres are frequently used to increase either seed quantity or reduce heterozygosity in many breeding schemes. The use of off-season sites is often restricted to annual spring crops, and there are only a few good examples where they have helped accelerate homozygosity in winter annuals, and virtually none in breeding biennials.

If off-season sites are to be incorporated into the breeding scheme, care must be taken to ensure that '*selectional adaptability*' of the off-season site does not have adverse effects on the segregating plant populations. For example, the spring barley breeding scheme at the Scottish Plant Breeding Station used to increase F_4 breeding selections to F_5 by growing these lines over winter in New Zealand. Although New Zealand has a climate that is very similar to that found in Scotland, there is a completely different spectrum of races of powdery mildew. As a result, mildew-resistant selections made in New Zealand were of no relevance when grown in Scotland, and so meant that all New Zealand trials needed careful spraying to avoid powdery mildew being confounded with other performance characters. Breeding companies developing varieties for the US market have developed large networks of off-season sites around the globe for crops such as maize and soybean.

The use of off-season sites benefits farmers as they significantly reduce the time required to take new varieties to the market, and so to respond to the ever-changing needs of humankind.

4.2.2 Breeding schemes for pure-line cultivars

There are probably as many different breeding schemes used by breeders of pure-line cultivars as there are breeders of inbreeding species. There are, however, three basic schemes: **bulk methods**, **pedigree methods** and **bulk/pedigree methods**. It should be noted that all the breeding schemes described involve more than a single cross at the *crossing stage*. A number of these crosses will be two-parent crosses (female parent × male parent, say $P_1 \times P_2$, although many breeders use three- and four-way parent cross combinations ($[P_1 \times P_2] \times P_3$, and $[P_1 \times P_2] \times [P_3 \times P_4]$, respectively).

Bulk method

The outline of a bulk scheme is illustrated in Figure 4.1. In this scheme, genetic variation is created by artificial hybridization between chosen parents.

The F_1 and several subsequent generations, in the illustration up to and including the F_5 generations,

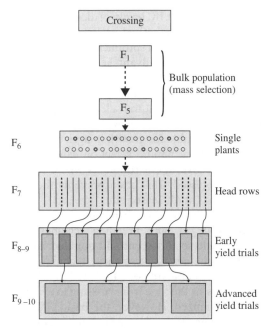

Figure 4.1 Outline of a bulk breeding scheme used for breeding pure-line crop species.

are grown as bulk populations. These bulks are left to set seed naturally which, with these species, will mean predominantly self-pollination. No conscious selection is imposed in these generation, and it is assumed that the genotypes most suited to the environment in which the bulk populations are grown will leave more offspring and hence predominate in future generations. Similarly, these bulk populations are usually grown under the stress and disease pressures common to the cultivated crop, and it is assumed that the frequency of adapted genotypes in the population increases. It is therefore very important that the bulks are grown in a suitable and representative environment. After a number of rounds of bulk increase, individual plants showing desirable characteristics are selected, often at the F_6 stage. From each selected plant, a head of seed is taken and grown as a row (a head-row). The produce from the best head-rows are bulk harvested, for initial yield trials. More advanced yield trials are grown from bulk harvest of desirable individuals.

The major advantage of the bulk method is that conscious selection is not attempted until plants have been selfed for a number of generations and hence the plants are nearly homozygous. This avoids the difficulty of selection among segregating populations where phenotypic expression will be greatly affected by levels of dominance in the heterozygotes. This method is also one of the least expensive methods of producing populations of inbred lines. The disadvantage of this scheme is the relatively long time from initial crossing until yield trials are grown. In addition, it has often been found that the natural selection, which occurs through bulk population growth, is not always that which is favourable for growth in agricultural practice. In addition, natural selection can, of course, only be effective in environments where the character is expressed. This often prevents the use of bulk methods at off-season sites.

Other methods have been used to produce homozygosity in bulk breeding schemes. These include single seed descent and doubled haploidy. Breeding schemes that use these techniques have increased the popularity of bulk breeding scheme in recent years, as the time from crossing to evaluation can be minimized. However, the basic philosophy is similar, being to produce near-homozygous lines, and thereafter select amongst these. Where rapid acceleration to homozygosity techniques are used,

it is essential to ensure that no negative selection occurs. For example, research has shown that creating homozygous breeding lines of canola (*B. napus*) through pollen culture produces a higher than random frequency of plants with low erucic acid in the seed oil. If low erucic acid content is desired, this poses a selection advantage. If, however, an industrial oilseed cultivar were desired (one with high erucic acid content), then using embryogenesis would be detrimental.

Pedigree method

The outline of a pedigree breeding scheme is shown in Figure 4.2. In a pedigree breeding scheme, single plant selection is carried out at the F_2 through to F_6 generations. The scheme begins by hybridization between chosen homozygous parental lines, and segregating F_2 families are obtained by selfing the heterozygous F_1s. Single plants are selected from amongst the segregating F_2 families. The produce from these selected plants is grown in plant/head rows at the F_3 generation. A number of the 'most desirable' single plants (in Figure 4.2, four plants)

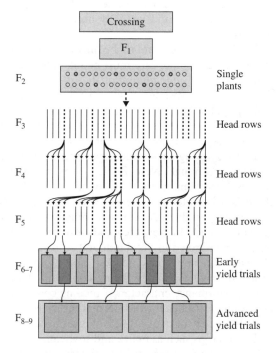

Figure 4.2 Outline of a pedigree breeding scheme used for breeding pure-line crop species.

are selected from the 'better' plant rows and these are grown in plant rows again at the F_4 stage. This process of single plant/head selection is repeated until plants are 'near' homozygous (i.e. F_6.) At this stage the most productive rows are bulk harvested and used as the seed source for initial yield trials at F_7.

In addition to being laborious (as a considerable amount of record-keeping is required) and relatively expensive, annual discarding may lead to the loss of valuable genotypes, particularly under the changing environmental conditions from year to year, making selection difficult. Other disadvantages of the pedigree method are that it requires more land and labour than other methods; experienced staff with a 'good breeders' eye' are necessary to make plant selections; selection is carried out on single plants where errors of observation may be great; while actual yield testing is not possible in the early generations.

If selection was effective on a single plant basis, pedigree breeding schemes would allow inferior genotypes to be discarded early in the breeding scheme, without the need for being tested in more extensive, and costly, yield trials. Unfortunately, pedigree breeding schemes offer little opportunity to select for quantitatively inherited characters, and even single gene traits can cause problems when selecting on a single plant basis in highly heterozygous populations.

Bulk/pedigree method

The outline of a bulk/pedigree breeding scheme is illustrated in Figure 4.3. This type of breeding scheme uses a combination of bulk population and single plant selection. An F_2 population is produced by selfing F_1 plants from controlled hybridizations. Individual plant selections from the segregating F_2 families are grown in plant progeny rows at the F_3 stage. Selected F_3 families are bulk harvested and preliminary yield trials are grown at the F_4 stage by planting the bulked F_3 seed. F_5 and F_6 bulk seed yield trials are grown, in each case by planting bulked seed from the previous year's trial. Selection of families is based upon performance in these trials. At the F_6 stage, single plant selections are once again made from the now near-homozygous lines. Progeny from these plant selections are grown then as plant rows at F_7; second-cycle initial

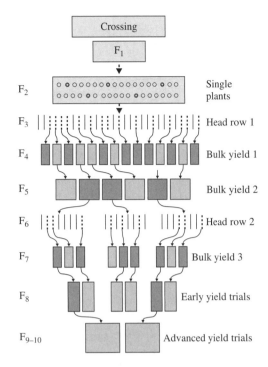

Figure 4.3 Outline of a bulk/pedigree breeding scheme used for breeding pure-line crop species.

yield trials at F_8; and more advanced yield trials at F_9.

The advantage of this combined breeding scheme is that inferior individuals, lines, families or populations are identified and discarded early in the breeding scheme. More than a single cultivar may be derived from a family or heterogeneous line identified as being superior by the earlier generation testing. Disadvantages will include, with fixed resources, the use of testing facilities for evaluation of individual lines in the early generations and so reducing the number of the later, more highly inbred lines that can be evaluated. Despite these disadvantages, bulk/pedigree schemes (or close derivatives of) are most commonly used to develop pure-line cultivars.

Modified pedigree method

Most breeding schemes have developed breeding schemes that are combinations of bulk and pedigree methods. For example, the breeding scheme used for developing winter barley cultivars at

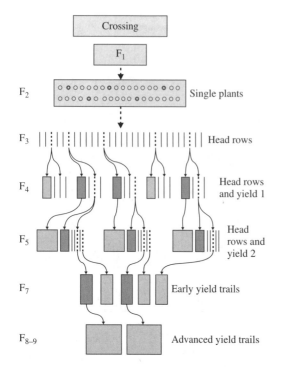

Figure 4.4 Outline of a modified pedigree breeding scheme used for breeding pure-line crop species.

the Scottish Plant Breeding Station was a ***modified pedigree breeding scheme*** (illustrated in Figure 4.4).

The modified pedigree breeding scheme enables yield trials to be grown simultaneously with pedigree selection. Single plants are selected from amongst segregating F_2 families. Seed from these selections are grown as plant progeny rows at F_3. One, or more, single plants are selected from each of the desirable F_3 plant rows, and the remainder of the row is bulk harvested. The single plant selections are grown as plant progeny rows at F_4. It is common in many crops, for example wheat and barley, to select single ears/heads rather than whole plants. In this case, the plant progeny rows are commonly referred to as 'head-rows' or 'ear-rows', as seed from a single ear/head is used to plant a single row. Simultaneously, the harvested F_3 bulk is planted in a preliminary yield trial. The seed from the bulk yield trial is used to plant a more extensive bulk yield evaluation trial at the F_5 stage. Based on the results from the F_4 bulk yield trial, the most productive populations are identified. Single plant selections are made from the corresponding

plant progeny rows and the remaining row is bulk harvested for a further yield trial the following year at the F_5 generation. This process is repeated at the F_5 to F_6 stage, by which time near homozygosity is achieved in the remaining lines.

The advantage of the modified pedigree breeding scheme is that it utilizes progeny bulk evaluation for yield and other quantitatively inherited characters, while single gene traits can be screened on a plant progeny row basis. In addition, this scheme allows for evaluation of quantitative characters while simultaneously inbreeding the selections.

4.2.3 Number of segregating families and selections

There have been numerous debates amongst plant breeders concerning the question of how many plants or families should be evaluated, and selected, at each stage in a breeding scheme. Unfortunately, there is no simple recipe to help new breeders, and the questions can only be addressed from an empirical standpoint. Plant breeding is a *numbers game* and the chance of success is often associated with screening many thousands of breeding lines. However, plant breeding programmes should only be as large as the specific breeding group can handle and as resource availability allows. Therefore, it is not productive to grow more lines at any stage in a breeding scheme than can be *effectively and accurately assessed*.

It is often easier to work backwards and ask how many lines can be handled at, say, the advanced yield trial stage in the breeding scheme, and then move backwards to the previous stage and predict how many lines are required at that stage to ensure that the required number are selected, and so on.

Similarly, the number of initial cross combinations that should be used differs markedly in different breeding programmes. Often a large number of crosses need to be screened, as the breeder cannot identify the most productive cross combinations. With experience of specific parents in cross combination and the benefit of 'cross prediction' techniques (see Chapter 7), it is possible to reduce the number of crosses screened on a large scale and hence allow breeders to put greater emphasis on cross combinations with the highest probability of producing desirable recombinants and hence cultivars.

4.2.4   Seed increases for cultivar release

At the other end of the breeding programme, once desired cultivars have been identified, it is necessary to produce a suitable quantity of seed that will be grown and increased for varietal release. This 'seed lot' is usually called **Breeders' seed**, as in most cases producing this seed is the responsibility of the breeder rather than a 'seedsman'. It is vital that breeders' seed lots are pure, free from variants, and that the cultivar that is to be released is 'true-to-type'. Breeders' seed is used in multiplication to produce '*foundation*' seed, which, in turn, is used to produce the various levels of '*registered*' or '*certified seed*', which is eventually sold to farmers.

Producing high quality breeders' seed is very similar to the breeding schemes (described above). In general there are two basic methods of producing breeders' seed, **mass bulk increase** and **progeny test increase.**

Mass bulk increase

In mass bulk increase schemes, a uniform sample of seed from the selected line is chosen and planted on only one occasion to result in the breeders' seed lot. The advantage of this simple method is that it is inexpensive and takes only a single season to obtain the required seed. The disadvantages are mainly related to the purity, homozygosity and homogeneity of the cultivar entering into commercialization.

Progeny test increase

The progeny test increase method is more expensive and takes longer to obtain the seed required. This method is very similar to the bulk/pedigree breeding scheme. A number of single plants are selected from the homozygous/near homozygous advanced breeding line. These are grown as plant progeny rows. Individual plant rows are discarded if they display off-types or are non-uniform. The remaining rows are harvested individually and the seed from each row is used to plant larger progeny plots the following season. These plots are again inspected and those that do not have the required homogeneity or show off-types are discarded. At harvest the progeny plots are bulk harvested as breeders'

seed. Breeders who wish to maintain a degree of heterozygosity in the released cultivar will include greater numbers of initial single plant selections in the scheme, or they may not be as restrictive in decisions to discard progeny rows or plots. The advantage of the progeny multiplication method is that it allows greater control by the breeder and results in greater homogeneity in the released cultivar.

4.3   Developing multiline cultivars

Multiline cultivars (multilines) are mixtures of a number of different genetic lines, often varying in their disease resistance genes. Multiline cultivars are almost exclusively composed of mixtures of lines from pure-line species. Multilines have been developed for a number of different crop species including barley, wheat, oats and peanuts. In turf grass, intraspecific and interspecific multiline cultivars are grown commercially.

Multilines have been suggested as one means to minimize yield or quality losses due to diseases or pests that have multiple races and where the race specificities can change from year to year. In this latter case there is a lower probability that all plants within the multiline (with a range of specific disease resistance genes or resistance mechanisms), as opposed to a pure-line cultivar, would be affected as severely as when the population carries only one resistance gene providing protection to just one pathogen race. It has also been suggested that the use of multilines would result in more durable mechanisms of disease resistance in crop species.

Research has also suggested that multiline cultivars are more stable over a range of different environments than are pure-line cultivars. The reason for this has been related to the heterogeneous nature of the mixture where some lines in the mix do well in some years or locations while others perform better under other conditions. Therefore multilines show fewer genotype by environment interactions, a primary reason for their popularity with peanut breeders. Similarly, such considerations have led to mixtures of rye grass and Kentucky bluegrass being sold commercially. Rye grass has rapid emergence and establishment and does better than bluegrass in shaded areas.

Marketing of multiline varieties in the US has advantages over other types of cultivar as seed companies can sell the seed without a common brand name if the seed sold is labelled with a '*cultivar not stated*' label. Multilines can also be sold under more than one name. For example, the same multiline can be sold with the brand names 'Browns Appeal', 'Browns Wonder' and 'Wonder Why' by the same or different seed sales groups. In other countries, however, multiline cultivars must comply with the set standards of DUS (Distinctiveness, Uniformity and Stability) required for other inbred cultivars, and this has limited their use because of the difficulties in obtaining such homogeneity standards in a mixture.

The same care needs to be taken when producing breeders' seed for a multiline cultivar as is the case with a pure-line cultivar. The individual lines forming the mixture are increased independently by either mass or progeny multiplication methods (above). The individual components are then mixed in the proportions required, the seed mixed to form the breeders' seed, from which foundation seed is produced. The prevalent diseases, yielding ability or other appropriate factors will determine the proportion of lines within the mix. It is important when calculating multiline mixture proportions to take into account the seed size (if mixing by weight) and also the germination potential of each line (which may be different for the different lines).

One major complication relating to seed mixture proportions is the reproductive potential or productivity of each genotype in the mixture. For example, if the given proportion of two-parent lines (A and B) in a very simple multiline is 1:1, but the reproductive potential of A is twice that of B, a 1:1 mix of breeders' seed will result in a 2:1 ratio of the lines being harvested from foundation seed and a 4:1 in registered seed, and finally a 8:1 ratio being sold to the farmer after one further round of multiplication, to obtain certified seed. Similarly, environmental conditions may affect the proportions of mixed lines in the multiline. These changes could reflect that foundation and certified seed were produced in an atypical environment or where a different disease spectrum exists.

Some multiline cultivars are mixtures of isogenic (or near-isogenic) lines that differ for a single gene (usually conferring resistance to a certain strain of a pathogen). The most common method used in developing isogenic lines (lines that only differ in their genotypes by specified genes) in plant breeding is through the use of backcrossing. The genetics

of **backcrossing** will be covered later in the qualitative genetics section. However, a brief description will be given here.

4.3.1 Backcrossing

Backcrossing is a commonly used technique in developing pure-line cultivars. This technique has been used in plant breeding (not only in inbred species) to transfer a small, valuable portion (often ideally a single gene) of the genome from a wild or unadapted genotype into the background genotype of an adapted and already improved cultivar.

Backcrossing is an operation that involves a ***recurrent parent*** and a ***non-recurrent parent***. In many cases the non-recurrent parent is an unadapted line or genotype, and hence is not expected to contribute characters other than the specific one that it is desired to transfer to the resulting selections. It is therefore usual to choose a recurrent parent which is already suited to the environment (i.e. the most adapted genotype or cultivar available). The process of isogenic line production will be identical and repeated for each line that is to be produced.

The process will be described for the case where the homozygous allele (RR) of interest from the non-recurrent parent is completely dominant in showing resistance to a disease and the recurrent parent has the recessive susceptible alleles (rr). First the recurrent parent is crossed to the non-recurrent parent producing F_1 seeds which are therefore heterozygous (Rr) for this locus and where each of the two parents contribute equally to the genotype. The F_1s (Rr) are crossed back to the recurrent parent ($rr \times Rr$), to produce backcross 1 (BC_1). The seeds from this '*backcross*' are, for this locus of the genotypes Rr or rr, in equal proportions, which can be screened to identify the disease-susceptible lines rr, as opposed to the disease-resistant Rr. The Rr lines can then be used to cross back again to the recurrent parent rr, to produce the second backcross or BC_2, which again comprises the same genotypes as the BC_1. This process of screening for the presence of the heterozygous resistant lines and backcrossing them to the recurrent parent is repeated a number of times with the aim of developing a line which comprises all the genes from the recurrent parent except at the '*resistance locus*', which will have the resistance allele (R) instead of the susceptible allele (r). In other words we effectively '*add*' this to the genotype. The number of backcrossing generations will depend on how closely the breeder wants the isogenic line to resemble the recurrent parent or how well the backcross genotypes are performing. The proportion of the recurrent parent genotype in each backcross family will increase with increased backcrossing, and can be calculated by the formula:

$$1 - (1/2)^g$$

where g is the number of backcrossing generations, including the original cross ($P_1 \times P_2$) to produce the F_1. The following are proportions of the genes that are theoretically recovered from the recurrent parent according to the number of backcrosses:

$$F_1 = 50.0\%$$
$$BC_1 = 75.0\%$$
$$BC_2 = 87.5\%$$
$$BC_3 = 93.8\%$$
$$BC_4 = 96.9\%$$

The above percentages of the recurrent parent genotype in the resulting progeny hold reasonably well in the early backcross generations. However, with increased backcrossing, the percentage of the genotype of the non-recurrent parent (the 'wild' type) that is retained will be increasingly influenced by linkage. The resulting backcross line can therefore often contain a higher proportion of wild-type genotype than desired, and more backcross generations will be required to obtain the desired proportion of the cultivated/adapted (recurrent) parent.

Once the desired number of backcrossing generations has been completed, then the heterozygous (*Rr*) disease-resistant lines are selected, self-pollinated or inter-mated, and lines homozygous for the resistance gene (*RR*) can be identified.

The backcrossing method where the gene of interest is recessive is a slightly more complex (and often more lengthy) process. The general theory is the same but in this case it is necessary to progeny test the backcrossed generations in order to separate the homozygous and heterozygous plants that need to be selected. Progeny testing can be avoided (or reduced) when tightly linked co-dominant molecular markers are available. Molecular markers also can be used to increase the frequency of the adapted (recombinant) parent genome in the backcross family. This is discussed in more detail in Chapter 8.

4.4 Development of open-pollinated population cultivars

Crops that are generally produced as open-pollinated population include alfalfa, forage legumes, herbage grasses, maize (some), oil palm, perennial ryegrass, red clover, rye and sugar beet.

Development of open-pollinated population cultivars is a process that changes the gene frequency of desirable alleles within a population of mixed genotypes while trying to maintain a high degree of heterozygosity. So it is really the properties of the population that are vital, not individual genotypes (as in pure-line cultivars of self-pollinating crops). Instead of ending with a cultivar for release that is a uniform genotype, the population will be a complex mixture of genotypes, which together give the desired performance.

It is not considered desirable (and it is often very difficult) to develop homozygous or near-homozygous breeding lines or cultivars from these outbreeding species as they suffer severe inbreeding depression, carry deleterious recessive alleles or have strong or partial self-incompatibility systems. There are basically two different types of outbreeding cultivars available: **open-pollinating populations** and **synthetic cultivars**.

4.4.1 Breeding schemes for open-pollinating population cultivars

In open-pollinating populations, selection of desirable cultivars is usually carried out by *mass selection*, *recurrent phenotypic selection* or *selection with progeny testing*. Open-pollinating (outbred or cross-pollinating) cultivars are maintained through open-pollinated populations resulting from random mating.

Mass selection

Mass selection is a very simple breeding scheme that uses natural environmental conditions to alter the allelic frequencies of an open-pollinating population. A new population is created by cross-pollinating two different existing open-pollinating populations. In this case a representative set of individuals from each population will be taken to be crossed – single plants will not of course be representative of the populations. So it is common (even if mistaken) to select individual plants from each population but to take a reasonably large sample of such plants. How they are crossed depends upon the choice of the breeder, but often Bi-Parental matings (BIPs) are performed where specific parents are selected and hybridized. Alternatively, it is common to collect pollen in bulk and use this to pollinate specifically selected female plants in another population, whilst in many other instances breeders allow random mating or open/cross pollination to occur naturally.

The seed that results from such a set of crosses is grown under field conditions over a number of seasons. The theory of the approach is that genotypes that are adapted to the conditions will predominate and be more productive than those that are '*less fit*'. It is also assumed that crossing will be basically at random and result in a population moving towards equilibrium.

Problems with mass selection are related to partial, or complete lack of, control of the environmental conditions other than by choosing suitable locations for the trials. In some instances it is possible to create disease stress by artificially inoculating susceptible plants with a pathogen to act as spreaders, or by growing very susceptible lines in close proximity to the bulk populations. However, the process is empirical and often subject to unexpected disturbances. It also assumes, as noted earlier, that natural selection is going to be in the direction that the breeder desires – an assumption that is not always justified. It should also be noted that care needs to be exercised in isolating this developing population from other crops of this species, which might happen to be growing within pollination distance.

Recurrent phenotypic selection

Recurrent phenotypic selection tends to be more effective than mass selection. The basic outline of this process is illustrated in Figure 4.5. A population is created by cross-pollination between two (or more) populations to create what is referred to as the **base population**. A large number of plants are grown from the base population and a subsample

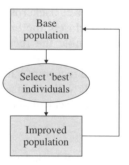

Figure 4.5 Basic recurrent phenotypic selection scheme.

of the most desirable phenotypes is identified and harvested as individual plants. The seeds of these selected plants are grown out, allowed to cross-pollinate at random, and thus to produce seed for a new population – an **improved population**.

This process is repeated a number of times, in other words, it is recurrent. The number of cycles performed will be determined by the desired level of improvement required over the base population, the initial allele frequencies of the base population, and the heritability of the traits of interest in the selection process. Recurrent phenotypic selection has been shown to be effective, but mainly in cases where there is high heritability of the characters being selected for, such as disease and pest resistances. The techniques are not nearly as effective where traits have a lower heritability, such as yield or quality traits.

It is common practice (and a good idea) to retain a sample of the base population so that the genetic changes due to selection can be evaluated in a later season.

Progeny testing

There are actually a variety of possibilities within the main heading of recurrent phenotypic selection, for example half- or full-sib selection with test crossing; and selection from S_1 progeny testing. All the schemes basically involve selecting individuals from within the population, and crossing or selfing these to produce seed. Part of the seed is sown for assessment and part is retained. Once the results of the assessments are available, the remnant seed from the progenies that have been shown to be superior are then sown as a composite population for plants again to be selected, and so on. At any stage seeds can be taken out for commercial exploitation.

4.4.2 Backcrossing in open-pollinated population cultivar development

Backcrossing is not as commonly used with open-pollinated population cultivar development, unlike pure-line cultivar development, but nevertheless the backcrossing technique is still used. When used, the main difference is that the recurrent and non-recurrent parents in the backcrosses are plant populations rather than homozygous lines. The basic assumption of any backcrossing system is that the technique is unlikely to result in a change in performance of the recurrent parent, other than for the single character being introduced. However, even when an allele has been introduced it is difficult to ensure its even distribution throughout the whole population.

Seed production

In most cases, seed production of open-pollinated cultivars simply involves taking a sample of seed from the population but under increasingly stringent conditions, to avoid the problems noted before, in avoiding contamination or cross-pollination from other populations or cultivars.

4.5 Developing synthetic cultivars

A synthetic cultivar basically gives rise to the same end result as an open-pollinated cultivar (i.e. population improvement), although a synthetic cultivar cannot be propagated by open-pollination without changing the genetic make-up of the population. This has perhaps been a primary reason for the rapid change-over from open-pollinated cultivars to synthetic ones, since it means that farmers need to return to the seed companies for new seed each year. It has been commercial seed companies that have been responsible for breeding almost all synthetic varieties. For example, before 1950 in the USA there were only two alfalfa breeders working in private seed companies while 23 were breeding in the public sector. By 1980 there were 17 public-sector alfalfa breeders, but now there are more than three times (52) the number of private breeders developing synthetic lines.

A synthetic cultivar must be reconstructed from its parental lines or clones. Within the US, maize is almost exclusively grown as hybrid cultivars whereas in many other countries maize crops are grown as synthetics. Synthetics have also been used almost exclusively in the development of alfalfa, forage grass and forage legume cultivars, and have also been used to develop varieties from other crop species (e.g. canola).

The breeding method used for the development of synthetic cultivars is dependent on the ability to develop either homozygous lines from a species or to propagate parental lines clonally. In the case of maize, for example, synthetic cultivars are developed using inbred lines as a three-stage process:

- develop a number of inbred lines;
- progeny-test the inbred lines for general combining ability;
- identify the 'best' parents and intercross these to produce the synthetic cultivar.

This process is almost identical to the procedure used for developing hybrid cultivars, and only differs in the last stage where many more parents will be included in a synthetic than in a hybrid cultivar. To avoid repetition, therefore, this section on synthetics will only cover the case of developing synthetic cultivars where inbred line development is not possible (for example, due to high inbreeding depression. So we will deal here with clonal synthetics. The process of developing a synthetic cultivar from clonal lines is illustrated in Figure 4.6.

Following Figure 4.6, clonal selections for use as parents can be added to the nursery *ad infinitum*, on the basis of continued phenotypic recurrent selection of the base population or by selections from those that have been produced by designed cross-pollinations. The second stage involves *clonal evaluation* and is conducted using replicated field trials of asexually reproduced plant units. The aim of the clonal testing is to identify which clonal populations are phenotypically most suited to the environments where they are to be grown. The clonal trials are often grown in two or more locations to include an assessment of environmental stability

A test cross or polycross technique will then be used to 'genetically' test the '*best*' clones identified from the clonal screen. The aim of this 'genetic' test is to determine the general combining ability of each

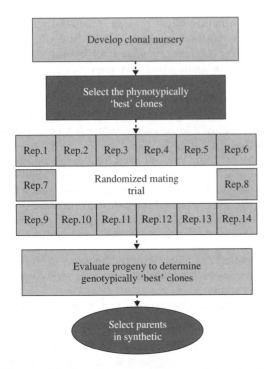

Figure 4.6 Breeding procedure used to develop synthetic cultivars from a clonally reproduced open-pollinated crop species (i.e. alfalfa) and using polycross progeny testing.

clonal line in cross-combinations with other genotypes in the selected group of clones.

If a test cross (often called a '*top cross*') is used, all the selected clones are hybridized to one (or more) test parent. The test parent will have been chosen because it is a desirable cultivar or it may be chosen because of past experience of the individual breeder. The test parent is a heterozygous clone that produces gametes of diverse genotypes. This diversity of gametes produced from the tester will help an assessment of the average ability of each clone to produce superior progeny when combined with alleles from many different individuals. Test cross evaluations are most useful when the variation that is observed within the different progeny is a result of differences between clones under evaluation and not due to only a small sample of genes coming from the test parent.

A polycross does not use a common test parent but rather a number of different parents. It therefore differs from a test cross as the seed progenies to be evaluated result from inter-crossing between

the clones that are under test (i.e. each clone under evaluation is used as female and randomly mated to all, or a good range, of other clonal selections). A polycross, like the test cross, is used to determine the general combining ability of the different potential clonal parents. The seed so produced from a polycross is then tested in randomized field trials. It is essential that the trials are randomized and that the level of replication is high enough to allow the possibility of hybrid seed being from as many other clones as possible.

Seeds from test crossing and polycrossing are grown in **progeny evaluation trials** to evaluate the genotypic worth or determine the general combining ability of each of the clones. Progeny evaluation trials are very similar to any other plot evaluation trial and are best grown in more than a single environment. Progeny evaluations are also often repeated over a number of seasons to obtain more representative evaluations of the likely performance over different years.

Depending on the results from the progeny evaluation trials, clones that show greatest general combining ability will be used as parents for the synthetic cultivar. These clonal parents are mated in a number of combinations to produce experimental synthetics. The number of clones that are selected and the number of parents that are to constitute the synthetic will determine how many possible combinations are possible. If there were only four clones selected, then there would be a total of 11 different synthetic cultivars (that is, the six possible 2-clone combinations, the four 3-clone combinations and the one 4-clone combination). With 6, 8, 10 and 12 parents, the number of possible synthetics would be 57, 247, 1,013 and 4,083, respectively. Therefore it is useful to try to predict the performance of synthetic cultivars without actually producing seed. One formula used is:

$$F_1 - [(F_1 - P)]/n$$

where F_1 is the mean performance of all possible single crosses among n parents, and P is the mean performance of the n parents. It should, however, be noted that there is an assumption of an absence of epistasis (interaction between alleles at different loci) in order to obtain a good estimate of synthetic performance, and so predictions are therefore often far removed from what is actually observed.

Such predictive methods should therefore be used with caution.

4.5.1 Seed production of a synthetic cultivar

The parents used to produce a synthetic cultivar can be intercrossed by hand-pollination to produce the first-generation synthetic (Syn.1). The aim is to cross every parent in the synthetic with all others (i.e. a half diallel cross). This can be difficult with some synthetics (e.g. alfalfa) where the number of parents included is often around 40.

It is therefore more common in situations where many parents are used in a synthetic cultivar to produce Syn.1 seed using a polycross procedure. The selected parents are grown in close proximity in randomized block designs with high replication. The crossing block is obviously grown in isolation for any other source of the crop to avoid cross-contamination. In cases where insect pollinators are necessary to achieve cross-pollination, attempts are made to ensure that these vectors are available in abundance. For example, alfalfa breeders introduce honey bees or leaf-cutter bees to pollinate synthetic lines.

Seed from Syn.1 is open-pollinated to produce Syn.2, which is subsequently open-pollinated to give the Syn.3 population, and so on. The classes of synthetic seed are categorized as breeders' seed, foundation seed and certified seed. In this case breeders' seed would be Syn.1, foundation seed Syn.2, and the earliest certified seed Syn.3.

In summary, the ***characteristics of a synthetic cultivar*** are:

- Synthetic cultivar species need to have potential to have parental lines, which reproduce from source material (either clonally or as an inbred line), and hence populations with the specific genetic makeup of the synthetic cultivar can be reproducibly produced from these base parents.
- To develop synthetic cultivars, the contributing parental material is tested for combining ability and/or progeny evaluated.
- Pollination of synthetic cultivar species cannot be controlled, and there has to be some natural method of random mating between parents (e.g. a suitable out-crossing mechanism such as a self-incompatibility system or other mechanism to promote cross pollination combined with an appropriate pollinator or wind pollination).

- The source parental material must be maintained for further use.
- Open-pollinated populations have limited life and are then reconstituted from the base parental lines on a cyclic basis.

4.6 Developing hybrid cultivars

Crops that are commonly produced and sold as hybrids include maize, Brussels sprouts, kale, onions, canola, sorghum, sunflower and tomato.

Although attempts have been made to develop hybrid cultivars from almost all annual crop species, it is unfair and incomplete to consider the evolution of hybrids without a brief history of the developments in hybrid corn. At the beginning of the 20th century it became apparent that genetic advances and yield increases achieved by corn breeders were markedly lower than those realized by other small-grain cereal breeders developing wheat and barley cultivars. Indeed, the pedigree selection schemes used by corn breeders, although considered suitable at that time, were essentially not effective. The knowledge that hybrid progeny produced by inter-mating two inbred lines often showed hybrid vigour or heterosis (i.e. trangressive segregation whereby the hybrid progeny produces higher yield than the better parent) suggested that hybrid cultivars could be exploited by corn breeders on an agricultural scale by manually detasseling female parents to produce a population that was entirely composed of F_1 hybrid seed that could be sold for commercial production.

The first suggestion of using controlled crosses was made by W.J. Beal in the late 19th century, based specifically on the concepts of Darwin on inbreeding and outbreeding. These ideas were then refined by G.H. Shull in 1909, on the basis of genetic studies, who put forward the idea of a hybrid cross being produced by first developing a series of inbred, or near-inbred, breeding lines and inter-mating these as inbred × inbred crosses (single cross hybrid) and using the hybrid seed for production. These hybrid progeny were indeed high-yielding and showed a high degree of crop uniformity. His basic concept, however, was not adopted at that time because even the most productive inbred lines had very poor seed yields (most likely due to inbreeding depression) and consequently, hybrid seed production was very expensive.

In the interim the traditional open-pollinated corn cultivars were quickly superseded by double cross hybrids ([Parent A × Parent B] × [Parent C × Parent D]) suggested by D.F. Jones in 1918. Double cross hybrids were not as high yielding or as uniform compared with the single cross hybrids proposed by Shull. However, hybrid seed production was less expensive than single cross hybrids, and as a result double cross hybrids completely dominated US corn production by the 1940s.

Initially all commercial hybrid corn seed was produced by detasseling *female* plants and growing them in close proximity to non-detasseled *males*, and harvesting seed only from the *female* plants. When this method of hybrid corn seed production was most prominent it was estimated that more than 125,000 people were employed in detasseling operations in the US in any growing season. Increased labour costs, combined with developments in using cytoplasmic male sterility (CMS) production systems in the 1960s, rendered detasseling of females in hybrid corn production obsolete for some time, but is now once again the method of choice, although some other biotechnology approaches to this are now being considered.

From the onset of hybrid corn breeding it was realized that there was a limit to production based on the inbred lines available, and a big effort was put into breeding superior inbred lines to use in hybrid combinations. Introduction of efficient and effective CMS hybrid production systems combined with the development of more productive inbred parents led to the return of single cross hybrids, which now dominate corn production in the US and many other developed countries. The relatively high cost of hybrid corn and other hybrid crop seeds does, however, limit the use of hybrid cultivars in many developing countries. Nevertheless, in several South American countries like Argentina, Chile, Paraguay, Uruguay and Brazil, nearly all the maize seed used by farmers is now hybrid because the superior yield they provide largely offsets their higher price.

The rapid increase in popularity and success achieved in hybrid maize could not have occurred without two very important factors. The first is that many countries, including until recently the US, do (or did) not have any Plant Variety Rights legislation or other means that breeders could use to protect proprietary ownership of the cultivars they bred. There was, therefore, little incentive for private companies to spend time, resources and effort in developing clonal, open-pollinating or inbred cultivars as individual farmers could increase seed stocks themselves, or they could be increased and sold by other seed companies. Hybrid varieties offered the potential for seed/breeding companies to have an in-built economic protection. The developing companies kept all the stocks of the parental lines and only sold hybrid seed to farmers. These hybrids, although uniform at the F_1 stage, would segregate if seed were retained and replanted (i.e. they would be F_2 progenies). Secondly, the introduction of hybrid maize occurred simultaneously with the transition from traditional to intensive technology-based agricultural systems. The new hybrids were indeed higher yielding, but were also more adapted to the increased plant populations, rising soil fertility levels and overall improved crop management of the times.

There are hardly any agricultural crops where hybrid production has not at least been considered, although hybrids are used in still relatively few crop species. The reasons behind this are firstly that not all crops show the same degree of heterosis found in maize, and secondly that it is not feasible in many crop species to find a commercial seed production system that is economically viable in producing commercial hybrid seed. Indeed, if maize had not had separate male and female reproductive organs and hence allowed easy female emasculation through detasseling, hybrid cultivar development might never have been developed, or acceptance would have been delayed by at least 20 years, until cytoplasmic male sterility systems were available.

Hybrid cultivars have been developed, however, in sorghum, onions and other vegetables using a cytoplasmic male sterility (CMS) seed production system; in sugar beet and some *Brassica* crops (mainly Brussels sprouts, kale and canola) using CMS and self-incompatibility to produce hybrid seed; and in tomato and potato using hand emasculation and pollination.

If hybrid cultivars are to be developed from a crop, then the species must:

- show a high degree of heterosis;
- be capable of being managed so as to produce inexpensive hybrid seed;
- not easily be produced uniformly, and have a high premium for crop uniformity.

There are many differing views regarding the exact contribution of hybrids in agriculture. In hybrid maize there would seem little doubt that there have been tremendous advances made. However, this has been the result of much research time and also large financial investments. In addition, it should be noted that the yielding ability of inbred parents in hybrid breeding programmes have been improved just as dramatically as their hybrid products. Most other hybrid crops (with the exception of sorghum) are also outbreeders. Almost all outbreeding crops show degrees of inbreeding depression and, therefore, its counterpart heterosis. In such cases there are strong arguments, certainly in practical terms, for exploiting heterozygosity to produce productive cultivars. This implies that hybrid cultivars can offer an attractive alternative over open-pollinated cultivars or even synthetic lines, although seed production costs will always be a major consideration. In inbreeding crops, hybrid cultivar production is much more difficult to justify on '*biological grounds*'.

Committed '*hybridists*', of whom there are many (especially within commercial seed companies), would argue that:

- Yield heterosis is there for the exploitation.
- Hybrid cultivars are economically attractive to breeding organizations and seed companies.
- Technical problems (usually associated with seed production) are simply challenges to be faced.
- The biological arguments are irrelevant.
- The rate of hybrid cultivar adoption is overwhelmingly rapid in countries where farmers can afford them.

Skeptics (of which there seem to be fewer, or who are less outspoken) argue on the basis of experimental data available to date that:

- There is no good evidence for over-dominance, and so it is definitely possible to develop pure-line, inbred cultivars that are as productive as hybrid lines.
- The economic attractions of seed companies should be weighed against high seed costs for the farmers; that technical skills could be put to better (more productive) use.
- The biological reality is all-important.

These latter sceptics usually, however, accept that in the case of outbreeding species, hybrids give a faster means of getting yield increases, while in the longer term inbred lines would match them, but in inbreeding crops this differential in speed is not present.

4.6.1 Heterosis

The performance of a hybrid is a function of the genes it receives from both its parents, but can be judged by its phenotypic performance in terms of the amount of heterosis it expresses. Many breeders (and geneticists) believe that the magnitude of heterosis is directly related to the degree of genetic diversity between the two parents. In other words, it is assumed that the more the parents are genetically different, the greater the heterosis will be. To this end, it is common in most hybrid breeding programmes to maintain two or more distinct germplasm sources (**heterotic groups**). Breeding and development is carried out within each source and the different genetic sources are only combined in the actual production of new hybrid cultivars or while testing experimental combinations. For example, maize breeders in the US observed significant heterosis by crossing Iowa Stiff Stalk breeding lines with Lancaster germplasm. Since this discovery, these two different heterotic groups have not been intercrossed to develop new parental lines, but rather have been kept genetically separated for parental development, so crossing and selection has been imposed on each heterotic group separately.

It has proved difficult to clearly and convincingly define the underlying causes of heterosis in crop plants. There are very few instances where heterozygous advantage *per se* has been shown to result from over-dominance. The counterpart to heterosis, inbreeding depression, is generally attributed to the fixation of unfavourable recessive alleles, and so it is argued that heterosis should simply reflect the converse effect. Therefore unfavourable recessive alleles in one line would be masked, in the cross between them, by dominant alleles from the other. If this is all there is to it, then heterosis should be fixable in true breeding lines by the selection of lines with only the favourable alleles. In general it has been found that this simple rationalization does not explain all the observed effects. Thus, the question is whether this breakdown in the explanation is related to a statistical problem of the behaviour of a large number of dominant/recessive alleles, each with small effect; whether the failure to detect over-dominance is simply a technical failure rather than a lack of biological reality; or whether a more complex explanation needs to be invoked. Dominance can be regarded as the interaction between alleles at the same locus, their interaction giving rise to only one of their products being observed (dominance is expressed) or a mixture of the products of the two (equal mixing giving no dominance, and inequality of mixing giving different levels of incomplete dominance). But another well-established type of interaction of alleles can occur, that between alleles at different loci (called non-allelic interaction or epistasis). In addition we cannot simply ignore the fact that linkage between loci is a recognized physical reality of the genetic system that we now regard as being the basis of inheritance. It has been shown that the combination of these two well-established genetic phenomena can produce effects that are capable of mimicking over-dominance.

To examine the effect of having many loci showing dominance and recessivity, determining the expression of a character differing between the parents of a single cross, let us examine the case of two genetically contrasting parental lines that differ by alleles at only five loci. Consider the two following cases, where we assume that: capital letters represent '*increasing*' alleles, lower case '*decreasing*' ones; each locus contributes in additive fashion to the expression of the character (i.e. the phenotype, let us consider yield); each increasing allele adds the same amount to the yield (2 units) while the decreasing allele adds nothing to yield;

and that dominance is complete and for increasing expression.

The level of heterosis can be considered in two forms: one is the F_1 *minus the mid-parent*, called mid-parent heterosis, while the second is F_1 *minus the best parent*, called best parent heterosis. The mid-parent heterosis need not detain us here, as it is of no direct interest. If best parent heterosis is positive (i.e. the F_1 exceeds the performance of the best parent), then generally heterosis has been ascribed to the presence of over-dominance, but this need not be the case.

In Case 1 the best parent (P_1) has a phenotype of 10 yield units; and the F_1 has a phenotype of 10 yield units; that is, they have identical yields and *no heterosis is detected*.

In Case 2 (with exactly the same alleles and effects but with different starting arrangement of alleles between the parents), the best parent (P_1) has a phenotype of 6 yield units; and the F_1 has a phenotype of 10 yield units; that is, *heterosis would be 4 units* and the F_1 is, in fact, 40% more productive than the better parent.

It is, however, clearly wrong to consider this to be over-dominance, since none of the individual loci show such an effect. It would certainly be possible (in fact, Parent 1 in Case 1 shows this) to fix this level of performance in homozygous, true breeding lines with no heterozygosity (i.e. AABBCCDDEE).

Now let us consider the statistical probability underlying these combinations. With five loci, assuming no linkage, no effects of selection and a random assortment of gametes in the F_1, after a

---

Case 1

Parent 1 = *AABBCCDDEE* **and** Parent 2 = *aabbccddee* **then:**

	Parent 1		**Parent 2**
Genotype	*AABBCCDDEE*	×	*aabbccddee*
Phenotype (yield)	$2+2+2+2+2 = 10$		$0+0+0+0+0 = 0$
	F_1		
Genotype	*AaBbCcDdEe*		
Phenotype (yield)	$2+2+2+2+2 = 10$		

Case 2
Parent 1 = *AAbbCCddEE* **and Parent 2** = *aaBBccDDee* **then:**

	Parent 1		Parent 2
Genotype	*AAbbCCddEE*	×	*aaBBccDDee*
Phenotype (yield)	$2 + 0 + 2 + 0 + 2 = 6$		$0 + 2 + 0 + 2 + 0 = 4$

F₁
Genotype *AaBbCcDdEe*
Phenotype $2 + 2 + 2 + 2 + 2 = 10$

number of rounds of selfing, the probability (which is equivalent to the frequency) of having a genotype that combines the five dominant alleles as homozygotes would be $(1/2)^5 = 1/32$ (or 0.03125, just over 3%). If, however, you wish to have some assurance that a breeding population contains at least one of these genotype recombinants, amongst all the possible inbred lines that can be produced, the number of lines needed to be grown and screened is given by the equation:

$$n = \ln(1-p)/\ln(1-x)$$

where n is the number of inbred lines that would need to be screened, p is the desired probability that at least one line with the desired genotype will exist in the population, and x is the frequency of the genotype of interest. In the example with five loci, it would be necessary to screen at least

145 inbred lines from the progeny derived from the cross between Parent 1 and Parent 2 (in both cases) to be 99% certain that at least one example of the genotype required would exist.

Of course, a quantitatively inherited trait like yield is not controlled by five loci but more likely 50, 500 or 5,000. To give some idea of the number of genotypes needed to obtain a specific combination of alleles, when the number of loci increases, see Table 4.1. The number of plants that need to be screened to be 90%, 95%, 99%, 99.5% and 99.9% sure that at least one genotype of the desired combination exists, is shown for cases with 5, 6, 7, 8, 9, 10 and 20 loci. With 20 loci, a modest number compared with the possible real situation, a breeder would need to evaluate almost 5 million inbred lines to be sure that the one he wants is present. So hybrids do offer a better probability of

Table 4.1 Number of recombinant inbred lines that would require evaluation for the breeder to be 90%, 95%, 99%, 99.5% and 99.9% sure that a homozygous lines will be a specific combination of alleles at each of 5, 6, 7, 8, 9, 10 and 20 loci.

	Probability of obtaining at least one individual of the desired genotype				
	0.900	0.950	0.990	0.995	0.999
5 loci	76	94	145	167	218
6 loci	146	190	292	336	439
7 loci	294	382	587	676	881
8 loci	588	765	1,177	1,354	1,765
9 loci	1,178	1,532	2,356	2,710	3,533
10 loci	2,357	3,066	4,713	5,423	7,070
20 loci	2,414,434	3,141,251	4,828,869	5,555,687	7,243,317

success in this instance, but not because they show over-dominance at their loci!

A second consideration with hybrid lines that has been postulated is that the heterozygosity of the cultivar makes it more stable over a range of different environments. This may be true, but there is no direct evidence for such a basic biological effect and it does not explain the extremely high genotype by environment interactions found to be exhibited by hybrid maize cultivars.

4.6.2 Types of hybrid

There are a number of different types of hybrid, apart from the single cross types concentrated on above. The different types of hybrid differ in the number of parents that are used in hybrid seed production. Consider four inbred parents (A, B, C and D) – types of hybrid that could be produced would include:

- Single cross hybrids (e.g. A × B, B × D, etc.)
- Three-way hybrids (e.g. [(A × B) × C, D × (C × A)], etc.)
- Double cross hybrids (e.g. [A × B] × [C × D, C × A] × [B × D], etc.)

Single cross hybrids are genetically uniform, whereas three-way or double cross hybrids are genetically heterogeneous (a three-way cross is less heterogeneous that a double cross). In general single cross hybrids have the highest level of heterozygosity and are more productive than the other two, but on the other hand are most expensive in terms of hybrid seed production.

4.6.3 Breeding system for F_1 hybrid cultivars

The three major steps in producing hybrids are therefore:

- development of inbred lines to be used as parents;
- test cross these lines to identify those that combine well;
- exploit the best single crosses as hybrid cultivars.

The system used to develop hybrid cultivars is illustrated in Figure 4.7. The scheme involves six stages:

- Produce two, or more, segregating populations.
- Develop inbred lines (parents) from each of the two populations.

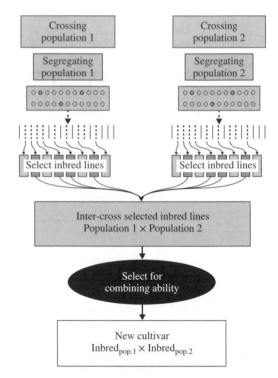

Figure 4.7 Outline of a hybrid breeding scheme.

- Evaluate performance of inbred lines phenotypically.
- Evaluate general combining ability of selected inbred lines.
- Evaluate hybrid cross combinations.
- Increase inbred parental lines, and commercialize hybrid cultivars.

The procedure used to develop inbred parent lines in hybrid cultivar development is similar to that used to breed pure-line cultivars, and the advantages and disadvantages of various approaches are the same. Breeders have used bulk methods, pedigree methods, bulk/pedigree methods, single seed descent and out-of-season extra generations (off-station sites) to achieve homozygosity. Regardless of the breeding approach of choice, these are independently imposed on the different heterotic groups from which inbreds are extracted. One of the most important objectives is to maintain high plant vigour and to ensure that the inbred lines have as high seed productivity as possible. This is not always easy, particularly in species where there is a high frequency of deleterious recessive alleles present in the segregating populations. Breeders must decide the level of homozygosity that is required. On one hand, the more homozygous (the extreme, of course, being 100% homozygosity) the inbred lines are, then the more uniform will be the resulting hybrid. The more heterozygous inbred lines may, however, be more productive as parents and hence help to reduce the cost of hybrid seed production.

Combining ability (or more relevantly, general combining ability, GCA) is evaluated with the aim of identifying parental lines which will produce productive progeny in a wide range of hybrid cross. Generally, it is not possible to cross all possible parental lines in pairwise combinations, as the number of crosses to be made and evaluated increases exponentially with the increased number of parents. It is therefore more usual to cross each parent under evaluation to a common **test parent** or **tester**. The tester used is common to a set of evaluations, and therefore, general combining ability is determined by comparing the performance of each progeny, assuming that the only difference between the different progenies can be attributed to the different inbred parents. Testers are usually highly developed inbred lines that have proved successful in hybrid combinations in the past. A far better prediction of general combining ability would be achieved if more than one tester were used. This, however, is not common practice, and breeders have tended to prefer to test more inbreds rather than to increase the number of test parents used. As the hybrid programme progresses, it is common, however, to replace older testers with newer ones.

Evaluation of specific combining ability (or actual individual hybrid combination performance) is carried out when the number of parents is reduced to a reasonable level. The number of possible cross combinations differs with the number of parents to be tested. The number of combinations is calculated from:

$[n(n-1)]/2$ for pairwise crosses

$[n(n-1)(n-2)]/2$ for three-way crosses

$[n(n-1)(n-2)(n-3)]/8$ for double crosses

where n is the number of parents to be evaluated. For example, if 20 parents are to be tested then there would be: 20 crosses to a single tester; 190

pairwise crosses, 3,420 three-way crosses and 14,535 double cross combinations possible. It is therefore common to predict the performance of three-way and double crosses from single cross performance rather than actually test them. The three-way cross $[(P_1 \times P_2) \times P_3]$ performance can be predicted from the equation:

$$1/2[(P_1 \times P_3) + (P_2 \times P_3)]$$

where $P_1 \times P_3$ is the performance of the F_1 progeny from the cross between P_1 and P_3. It is noted that the actual single cross in the hybrid predicted $(P_1 \times P_2)$ is not used in the prediction.

To predict the performance of a double cross $[(P_1 \times P_2)(P_3 \times P_4)]$ the following equation is used:

$$1/4[(P_1 \times P_3) + (P_1 \times P_4) + (P_2 \times P_3) + (P_2 \times P_4)]$$

where $P_1 \times P_3$ is the performance of the F_1 progeny from the cross between P_1 and P_3, and so on. Note again that the two single crosses used in the double cross do not appear in the prediction equation.

The assumptions underlying this will not be discussed here, but we simply note that this is what is carried out quite often in practical breeding.

4.6.4 Backcrossing in hybrid cultivar development

Backcrossing has featured quite highly in hybrid breeding schemes. Backcrossing is used in hybrid development for two purposes:

- To introduce a single gene into an already desirable inbred parent.
- To produce near isogenic lines, which can reduce the cost of seed production, by *convergence*, called *convergent improvement*. This is used to make slight improvements to specific hybrid cross combinations and involves recurrent backcrossing between a single cross hybrid ($F_1 = P_1 \times P_2$) and both of its two parents (P_1 and P_2). The result will be to produce, from P_1, P_{1*} (where P_{1*} is a near-isogenic line of P_1). Similarly, near-isogenic lines are developed for P_2. These can be used in a modified single cross ($[P_1 \times P_{1*}] \times P_2$; see also A' maintainer lines in cytoplasmic male sterile systems, below), or a double modified single cross ($[P_1 \times P_{1*}] \times [P_2 \times P_{2*}]$). The aim of this is to increase the efficiency and reduce the cost of seed production.

4.6.5 Hybrid seed production and cultivar release

The inbred lines used as parents are increased in exactly the same way as pure-line cultivars, and hence no further description is needed here.

The first and highest priority of hybrid seed production is to complete the task as cheaply as possible with the maximum proportion of hybrid offspring. There are four basic means that have been used to produce commercial amounts of hybrid seed:

Mechanical production In hermaphroditic plants, females are emasculated and pollination is achieved either by hand or naturally. In diclinous plants simple physical protection (such as bagging) and hand-pollination are used. The problems with mechanical hybrid production are related to cost, labour inputs and the number of seeds produced by each pollination operation. Examples include tomatoes and potatoes, along with detasseling (as described earlier) in maize.

Nuclear male sterility Nuclear male sterility (NMS) is found in many crop species, and sterility is usually determined by the expression of recessive alleles *(ms ms)* while fertility is present with the dominant homozygote *(Ms Ms)* and the heterozygotes *(Ms ms)*. Clearly it is not possible to maintain a pure male sterile line by simple sexual reproduction. However, if seeds are collected from male sterile plants they will be heterozygous for the sterility locus and on selfing will segregate 75% fertile *(Ms ms* or *Ms Ms)* and 25% sterile *(ms ms)*. The major problems with NMS are related to introducing the necessary genes into commercially desirable genotypes, and the expression being environmentally dependent. Success has been achieved in castor and attempts have been made with cotton, carrot, tomato, sunflower and barley.

Self-incompatibility To produce hybrids one simply interplants two self-incompatible, but cross compatible, inbred lines and harvests all the seed, all of which will be hybrid. Problems have been related to overcoming the self-incompatibility of parental lines in order to maintain them, which can be costly (e.g. bud pollination by hand is often necessary to overcome incompatibility in some *Brassica* species). Incompatibility systems are also often influenced by environmental conditions. Examples of hybrid crops include several vegetables, canola and sunflower.

Cytoplasmic male sterility Cytoplasmic male sterility (CMS) is of wide natural occurrence and can sometimes be induced by backcrossing a desired, adapted genotype (i.e. its nucleus) into a foreign cytoplasm such that the nuclear–cytoplasmic interaction results in male sterility. A practical hybrid system also requires, of course, the restoration of male fertility, usually achieved with 'restorer genes'. These, fortunately, are mostly dominant. If the hybrid product need not produce seeds (e.g. onions or beets) the restorer gene is then not needed. The combination of effective sterility and reliable restorer genes are not often available. Another problem is that the cytoplasm itself (which is always a component of the resulting hybrid) often has deleterious effects on plant vigour. The lack of pollen in the female plants can also have an effect on hybrid production. For example, if the pollinating vectors are insects they may well prefer to visit only the fertile males. Examples of success are onions, carrot, sugar beet, maize and sorghum.

To produce hybrid cultivars using cytoplasmic male sterility requires three types of genotype: male fertile lines (called B lines) with no cytoplasmic male sterility genes (Figure 4.8), which are homozygous for a dominant restorer gene (i.e. normal *(n)* cytoplasm, *Rf Rf*); cytoplasmic sterile female lines with male sterile cytoplasm *(cms)* with no restorer genes, *rf rf* (called A lines); and 'male-fertile' female lines (called A' lines or A' maintainer lines) with normal *(n)* cytoplasm and no

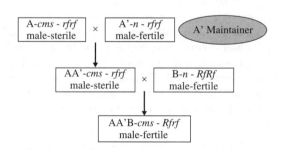

Figure 4.8 Hybrid seed production system using cytoplasmic male sterility requires three types of genotype: male fertile lines (called B lines) with no cytoplasmic male sterility genes and which are homozygous for a dominant restorer gene; cytoplasmic sterile female lines with male sterile cytoplasm but with no restorer genes; and 'male-fertile' female lines (called A' lines or A' maintainer lines) with normal cytoplasm and no restorer genes.

restorer genes, *rf rf*. A' lines are usually isogenic lines of the cytoplasmic male sterile A-female parents, produced by recurrent backcrossing, and are used to maintain and increase seed of the A cytoplasmic male sterile female parent.

In addition, several companies have used biotechnology to develop systems that enable inbred lines to become male sterile through the expression of genes specifically in male reproductive tissues. These will become commercially available in the years to come.

4.7 Development of clonal cultivars

Crops that are generally produced as clonal cultivars include bananas, cassava, citrus, potatoes, rubber trees, soft fruit (raspberry, blackberry and strawberry), sugarcane, sweet potatoes, and top fruit (apples, pears, plums, etc.).

Clonal crops are perennial plant species, although a few clonally propagated crops (e.g. potato and sweet potato) are grown as annual crops. Some clonally propagated crops are very long-lived (e.g. rubber, mango and rosaceous top fruits) and can produce crops for many years after being established. Indeed, there are instances whereby fig and palm cultivars have survived over a thousand years, and are still in commercial production. Other clonally propagated crops have a shorter lifespan yet remain in commercial production for several years after being propagated (e.g. sugarcane, bananas, pineapples, strawberries and *Rubus* spp).

There are many methods of propagation used in clonal crop production. Apples, pears, cherries, various citrus, avocados and grapes are propagated through budding and grafting onto various root stocks. Leafy cuttings are used to propagate pineapple, sweet potato and strawberry. Leafless stem cuttings are used to propagate sugarcane and lateral shoots are used for banana and some palms. There also is, for a number of species, the potential for clonal reproduction via tubers (swollen stems), as is the case for potatoes.

In general, clonal crop species are out-breeders, which are intolerant to inbreeding. Individual clones are highly heterozygous and so it is easy to exploit the presence of any heterosis that is exhibited. Imagine, for example, that corn could be easily reproduced asexually (say through apomixis); then there would be no need to develop hybrid corn

cultivars because the highly heterozygous nature of a hybrid line could be 'genetically fixed' and exploited through asexual reproduction.

The process of developing a clonal cultivar is, in principle, very simple. Breeders generate segregating progenies of seedlings, select the most productive genotypic combination and simply multiply asexually; this also stabilizes the genetic makeup (i.e. avoids problems relating to genetic segregation arising from meiosis). Despite the apparent simplicity of clonal breeding, it should be noted that while clonal breeders have shared in some outstanding successes, it has rarely been due to such a simple process, as will be noted from the example below.

4.7.1 Outline of a potato breeding scheme

The breeding scheme that was in use at the Scottish Crop Research Institute prior to 1987 is illustrated in Figure 4.9. This breeding scheme is similar to what many potato breeders still use today. It should be noted that the programme used two contrasting growing environments. The **seed site** (indicated by grey boxes) was located at high altitude and

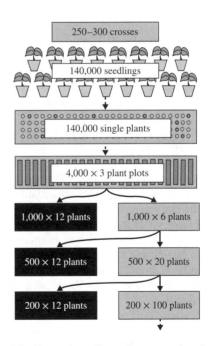

Figure 4.9 Potato breeding scheme used at the Scottish Crop Research Institute prior to 1987.

was always planted later and harvested earlier than would be considered normal for a typical **ware crop** (indicated by black boxes), in order to minimize problems of insect-borne virus disease infection. A ware crop is the crop that is produced for consumption rather than for replanting.

Each year, between 250–300 cross pollinations were carried out between chosen parents. From each cross combination the aim was to produce around 500 seeds, leading to about 140,000 seedlings being raised in small pots grown in a greenhouse (two greenhouse seasons were needed to accommodate the 140,000 total). At harvest, the soil from each pot was removed and the tubers produced by each seedling were placed back into the now empty pots. At this stage a breeder would visually inspect the small tubers in each pot (each seedling being a unique genotype) and either select or reject the produce of each seedling. One tuber was taken from amongst the tubers produced by the selected seedlings, while all tubers from rejected clones were discarded. The seedling generation in the greenhouse, as in most clonal crops, was the one and only generation that derived directly from *true botanical seeds* (i.e. from sexual reproduction). All other generations in the programme were derived by vegetative (clonal) reproduction, in other words from tubers.

The following year the single selected tubers (approximately 40,000) were planted in the field at the *'seed site'* as single plants within progeny blocks. This stage was referred to as the first clonal year. At harvest each plant was hand-dug, and the tubers exposed on the soil surface in a separate group for each individual plant. A breeder would then visually inspect the produce from each plant and decide, on that basis, to reject or select each group of tubers (i.e. each clone). Three tubers were retained from selected plants and planted in the field at the same seed site in the following year (the second clonal year) as a three-plant, unreplicated plot. First and second clonal year evaluations were therefore carried out under seed site conditions to reduce as far as possible the chances of contamination of the plants and hence the tubers, especially by virus diseases.

Second clonal year plots were harvested mechanically and, again, tubers from each plot exposed on the soil surface. A breeder examined the tubers produced and decided to select or reject each clone,

again on the basis of visual inspection. Tubers from selected clones were retained and grown in the subsequent year for the first time under *'ware growing conditions'* in unreplicated trails at the third clonal year stage. Tubers from each selection were also retained at the seed site and used to grow a six-plant plot in the following year. After the second clonal year, however, the seed site was only used to increase clonal tubers and no selection was carried out on the basis of the performance at this site.

At the ware site, measurements of a variety of characters were taken and yield was recorded (Figure 4.10). Thus the third clonal year was the first one where selection was based upon objective measurements, principally yield but also other performance characters and disease reaction. The fourth and fifth clonal generations were repeats of the third year with reduced numbers of entries after each successive round of selection, but with more replicates and larger plots; this was paralleled by larger multiplication plots of 20 plants and 100 plants, respectively, at the seed site.

In the sixth, seventh and eighth clonal generations, surviving clones were evaluated at a number of different locations (*'regional trials'*) in the UK. After each round of trials, the most desirable clones were advanced (i.e. re-trialled) and less attractive clones discarded.

Clones that were selected in each of the three year's regional trials were entered into the UK National List Trials (a statutory government-organized national testing scheme). Depending on performance in these trials a decision was made regarding cultivar release and initial foundation seed lots were initiated. If all went well, farmers could be growing newly developed cultivars within 17 years of the initial cross being made!

4.7.2 Time to develop clonal cultivars

Despite the lengthy time period between crossing and farmers growing a new potato cultivar, this is a short time period in comparison with some of the other asexually propagated crops. In potato the long selection process is related to the difficulty in evaluating a crop where the phenotype is greatly affected by the environment (both where the seed and ware crops were grown). In the case of potato, some of the length of the process is related to a slow multiplication rate, around 10:1 per generation.

Figure 4.10 Potato yield assessment trials produce a vast volume of product (tubers) from a relatively small area.

In addition, seed tubers are bulky and require large amounts of storage space. Planting material for one acre of potatoes will require approximately 2,000 lbs of seed tubers. To some extent this lengthy period can be reduced by *in vitro* multiplication of selected genotypes, but this does not reduce the need for adequate testing of genotypes over a reasonable number of seasons and sites.

With many other clonal species the process from crossing to cultivar release can be very lengthy. In apple breeding, for example, it is often said that if a breeder is successful with the very first parent cross combination, then it is still unlikely that the cultivar will be released (from that cross) by the time the breeder retires! In this case there is the obvious difficulty in the time taken from planting an apple seed to the time that the first fruit can be evaluated. With several other clonal crops, even the time to develop a new apple cultivar can be considered a relatively short time period in terms of the life of a cultivar. For example, the most common date palm cultivar in commercial production is a clonal line derived from the Middle-east called 'Siguel'. To our knowledge, it is so old that no one knows the derivation of this line, but it still predominates as the leading cultivar around the world.

Clonal crop species have shown a high frequency of exploited natural mutations compared with other crop types, probably simply as a result of their clonal propagation making any 'variant' obvious and easily multiplied. As a result many clonal cultivars are the result of natural mutations rather than arising from selection following a specific hybridization between parental lines. For example, the potato cultivar 'Russet Burbank' is a mutation from 'Burbank', and similarly 'Red Pontiac' is a natural mutation of 'Pontiac', and many apple cultivars are simply fruit colour mutants.

4.7.3 Sexual reproduction in clonal crops

In crops that are reproduced from true botanical seed, there is definite selection for reproductive normality and high productivity of sexual reproduction. In clonal crops this has not always been the case, and the result can hinder the ability of breeders to generate variation by sexual reproduction. There are two main types of clonally reproduced crops (excluding apomicts):

- those that produce a vegetative product;
- those producing a reproductive product (i.e. fruit).

Those that produce a vegetative product have almost all been selected to have reduced sexual reproductive capacity and can exhibit problems in relation to sexual crosses. This is probably the result of conscious or subconscious selection for plants that do not *waste energy* on aspects of sexual reproduction and will therefore put more energy into the vegetative parts. Extremes are found in yams and sweet potatoes in which many cultivars never flower and in many cases they cannot be stimulated to reproduce sexually. Modern potato cultivars have far less flowering than their wild relatives, and of those that do flower, many have very poor pollen viability or are pollen sterile. In the case of potato this has been the result of conscious human selection, based on the argument noted above and where sexual seed development can also cause a problem of generating 'volunteers' in subsequent crops.

In clonal crops where the reproductive product is used, there is of course no question about reducing flowering, but nonetheless, reproductive peculiarities and sterility problems are still very common. In general, selection has favoured the vegetative part of fruit development at the expense of seed production. In the extreme, bananas are vegetatively parthenocarpic (i.e. formation of fruit without seeds). Wild bananas are diploid ($n = 11$) and reproduce normally, producing fruit with large seeds. Commercially grown bananas are triploid and hence sterile, so many banana cultivars cannot be used as parents in a breeding scheme. Pineapples are also parthenocarpic but self-incompatible, so that clonal plantings give seedless fruits, even though fruit would be seedy if pollinated by another genotype. In addition, mangoes and some citrus sometimes produce polyembryonic plants (where multiple embryos are formed from a zygote by its fission at an early development stage, and hence in effect, results in clonal pseudo-seedlings identical to the mother genotype), which can be a great nuisance to breeders when they are selecting amongst segregating populations of sexually generated progeny.

In conclusion, reproductive derangement in clonal species can result in potential sexual crosses between particular parent combinations not being possible, or that individual parental lines cannot be used for sexual reproduction. This limits the options open to breeders of clonally reproduced crops.

4.7.4   Maintaining disease-free parental lines and breeding selections

Several decades ago it was thought that clonal cultivars were subject to **clonal degeneration**, seen as a reduction in productivity with time. It is now known that this degeneration is primarily the result of clonal stocks becoming infected with bacteria or viruses. For example, both potatoes and strawberries can be affected in this detrimental way by infection by bacterial or viral diseases. As an additional example, a bacterial disease causes stunting disease in sugarcane ratoons, which is not, as was postulated, caused by genetic '*drift*' bringing degeneration. It should, however, be noted that the cause of reduction in performance is often not solely due to disease buildup, and some degeneration of clonal lines can be the result of accumulation of deleterious natural somatic mutations.

Nevertheless, a primary problem in breeding clonal crops is to keep parental lines and breeding stocks free from viral and bacterial diseases that are transmitted through vegetative propagation. Often it is necessary to maintain disease-free breeding stocks that are separated from those being tested for adaptation potential, as was the case with the potato breeding scheme described above.

4.7.5   Seed increase of clonal cultivars

One of the primary concerns associated with the increase of new clonal cultivars is to prevent the stocks becoming infected by viral or bacterial diseases during commercial increase. Biotechnology approaches such as *in vitro* culture can often help in providing rapid increases of clonal stocks, which can be readily maintained true to type. They also enable the elimination, or at least a very significant reduction, of the pathogen load that builds up through continuous clonal propagation in the field. Such methods, however, cannot always be easily applied to many (particularly woody plant types) species, where rates of multiplication *in vitro* are often low and re-establishment under normal field conditions is not always possible. *In vitro* multiplication and applications for disease-free clonal material are discussed later.

4.8   Developing apomictic cultivars

It is possible to develop clonal cultivars through obligate apomixis (e.g. as with buffel grass, *Cenchrus ciliaris*). In species that are grown from obligate apomictic seeds, there tends to be a positive relationship between the productivity of a clone and its level of heterozygosity. It is therefore desirable to develop sources of genetic variability to maximize productivity within the population and this is a continual challenge to breeders of apomictic crops. Essentially all the seed produced by an obligate apomict are produced as a result of asexual reproduction. By far the most common means of producing genetic variation and/or increasing heterozygosity, however, is by sexual reproduction or hybridization between two chosen lines or populations. In developing apomictic cultivars it is important that some seed can be sexually produced so that this type of variation can be exploited.

There are basically two methods of producing apomictic cultivars.

- Selfing a sexual clone to produce a segregating S_1 (due to the clone being highly heterozygous). The progeny can be screened for 'sexual' and 'apomicts'. If apomicts are obligate they have the potential of being a clonal cultivar.
- Crossing a sexual clone to an apomict clone to produce an 'F_1' and selecting for sexual or apomicts reproducing clones within the segregating population. Again, obligate apomicts have the potential of being new cultivars.

The method is illustrated in Figure 4.11. Of course, a breeding scheme can use a combination of both

methods of selfing and intercrossing to develop a breeding system. In many cases lines have been developed by first selfing clones to achieve some degree of inbreeding, which is aimed at reducing the frequency of deleterious alleles, and then intercrossing the partial inbred lines to develop the apomict cultivar. The scheme shown in Figure 4.11 also could begin by selfing a semi-apomictic selection rather than a cross between an apomict and a sexual parent.

Once the apomicts have been identified, the most adapted lines are selected through repeated rounds of evaluation in different environments and years.

4.9 Summary

In summary, different crop species are more or less suited for developing different cultivar types. Individual plant breeders have developed specific breeding schemes that best suit their situation according to the type of cultivar being developed, the crop species and the resources (financial and others) that are available.

Irrespective of which exact scheme is used, breeding programmes also differ in the number of individual phenotypes that are evaluated at each stage and in the characters that are used for selection at each of the breeding stages. To address the issue of numbers in the breeding programme, it is necessary to consider the mode of inheritance of the factors of interest, and to have an understanding of the genetics underlying their inheritance.

Think questions

(1) Disease resistance to powdery mildew has been identified in a primitive, uncultivated species related to cultivated barley (*Hordeum vulgare*). Research has shown that the resistance is controlled by a single dominant allele (notation = *RR*). There is a resistance screen test available which shows 100% reliability. Outline a breeding scheme that could be used to introgress this character into an existing cultivar ('Golden Promise') which is susceptible to powdery mildew (notation = *rr*). The aim is for the resulting resistant lines to comprise at least 96% of the Golden Promise genotype.

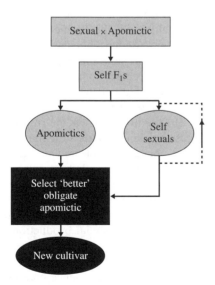

Figure 4.11 Breeding scheme for developing near obligate apomict cultivars.

Consider now that the disease resistance is controlled by a single recessive allele – how would this affect the breeding method you have described above?

(2) Crop cultivars can be divided into pure-line cultivars, open-pollinated populations, hybrid cultivars or clonal cultivars. Briefly, describe the major problems that can be encountered or the attributes of the crop types that can be utilized, when breeding: a pure-line cultivar; an open-pollinated population; a hybrid cultivar; and a clonal cultivar.

(3) You have been appointed as alfalfa (*Medicago sativa* L.) breeder to develop crops for the Alfred Alfoncia Seed Company Ltd. Outline a suitable breeding scheme you will use to develop superior synthetic cultivars for your new employer. Show each stage of the scheme on a year-by-year basis.

Now, after several rounds of clonal selection, you have identified four potential highly productive parent lines for developing a new synthetic (lines are coded as A, B, C and D). Hybrid seed was produced from all two-parent cross combinations possible and these were evaluated in yield trials along with the parent lines. From the yield (kg) results (below), indicate the three parent synthetic most likely to be highest yielding and state the expected yield.

$$\text{Parent A} = 18; \text{Parent B} = 22;$$
$$\text{Parent C} = 21; \text{Parent D} = 25$$
$$A \times B = 25 \quad A \times C = 26 \quad A \times D = 24$$
$$B \times C = 30 \quad B \times D = 21 \quad C \times D = 24$$

(4) Outline three characteristics of a crop species that would merit consideration before developing hybrid cultivars from that species.

(5) Outline the advantages and disadvantages of a bulk breeding scheme and a pedigree breeding scheme to develop pure-line cultivars

Many breeding schemes for pure-lines are a combination of bulk and pedigree schemes. Design a suitable bulk/pedigree breeding scheme (it is not necessary to make notes on when specific characters are selected) that could be used to develop superior pure-line cultivars. Outline any advantages or disadvantages of your breeding scheme.

(6) A hybrid breeding programme has identified six superior inbred lines (A, B, C, D, E and F) which were intercrossed in all combinations and the F$_1$ progenies evaluated for productivity in a field yield trial. The results from each single cross in the trial were:

A × B = 61	A × C = 65	A × D = 51	A × E = 56
A × F = 62	B × C = 63	B × D = 65	B × E = 59
B × F = 60	C × D = 59	C × E = 58	C × F = 60
D × E = 57	D × F = 68	E × F = 54	

Which parent combination will provide the most productive **single cross** hybrid and what would the expected yield be? Which parent combination will provide the most productive **three-way cross** hybrid and what would the expected yield be? Which parent combination will provide the most productive **double cross** hybrid and what would the expected yield be?

Explain why the most productive double cross combination (from the prediction equation) may not result in the most commercially suited hybrid cultivar.

(7) In order to justify development of hybrid cultivars it is necessary to have some method of producing inexpensive hybrid seed. Outline three systems that have been used to produce hybrid seed and indicate the advantages or disadvantages of each system.

(8) 'Russet Burbank' has been the predominant potato cultivar in the Pacific Northwest for many years. You have been asked to design a breeding programme that will produce cultivars that can replace some of the Russet Burbank acreage. Outline the breeding design you would use, indicating: (a) five breeding objectives; (b) the breeding scheme (use diagrams if necessary); and (c) indicating the number of clones evaluated and selected at each stage.

(9) In order to produce a hybrid cultivar using cytoplasmic male sterility you require three types of genotypes. Describe the differences between the three types required, and describe how commercial production will be conducted.

(10) You are to embark on a backcrossing study designed towards convergence breeding. You have at your disposal two genotypes coded C and M. Genotype C is a cytoplasmic male

sterile line, but one that does not have very good hybrid combining ability; genotype M has great hybrid combining ability and is male fertile (without a male fertile restorer gene). Design a scheme (listing each operation necessary) that will result in a genotype which is 95% the nuclear genotype of the M line but is cytoplasmically male sterile.

(11) You are walking along a deserted beach in southern California (are there any of these left?) and you inadvertently kick a brass oil lamp. On picking up the lamp you gently rub the side and – ***poof!*** – a 20 foot tall genie appears! '*Oh master!* ' says the genie, '*I can grant you two wishes*' (well, times are hard and the old three wishes rule does not apply any more). You think for a few minutes, then say '*I would like you to conjure me up the most wonderful plant that exists within the Universe,* *so that I can grow it on my farm in Idaho and make millions of dollars in profits*'. ***Poof!*** This plant appears before you – a plant species that has never been seen on Earth before.

As a plant breeder, list five questions you would ask the genie about the biology of this plant that would help you to design a cultivar development programme to increase the value of it as a crop.

Having sorted that out, you begin to think of your other wish. '*I would like you to tell me the formula for a chemical apomict*' (a chemical that, when applied to a crop, will result in 100% apomictic seed from the crop). '*Well*' says the genie. '*I can do this, but you must specify which crop species the chemical will work on*' (it can only work on one chosen species). What crop species would you choose to have apomictic seeds, and why?

5
Genetics and Plant Breeding

5.1 Introduction

Early plant breeders, basically farmers, did not have any knowledge of the inheritance of characters in which they were interested. The only knowledge they possessed was that the most productive offspring tended to originate from the most productive plants, and that the better flavour types tended to be derived from parent plants which were, themselves, of better flavour. None the less, the achievements of these breeders were remarkable and should never be underestimated. They moulded most of the crops as we recognize them today from their wild and weedy ancestral types.

In modern plant breeding schemes it is recognized, however, that it is very much more effective and efficient (or indeed essential) to have a basic knowledge of the inheritance or genetics of traits for which selection is to be carried out.

There are generally five different areas of genetics that have been applied to plant breeding:

- **Qualitative genetics**, where inheritance is controlled by alleles at a single locus, or at very few loci.
- **Population genetics**, which deals with the behaviour (or frequency) of alleles in populations and the conditions under which they remain in equilibrium or change, thus allowing predictions to be made about the properties and changes expected in populations.
- **Quantitative genetics**, for traits where the variation is determined by alleles at more than a few loci; traits that are said to be controlled by polygenic systems. Quantitative genetics is concerned with describing the variation present in terms of statistical parameters such as progeny means, variances and covariances.
- **Cytogenetics**, the study of the behaviour and properties of chromosomes, being the structural units that carry the genes that govern expression of all the traits.
- **Molecular genetics**, where studies are carried out at the molecular level. Molecular techniques have been developed to investigate and handle both qualitative and quantitative characters. Although the details of molecular genetics are generally outside the scope of this book, the impact of molecular genetic techniques is increasing, important and relevant.

5.2 Qualitative genetics

Few sciences have as clear-cut a beginning as modern genetics. As mentioned previously, early plant breeders were aware of some associations between parent plant and offspring, and at various times in history researchers had carried out experiments to study such associations. However, experimental genetics with real meaning began in the middle of the 19th century with the work of Gregor Johann Mendel, and was only fully appreciated after the turn of the 20th century.

Plant Breeding, Second Edition. Jack Brown, Peter D.S. Caligari and Hugo A. Campos.
© 2014 John Wiley & Sons, Ltd. Published 2014 by John Wiley & Sons, Ltd.
Companion Website: www.wiley.com/go/brown/plantbreeding

Mendel's definitive experiments were carried out in a monastery garden on pea lines. The flowers of pea plants are so constructed as to favour self-pollination, and as a result the majority of lines used by Mendel were either homozygous or near-homozygous genotypes. Mendel's choice of peas as an experimental plant species offered a tremendous advantage over many other plant species he might have chosen. The differences in characters he chose were also fortuitous. Therefore many present-day scientists have argued that Mendel had a great deal of luck associated with his findings because of the choices he made over what to study. When this is combined with segregation ratios that are better than might be expected by chance, many have concluded that he must 'have already foreseen the results he expected to obtain'.

Before considering an example of one of Mendel's experiments, there are a few general points to be made about his experiments. Others, before Mendel, had made controlled hybridizations or crosses within various species. So why did Mendel's crosses, rather than those of earlier workers, provide the basis for the modern science of genetics? First and foremost, Mendel had a brilliant analytical mind that enabled him to interpret his results in ways that defined the principles of heredity. Secondly, Mendel was a proficient experimentalist. He knew how to carry out experiments in such a way as to maximize the chances of obtaining meaningful results. He knew how to simplify data in a meaningful way. As parents in his crosses, he chose individuals that differed by sharply contrasting characters (now known to be controlled by single genes). Finally, as noted above, he used true breeding (homozygous) lines as parents in the crosses he studied.

Among other elements in Mendel's success were the simple, logical sequence of crosses that he made and the careful numerical counts of his progeny that

he recorded with reference to the easily definable characteristics on whose inheritance he focused his attention. It should be noted that many of the features listed as reasons for Mendel's success are very similar to the criteria necessary to carry out a successful plant breeding programme (including the '*luck element*').

Let us consider some of Mendel's experiments as an introduction to qualitative inheritance. These results are presented in Table 5.1. When Mendel crossed plants from a round-seeded line with plants from a wrinkled-seeded line, all of the first generation (F_1) had round seeds. The characteristic of only one of the parental types was therefore represented in the progeny. In the next generation (F_2), achieved by selfing the F_1, both round-seed and wrinkled-seed were found in the progeny. Mendel's count of the two types in the F_2 was 5,474 round-seed and 1,850 wrinkled-seed, a ratio of 2.96:1.

Mendel found it notable that the same general result occurred when he made crosses between plants from lines differing for other characters. Another example is when he crossed peas with yellow cotyledons with ones with green cotyledons; the F_1s all had yellow cotyledons, while he found a ratio of 3.01 yellows to 1 green in the F_2. Almost identical results were obtained when long-stemmed plants were crossed with short-stemmed plants, and when plants with axial inflorescence were crossed with plants with terminal inflorescence.

How did Mendel interpret his generalized findings? One of the keys to his solution was his recognition that in F_1 the heredity basis for the character that fails to be expressed was not lost. This expression of the character appears again in the F_2 generation. Recognizing the idea of dominance and recessiveness in heterozygous genotypes and the particulate nature of the heritable factors was the overwhelming genius of Gregor Mendel. This laid the foundation of genetics and hence the

Table 5.1 Results from some of Mendel's crossing experiments with peas.

Phenotype of parents	F_1 progeny	No. in F_2 progeny	F_2 ratio, dominant : recessive by phenotype
round (r) × wrinkled (w), seed	round	5,474 r : 1,850 w	2.96 : 1
yellow (y) × green (g), coty.	yellow	6,022 y : 2,001 g	3.01 : 1
long (lo) × short (sh), stem	long	787 lo : 277 sh	2.84 : 1
axil (ax) × terminal (ter), inf	axil	651 ax : 207 ter	3.15 : 1

explanation underlying the most important features of qualitative genetics.

5.2.1 Genotype/phenotype relationships

Within genetic studies there are two interrelated points. The first is concerned with the actual genetic makeup of individual plants or segregating populations and is referred to as the genotype. The second is related to what is actually expressed or observed in individual plants or segregating populations, and is termed the phenotype.

In the absence of any environmental variation (which can often be assumed with qualitative, single, major gene inheritance), the most frequent cause of difference between genotype and phenotype is due to **dominance** effects. For example, consider a single locus and two alleles (A and a). If two diploid homozygous lines are crossed where one has the genotype AA and the other aa, then the F_1 will be heterozygous at this locus (i.e. Aa). If this resembles exactly the AA parent, then allele A is said to be dominant over allele a. On selfing the F_1, genotypes will occur in the ratio $1/4 AA : 1/2 Aa : 1/4 aa$. If, however, the allele A is completely dominant to a, the phenotype of the F_2 will be $3/4 A : 1/4 a$. Therefore, **complete dominance** occurs when the F_1 shows exactly the same phenotype as one of the two parents in the cross.

Compare these ratios to those observed by Mendel (Table 5.1). You will notice that Mendel's data do not fit exactly to a 3:1 ratio that would be expected, given that each trait is controlled by a single completely dominant gene. It should, however, be remembered that gamete segregation in meiosis and pairing in fertilization are **random events**. Therefore it is highly unlikely, because of sampling variation, that exact ratios are found in such experiments. Indeed, applying statistical considerations, many modern researchers have claimed that Mendel's data appears to be 'too good a fit' for purely random events. Whether these geneticists are correct or not does not, however, detract in any way from the remarkable achievements of Mendel.

5.2.2 Segregation of qualitative genes in diploid species

To illustrate segregation of qualitative genes, and indeed many following concepts, consider a series of simple examples. Firstly, consider two pairs of single genes in spring barley (*H. vulgare* L.). Say that dwarf barley plants differ in plant height from tall types at the t-locus, with tall types given the genetic constitution TT, and dwarf lines tt. Barley lines with 6-row ears differ from lines with 2-row ears, where the central florets do not set, at the S-locus, with 6-row SS types being dominant over 2-row *ss types*. Let us consider the case where two homozygous lines are artificially hybridized, where one parent is homozygous tall and 6-row and the other is dwarf and 2-row. First consider the phenotypes expected for each trait.

Parents	Tall × Dwarf	6-row × 2-row
F_1	Tall	6-row

When the tall F_1 is backcrossed to the dwarf parent, a ratio of 1 tall : 1 dwarf is obtained in the resulting progeny. Similarly, when the six-row F_1 is crossed with a homozygous two-row, a ratio of 1 six-row : 1 two-row is obtained. Evidently, the difference between tall and dwarf behaves as a single major gene with the tall allele showing complete dominance to the dwarf allele; and the difference between six-row and two-row is also a single gene with six-row showing complete dominance to two-row. In terms of segregating alleles (i.e. genotypes), the above example would be:

Parents	$TT \times tt$	$SS \times ss$
F_1	Tt	Ss
F_2	$TT:Tt:tt$	$SS:Ss:ss$
	1:2:1	1:2:1

Where TT and Tt and similarly SS and Sc have the same phenotype, we get the 3:1 phenotypic segregation ratio. If each allele exerts equal effect (additive, and no dominance) then we would have three expected phenotypes in the Mendelian ratio of one tall, two intermediate and one short, according to the 1 TT : 2 Tt : 1 tt, or one 6-row, two intermediate and one 2-row.

Now assume that a large number of the F_2 plants were allowed to set selfed seed: what would be the ratio of phenotypes and genotypes in the F_3 generation? To answer this, consider that only the heterozygous F_2 plants will segregate (i.e. *TT* and *tt* types are now homozygous for that trait and so will breed true for that character). Of course the heterozygous F_2 has the same genotype as the F_1, and so it is no surprise that it segregates in the same ratio. At F_3 we therefore have $(1/4\ TT + 1/2 \times 1/4\ TT) : 1/2 \times 1/2\ Tt : (1/2 \times 1/4\ tt + 1/4\ tt)$ which results in 3/8 *TT* : 2/8 *Tt* : 3/8 *tt*. Similar results of 3/8 *SS* : 2/8 *Ss* : 3/8 *ss* would be obtained for 6-row versus 2-row. Applying these simple rules and expanding this to later generations, we get:

F_4	7/16 *TT*	:	2/16 *Tt*	:	7/16 *tt*
F_5	15/32 *TT*	:	2/32 *Tt*	:	15/32 *tt*
F_6	31/64 *TT*	:	2/64 *Tt*	:	31/64 *tt*
$\vdots$					
F_∞	1/2 *TT*	:	0 *Tt*	:	1/2 *tt*

Note that each successive generation of selfing reduces the proportion of heterozygotes by half (i.e. 1/2 *Tt* at F_2, 2/8 *Tt* at F_3, 2/16 *Tt* at F_4 and 2/32 *Tt* at F_5, etc.

Let us now consider how these two different pairs of alleles behave with respect to each other in inheritance. One way to study this is to cross individuals that differ in both characteristics:

tall, 6-row   ×   dwarf, 2-row

(*TTSS*   ×   *ttss*)

When this cross is made, the F_1 shows both dominant characteristics, tall and 6-row. Plant breeders' interests in genetics mainly relate to selection, and as no selection takes place at the F_1 stage, so the interest begins when the self-pollinated progeny of the F_1 is considered. In this case the F_1 individuals are assumed to produce equal frequencies of four kinds of gametes during meiosis (*TS, Ts, tS* and *ts*). An easy way to illustrate the possible combination of F_2 progeny is using a **Punnett square**, where the four gamete types from the male parent are listed in a row along the top, and the four kinds from the female parent are listed in a column down the left-hand side. The 16 possible genotype combinations are then obtained by filling in the square, that is:

Gametes from female parent	Gametes from male parent			
	TS	*Ts*	*tS*	*ts*
TS	*TSTS*	*TSTs*	*TStS*	*TSts*
Ts	*TsTS*	*TsTs*	*TstS*	*Tsts*
tS	*tSTS*	*tSTs*	*tStS*	*tSts*
ts	*tsTS*	*tsTs*	*tstS*	*tsts*

This also can be written, putting the allele representations for the same locus together, as:

Gametes from female parent	Gametes from male parent			
	TS	*Ts*	*tS*	*ts*
TS	*TTSS*	*TTSs*	*TtSS*	*TtSs*
Ts	*TTSs*	*TTss*	*TtSs*	*Ttss*
tS	*TtSS*	*TtSs*	*ttSS*	*ttSs*
ts	*TtSs*	*Ttss*	*ttSs*	*ttss*

Collecting the like genotypes we get the following frequency of **genotypes**:

1/16 *TTSS*;	2/16 *TTSs*;	1/16 *TTss*
2/16 *TtSS*;	4/16 *TtSs*;	2/16 *Ttss*
1/16 *ttSS*;	2/16 *ttSs*;	1/16 *ttss*

and frequency of phenotypes:

9/16 *T_S_* :	= tall and 6-row
3/16 *T_Ss* :	= tall and 2-row
3/16 *ttS_* :	= short and 6-row
1/16 *ttss* :	= short and 2-row

As with the single gene case above, if the alleles do not show dominance then there would indeed be nine different phenotypes in the ratio:

1 *TTSS* : 2 *TTSs* : 1 *TTss* :

2 *TtSS* : 4 *TtSs* : 2 *Ttss* :

1 *ttSS* : 2 *ttSs* : 1 *ttss*

In most cases when developing inbred cultivars, plant breeders carry out selection based on plant phenotype amongst early generation segregating populations (i.e. F_2, F_3 and F_4). It is obvious that

75% of the F_2 plants will be heterozygous at one or both loci, and dominance effects can mask the actual genotypes that are to be selected. Single F_2 plant selections for the recessive traits (short stature and 2-row) allow for identification of the desired genotype (*ttss*) but only 1/16th of F_2 plants will be of this type, while the recessive expression of the trait in most plants is hidden due to dominance.

The effects of heterozygosity on selection can be reduced through successive rounds of self-pollination. Plant breeders, therefore, do not only select for single gene characters at the F_2 stage. Consider now what would happen if a sample of F_2 plants from the above example were selfed: what would be the resulting segregation expected at the F_3 stage?

There are nine different genotypes at the F_2 stage, *TTSS*, *TTSs*, *TTss*, *TtSS*, *TtSs*, *Ttss*, *ttSS*, *ttSs* and *ttss*, and they occur in the ratio $1:2:1:2:4:2:1:2:1$, respectively. Obviously if any genotype is homozygous at a locus then these plants will not segregate at that locus. For example, plants with the genetic constitution of *TTSS* will always produce plants with the *TTSS* genotype. F_2 plants with a genotype of *TTSs* will not segregate for the *Tt* locus and will only segregate for the *Ss* locus. Similarly, F_2 plants with a genotype of *TtSs* will segregate for both loci, with the same segregation frequencies as the F_1.

From this we have:

	F₂ parents									
F_3	*TTSS*	*TTSs*	*TTss*	*TtSS*	*TtSs*	*Ttss*	*ttSS*	*ttSs*	*ttss*	Total
	4/64	8/64	4/64	8/64	16/64	8/64	4/64	8/64	4/64	
TTss	4	2	–	2	1	–	–	–	–	9
TTSs	–	4	–	–	2	–	–	–	–	6
TTss	–	2	4	–	1	2	–	–	–	9
TtSS	–	–	–	4	2	–	–	–	–	6
TtSs	–	–	–	–	4	–	–	–	–	4
Ttss	–	–	–	–	2	4	–	–	–	6
ttSS	–	–	–	2	1	–	4	2	–	9
ttSs	–	–	–	–	2	–	–	4	–	6
ttss	–	–	–	–	1	2	–	2	4	9
										64

Summing over rows, this results in a genotypic segregation ratio of 9/64 *TTSS* : 6/64 *TTSs* : 9/64 *TTss* : 6/64 *TtSS* : 4/64 *TtSs* : 6/64 *Ttss* : 9/64 *ttSS* : 6/64 *ttSs* : 9/64 *ttss*.

The phenotypic expectation at the F_3 would therefore be:

$$T\_S\_ \; = \; 25/64 \; = \; \text{tall and 6-row}$$
$$T\_Ss \; = \; 15/64 \; = \; \text{tall and 2-row}$$
$$ttS\_ \; = \; 15/64 \; = \; \text{short and 6-row}$$
$$ttss \; = \; 9/64 \; = \; \text{short and 2-row}$$

This result could be obtained in a simpler manner by considering the segregation ratio of each trait separately and using these to form a Punnett square. For example, the frequency of homozygous tall (*TT*), heterozygous tall (*Tt*) and homozygous short (*tt*) at the F_2 is 1:2:1, respectively. Similarly, the frequency of *SS*, *Ss* and *ss* is also 1:2:1. We can use these frequencies to construct a Punnett square such as:

Gametes	$TT - 1/4$	$Tt - 1/2$	$tt - 1/4$
$SS - 1/4$	1/16 *TTSS*	2/16 *TtSS*	1/16 *ttSS*
$Ss - 1/2$	2/16 *TTSs*	4/16 *TtSs*	2/16 *ttSs*
$ss - 1/4$	1/16 *TTss*	2/16 *Ttss*	1/16 *ttss*

This would give the same genotypic and phenotypic frequencies as shown above. It is, however, necessary to be familiar with the more direct method for situations where segregation is not independent (i.e. in cases of linkage).

It should be noted, as in the single gene case previously described, that increased selfing results in increased homozygosity. Therefore with increased generations of selfing there is an increase in the frequency of expression of recessive traits. If, in this case, a breeder wished to retain the 6-row short plant type, then it would be expected that 3/16 (approximately 19%) of all F_2 plants will be of this type (*ttS_*), and only 1/3 of these would at this stage be homozygous for both traits. If selection were delayed until the F_3 generation, then 15/64 (or just over 23%) of the population would be of the desired type, and now 60% of the selections would be homozygous for both genes. Obviously, if selection is delayed until after an infinite amount of selfing, then the frequency of the desired types would be 25% (and all homozygous).

5.2.3 Qualitative loci linkage

The principle of independent assortment of alleles at different loci is one of the cornerstones on which an understanding of qualitative genetics is based. Independent assortment of alleles does not, however, always occur. When certain different allelic pairs are involved in crosses, deviation from independent assortment regularly occurs if the loci are located on the same arm of the same chromosome. This will mean that there will be a tendency of parental combinations to remain together, which is expressed in the relative frequency of new combinations, and is the phenomenon of **linkage**.

It is often desirable to have knowledge of linkage in plant breeding to help predict segregation patterns in various generations and to help in selection decisions. All genetic linkages can be broken by successive generations of sexual reproduction, although it may take many rounds of recombination before tight linkages are broken. In general, however, if two traits of interest are adversely linked (i.e. the desired combination appears with lower than expected frequency), then increased opportunities for recombination need to be given before selection takes place.

The ratio of F_2 progeny after selfing F_1 individuals (i.e. say $AaBb$), of equal numbers of gametes of the four possible genotypic combinations (AB, Ab, aB, ab), leads to the ratio of $9A\_B\_ : 3A\_bb : 3aaB\_ : 1aabb$. Similarly, we find the genetic ratio of $1A\_B\_ : 1A\_bb : 1aaB\_ : 1aabb$, on test-crossing the F_1 to a complete recessive ($aabb$). Any significant deviation from these expected ratios is an indication of linkage between the two gene loci.

Consider a second simple example of recombination frequencies derived from a test cross, how the test cross can be used to estimate recombination ratios, and hence ascertain linkage between characters. Consider again the dwarfing gene in spring barley, but this time with a third single gene that confers resistance to barley powdery mildew. Mildew-resistant genotypes are designated as RR and susceptible genotypes as rr. A cross is made between two barley genotypes where one parent is homozygous tall and mildew resistant (i.e. $TTRR$); and the other is short and mildew susceptible (i.e. $ttrr$). We would expect the F_1 to be tall and resistant (i.e. $TtRr$). When the heterozygous F_1 is crossed with a completely recessive genotype (i.e. $ttrr$) we would expect to have a 1:1:1:1 ratio of phenotypes, tall and resistant: tall and susceptible: short and resistant: short and susceptible. When this cross was carried out and the progeny from the test cross was

examined, the following frequencies of phenotypes and genotypes were the ones actually observed.

Tall and resistant ($TtRr$)	=	79
Short and resistant ($ttRr$)	=	18
Tall and susceptible ($Ttrr$)	=	22
Short and susceptible ($ttrr$)	=	81
Total	=	200

Linkage is obviously indicated by these results, as the observed frequencies are markedly different from those expected at 50:50:50:50, on the basis of independent assortment. The two phenotypes, *tall and resistant* and *short and susceptible*, occur at a considerably higher frequency (79 and 81, respectively) than we expected. You will readily note that these are the same phenotypes as the two parents. The other two phenotypes (the non-parental, or recombinant, types) were observed with much lower frequency (18 and 22, respectively). Added together, the two recombinant types (*short and resistant*, and *tall and susceptible*) only account for 40 (20%) out of the 200 F_2 progeny examined, while we expected their contribution to collectively make up 50% of the progeny. Therefore the F_1 plants are producing TR gametes or tr gametes in 80% of meiotic events, or a frequency of 0.4 for each type. Similarly, recombinant Tr or tR gametes are produced in only 20% of meiotic events, or a frequency of 0.1 for each type. Knowing the frequency of the four gamete types, we can now proceed to predict the frequency of genotypes and phenotypes we would expect in the F_2 generation using a Punnet square.

Gametes from female parent	Gametes from male parent			
	$TR - 0.4$	$Tr - 0.1$	$tR - 0.1$	$tr - 0.4$
$TR - 0.4$	$TTRR$ 0.16	$TTRr$ 0.04	$TtRR$ 0.04	$TtRr$ 0.16
$Tr - 0.1$	$TTRr$ 0.04	$TTrr$ 0.01	$TtRr$ 0.01	$Ttrr$ 0.04
$tR - 0.1$	$TtRR$ 0.04	$TtRr$ 0.01	$ttRR$ 0.01	$ttRr$ 0.04
$tr - 0.4$	$TtRr$ 0.16	$Ttrr$ 0.04	$ttRr$ 0.04	$ttrr$ 0.16

Collecting like phenotypes together results in:

Tall and resistant	$= T\_R\_$	$= 0.66$
Short and resistant	$= ttR\_$	$= 0.09$
Tall and susceptible	$= T\_rr$	$= 0.09$
Short and susceptible	$= ttrr$	$= 0.16$

From a breeding standpoint, compare these phenotypic ratios to those that would have been expected with no linkage (i.e. $0.56 \ T\_R\_ : 0.19 \ ttR\_ : 0.19 \ T\_rr : 0.06 \ ttrr$). If, as might be expected, the aim was to identify individual plants which were short and resistant, then the actual occurrence of these types would drop from 19% to 9%, by more than half. It should also be noted that 20% recombination is not particularly high.

Overall, therefore, genes on the same chromosome (particularly on the same chromosome arm) are linked and do not segregate independently as they would when they are located on different chromosomes. Two heterozygous loci may be linked in coupling or repulsion. Coupling is present when desirable alleles at two loci are present together on the same chromosome and the unfavourable alleles are on another (e.g. *TR/tr*). Repulsion is present when a desirable allele at one locus is on the same chromosome as an unfavourable allele at another locus (e.g. *Tr/tR*).

In the test cross **TtRr × ttrr** the situations shown in Table 5.2 can arise.

The percentage recombination, expressed as the number of map units between loci, can be calculated from the equation:

$$\text{Recombination \%}$$
$$= \frac{\text{sum of non-parental classes}}{\text{total number of individuals}} \times \frac{100}{1}$$

Depending on what the initial parents were, this will be estimated as:

$$= \frac{TR + tr}{TtRr + Ttrr + ttRr + ttrr}$$

or

$$= \frac{Tr + tR}{TtRr + Ttrr + ttRr + ttrr}$$

Linkage can also be detected, but less efficiently, by selfing the F_1 and observing the segregation ratio of the F_2 to determine whether there is any deviation from the expectation, based on independent

Table 5.2 Expected frequency of genotypes resulting from a test cross where *TrRr* is crossed to *ttrr* with independent assortment, and with coupling and repulsion linkage.

Independent assortment	Coupling linkage	Repulsion linkage
(1/4) *TR*	> (1/4) *TR*	< (1/4) *TR*
(1/4) *Tr*	< (1/4) *Tr*	> (1/4) *Tr*
(1/4) *tR*	< (1/4) *tR*	> (1/4) *tR*
(1/4) *tr*	> (1/4) *tr*	< (1/4) *tr*

assortment (e.g. 1 *TTRR* : 2 *TTRr* : 1 *TTrr* : 2 *TtRR* : 4 *TtRr* : 2 *Ttrr* : 1 *ttRR* : 2 *ttRr* : 1 *ttrr*).

The relative positions of three loci can be **mapped** by considering the frequency of progeny from a **three gene test cross**. To examine three gene test crosses, consider the following cross between two parents where one parent has the genotype *AABBCC* and the other has the genotype *aabbcc*. The F_1 would have the genotype *AaBbCc*. In order to perform the test cross it is necessary to cross the F_1 family with a completely recessive genotype (i.e. *aabbcc*, in this case the recessive parent) and observe the frequency of the eight possible phenotypes. In our example the frequencies are as shown in Table 5.3.

If all three loci segregated independently, for example if they were on different chromosomes, we would expect the phenotype frequencies of the eight genotypes to be the same at 149.375 (or the total number of observations divided by the expected number of phenotype classes, $1195/8 = 149.375$).

Table 5.3 Observed and expected genotypes obtained from backcrossing the F_1 progeny from a cross between two parents where one parent has the genotype *AABBCC* and the other has the genotype *aabbcc*, to the recessive (*aabbcc*) parent.

Phenotype	Observed	Expected
A_B_C_	500	149.375
aabbcc	510	149.375
aaB_C_	50	149.375
A_bbcc	55	149.375
A_bbC_	35	149.375
aaB_cc	38	149.375
A_B_cc	4	149.375
aabbC_	3	149.375
Total	1195	

Obviously, the observed frequencies are very different from those expected.

The first point to note is that the frequency of the parental genotypes is considerably higher than expected, while all other, **recombinant,** types are less than expected.

Next, it is necessary to know the relative order that the loci appear on the chromosome. Such ordering or arrangement of loci along a chromosome is known as a genetic map. The middle locus can usually be determined from the phenotype that is observed at the lowest frequency. In this example the lowest observed phenotypes are A_B_cc and $aabbC\_$, both of which are effectively parental types at the A and B locus but involves non-parental combinations with C. As recombination of only the centre gene will involve a **double crossover** event, the frequency of occurrence will be lowest. Conversely, the pair with most phenotypic observed classes for recombinants (not counting double crossovers) involves the outside genes. From this, the C-locus would appear to be the middle one (i.e. requires a recombination event between A- and C-locus, plus recombination between the B- and C-locus).

Consider the possible combinations of alleles and the frequency at which they occur. For the $A - B$ loci we have:

Parental types	$A\_B\_$	=	$500 + 4$	$= 504$
	$aabb$	=	$510 + 3$	$= 513$
Recombinants	$aaB\_$	=	$50 + 38$	$= 88$
	A_bb	=	$55 + 35$	$= 90$

So percentage of recombinants = $(178/1195) \times 100 = 14.9\%$.

For $A–C$ loci we have:

Parental types	$A\_C\_$	=	$500 + 35$	$= 535$
	$aacc$	=	$510 + 38$	$= 548$
Recombinants	$aaC\_$	=	$50 + 3$	$= 53$
	A_cc	=	$55 + 4$	$= 59$

So percentage of recombinants = $(112/1195) \times 100 = 9.4\%$.

For the $B–C$ loci we have:

Parental types	$B\_C\_$	=	$500 + 50$	$= 550$
	$bbcc$	=	$510 + 55$	$= 565$
Recombinants	$bbC\_$	=	$35 + 3$	$= 38$
	B_cc	=	$38 + 4$	$= 42$

So percentage of recombinants = $(80/1195) \times 100 = 6.7\%$.

We obtain the **_map distances_** as being equivalent to the percentage recombinants and so we have:

$$A\text{-----------}14.9m\mu\text{------------}B$$
$$A\text{------}9.4m\mu\text{-------}C\text{---}6.7m\mu\text{---}B$$

From the map it is clear that the map distance A to B ($14.9\,m\mu$) is less than the added distance A–C plus C–B. The method of calculation used to estimate these distances assumes that only the non-parental types (i.e. recombinants) are included as the results of recombination, as would be the case where the linkage between two loci is considered. Where three loci are involved it is necessary to include the double crossover recombination events in estimating the distance between the furthest two loci. In this case we could calculate the $A–B$ distance from:

Non-recombinant types	$A\_B\_C\_$	500	$= 500$
	$aabb$	510	$= 510$
Single recombinant types	$aaB\_$	$50 + 38$	$= 88$
	A_bb	$55 + 35$	$= 90$
Twice × double crossover recombinants		$2(3 + 4)$	$= 14$

Therefore, the percentage of recombinants = $(192/1195) \times 100 = 16.1\%$, giving an estimated genetic distance between A–B of $16.1\,m\mu$, which is now the same distance as estimated by simply adding $A–C$ to $C–B$.

This indicates a general flaw in linkage map distance estimation, in that where there is no 'centre gene' locus, it is impossible to detect double recombination events, and as a result map distances will always provide underestimates of actual recombination frequencies.

To avoid such discrepancies, recombination frequencies, which are not additive, are usually converted to a cM scale using the function published by J.B.S. Haldane in 1919, being:

$$cM = (-\ln(1 - 2R)) \times 50$$

where R is the recombination frequency. From this we see that the map distances transform to:

$$A\text{-----------}17.6\,cM\text{------------}B$$
$$A\text{-------}10.4\,cM\text{-----}C\text{---}7.2\,cM\text{---}B$$

The standard error of these map distances can be calculated using the equation:

$$s.e. = [R/(1 - R)]/n$$

where R is the recombination frequency, and n is the number of plants observed in the three-way test cross.

5.2.4 Pleiotropy

Very tight linkage between two loci can be confused with pleiotropy, the control of two or more characters by a single gene. For example, the linkage of resistance to the soybean cyst nematode and seed coat colour seems to be a case of pleiotropy because the two characters are always inherited together. The only way linkage and pleiotropy can be distinguished is effectively the negative way, that is, to find a crossover product such as a progeny homozygous for resistance and yellow seed coat. Resistance and yellow seed coat could never occur in a true breeding individual if true pleiotropy was present. Thousands of individuals may have to be grown to break a tight linkage and prove that it is not actually pleiotropy.

5.2.5 Epistasis

Different loci may be independent of each other in their segregation and recombination patterns, but independence of gene transmission, however, does not necessarily imply independence of gene action or expression. In fact, in terms of its final expression in the phenotype of the individual, *no gene acts by itself*.

A character can be controlled by genes that are inherited independently but that interact to form the final phenotype. The interaction of genes at different loci that affect the same character is called either non-allelic interaction or *epistasis*. Epistasis was originally used to describe two different genes that affect the same character, one that masks the expression of the other. The gene that masks the other is said to be epistatic to it. The gene that masked was termed hypostatic. Epistasis causes deviations from the common phenotypic ratios in F_2 such as 9:3:3:1, which indicates segregation of two independent genes, each with complete dominance.

To examine epistasis, consider the following simple examples. A white kernel wheat cultivar is crossed to one with coloured kernels. The F_1 progeny all have coloured kernels. This situation could easily be interpreted as indicating that coloured kernels (CC) are dominant to white (cc). However, when the F_2 progeny is examined, the ratio observed is 15 coloured kernels to 1 white kernel, a much higher proportion of coloured kernels than expected (3:1) if a single dominant gene is responsible. In fact, two genes control kernel colour in wheat (gene A and gene B). Coloured wheat kernels are produced from a precursor product being acted on by an enzyme, or in this case either one of two enzymes, enzyme A or enzyme B. So when a homozygous wheat cultivar with coloured kernels ($AABB$) is crossed to one with white kernels ($aabb$), the F_1 ($AaBb$) produces both the A enzyme and the B enzyme, and hence all kernels are coloured. Now phenotypes from the F_2 progeny should have a ratio of 9 $A\_B\_$: 3 A_bb : 3 $aaB\_$: 1 $aabb$. However, coloured kernels can result from either the A-gene (enzyme A) or the B-gene (enzyme B), acting alone or together, so then the first 3 types ($A\_B\_$, A_bb, and $aaB\_$) all have coloured kernels and only the $aabb$ types have white kernels. This is called duplicate dominant epistasis where A is epistatic to B and bb, and B epistatic to A and aa.

Consider now a cross between two sweet pea cultivars, both homozygous, one with coloured flowers (anthocyanin) and the other with white flowers (no anthocyanin). All F_1 plants have coloured flowers. Again this could be interpreted as a situation whereby flower colour is controlled by a single dominant gene (CC). However, when the F_2 progeny is examined the ratio observed is 9 coloured flowers: 7 white flowers, certainly not a simple 3:1 ratio, or indeed a 9:3:3:1 ratio, expected in a simple dominant case of one and two genes, respectively. Once again it is known that two genes control flower colour in sweet pea (gene C and gene P). Anthocyanin production (and hence coloured flowers) in sweet pea requires two operations whereby a precursor is acted on by an enzyme C, produced by the presence of the dominant C-allele, to produce an intermediate product, which is then acted on by a different enzyme P, produced by the dominant P-allele. Therefore anthocyanin and coloured flowers only occur when both the C- and P-alleles are present. So coloured sweet pea flowers can only result from both the enzyme C (C-gene) and enzyme P (P-gene) produced together (i.e. 9/16 $C\_P\_$). From the other possible types: C_pp produces enzyme C, but not enzyme P; $ccP\_$ produces enzyme P but not enzyme C; while $ccpp$ produces neither enzyme, and so all these types have white flowers. This is termed duplicate recessive epistasis whereby *cc* is epistatic to *P*, and *pp* is epistatic to *C*.

One final example relates to fruit colour in squash where two loci are involved. At one locus, fruit colour is recessive to no colour. The first locus must be homozygous for the recessive allele before the second colour gene at another locus is expressed. At the first locus, white fruit colour is dominant (*W*) over coloured fruit (*w*). At the second locus yellow fruit colour (*G*) is dominant over green colour (*g*), which is recessive. When a heterozygous F$_1$(*WwGg*) is self-pollinated, three different fruit colours white:yellow:green are observed in a 12:3:1 ratio. All *W_* phenotypes (i.e. 9/16 *W_G_* and 3/16 *W_gg*) are white, because of dominant epistasis where *W* is epistatic to *G* and *g*. Phenotypes with *wwG_* (3/16) have yellow fruit, while *wwgg* genotypes (1/16) have green fruit.

All possible phenotypic ratios in F$_2$ for two unlinked genes, as influenced by dominance at each locus and epistasis between loci, are shown in Table 5.4.

In some ways the interaction between two different loci, in which the allele at one locus affects the expression of the alleles at another, will remind you of the phenomenon of dominance. The two phenomena are essentially different, however. **Dominance** always refers to the expression of one member of a pair of alleles relative to the other at the same locus, as opposed to another locus; **epistasis** is the term generally used to describe effects of non-allelic genes (i.e alleles at different loci) on each other's expression, in other words their interaction.

5.2.6 Qualitative inheritance in tetraploid species

Compared with crop species that are cultivated as pure inbred lines, there has been comparatively little research carried out on inheritance in auto-tetraploid species. This has been primarily due to the fact that the major autotetraploid crop species (e.g. potato) are clonally reproduced or are outbreeding species that suffer severe inbreeding depression (i.e. alfalfa). Many (or all) of these cultivars are highly heterozygous, and hence it is not as easy to carry out simple genetic experiments.

However, major gene inheritance in autotetraploids has been the topic of research/breeding

Table 5.4 Phenotypic ratios of progeny in the F$_2$ generation for two unlinked genes (where *A* is dominant to *a*, and *B* is dominant to *b*), with epistasis between loci.

F$_1$ phenotype				Explanation
*A_B_*	*A_bb*	*aaB_*	*aabb*	
9	3	3	1	No epistasis.
9	3	4	0	Recessive epistasis: *aa* epistatic to *B* and *bb*.
12	0	3	1	Dominant epistasis: *A* epistatic to *B*, or *bb*.
13	0	3	0	Dominant and recessive epistasis: *A* epistatic to *B* and *bb*; *bb* epistatic to *A* and *aa*. *A_* and *bb* produce identical phenotypes.
9	0	7	0	Duplicate recessive epistasis: *aa* epistatic to *B*, and *bb*; and *bb* epistatic to *A* and *aa*.
15	0	0	1	Duplicate dominant epistasis: *A* epistatic to *B* and *bb*; *B* epistatic to *A* and *aa*.

groups with the aim of parental development. Consider, for example, the potato crop, where there are several single traits that are controlled by major genes such as resistance to Potato Virus X, Potato Virus Y, Potato Cyst Nematode (*G. rostochiensis*) and late blight (*Phytophthora infestans*). All these qualitative traits show complete dominance. The technique used to develop parents in a breeding programme is aimed at increasing the proportion of desirable offspring in sexual crosses, and in the extreme to avoid the need to test breeding lines for the presence of the allele of interest. The technique is called ***multiplex breeding***.

In tetraploid crops any genotype may be ***nulliplex*** (*aaaa*), having no copies of the desired allele at the specific (*A*) locus; ***simplex*** (*Aaaa*), having only one copy of the desirable resistance allele (*A*); ***duplex*** (*AAaa*), having two copies of the allele; ***triplex*** (*AAAa*), having three copies of the allele; or ***quadruplex*** (*AAAA*) having four copies of the allele (i.e. homozygous at that locus). If the alleles show dominance, then genotypes that have at least one copy of the gene will, phenotypically, appear identical in terms of their resistance. But they will differ in their effectiveness as parents in a breeding programme. To determine the genotype of a clonal line, test crossing is necessary.

Table 5.5   The ratios of resistant to susceptible progeny amongst the genotypes derived by crossing a simplex, duplex, triplex and quadruplex resistant gene parent to a nulliplex parent.

Cross type	Phenotype Resistant : Susceptible	% resistant in progeny
Simplex (*Rrrr*) × nulliplex (*rrrr*)	1 : 1	50
Duplex (*RRrr*) × nulliplex (*rrrr*)	5 : 1	83
Triplex (*RRRr*) × nulliplex (*rrrr*)	1 : 0	100
Quadruplex (*RRRR*) × nulliplex (*rrrr*)	1 : 0	100

To illustrate the usefulness of multiplex breeding in potato, consider the problem of developing a parental line which, when crossed to any other line (irrespective of the genotype of the second parent), will give progeny all of which will be resistant to potato cyst nematode by having at least one copy of the H_1 gene (a qualitative resistance gene conferring resistance to all UK populations of the damaging nematode *Globodera rostochiensis*, and which has been shown to give relatively durable resistance).

The aim of multiplex breeding is therefore to develop parental lines that are either triplex or quadruplex (three or four copies of the desirable allele, respectively) for the H_1 allele. When crossed to any other parent, these multiplex lines will produce progeny that have at least a single copy of the H_1 gene and, due to dominance, all will be phenotypically resistant to *G. rostochiensis* nematodes. It will therefore not be necessary to screen for *G.*

rostochiensis nematode resistance in these progeny. The ratios of resistant to susceptible amongst the progeny of genotypes derived by crossing a simplex, duplex, triplex and quadruplex to a nulliplex are shown in Table 5.5.

Genotype ratios from all possible cross-combinations between nulliplex, simplex, duplex, triplex or quadruplex parents are given in Table 5.6.

It was at first thought that developing multiplex parents would be very difficult; instead it proved simply to be a matter of effort and application. The main difficulty lies in the fact that progeny tests need to be carried out to test the genetic makeup of the dominant parental lines (very similar to backcrossing where the non-recurrent parent has a single recessive gene of interest). Consider one of the worst situations, where both starting parents are simplex. Three-quarters of the progeny will be resistant, but only a quarter will be duplex. So a quarter of the progeny will be nulliplex, hence susceptible, and so upon testing can be immediately discarded. The resistant progeny, however, need to be testcrossed to a nulliplex to distinguish the duplex from simplex genotypes. Once identified, the selected duplex lines are intercrossed or selfed. In their progeny, 1/36 will be quadruplex and 8/36 will be triplex. A single round test cross with a nulliplex will be necessary to distinguish, on the one hand, the triplex and quadruplex lines from, on the other, those that are either duplex or simplex. A second round test cross using the progeny of the first test cross will be necessary (i.e. a second backcross to a nulliplex line) in order to distinguish the quadruplex from triplex lines.

Table 5.6   Genotype ratios from all possible cross-combinations between nulliplex (N), simplex (S), duplex (D), triplex (T) or quadruplex parents (Q).

Cross	Nulliplex (N)	Simplex (S)	Duplex (D)	Triplex (T)	Quadruplex (Q)
Nulliplex (*aaaa*)	All N				
Simplex (*Aaaa*)	1S : 1N	1D : 2S:1N			
Duplex (*AAaa*)	1D : 4S : 1N	1T : 5D : 5S : 1N	1Q : 8T : 18D : 8S : 1N		
Triplex (*AAAa*)	1D : 1S	1T : 2D : 1S	1Q : 5T : 5D : 1S	1Q : 2T : 1D	
Quadruplex (*AAAA*)	All D	1T : 1D	1Q : 4T : 1D	1Q : 1T	All Q

Once quadruplex lines have been identified, these can be continually intercrossed, or selfed, without further need to test. Similarly, quadruplex or triplex parental lines can be used in cross combination with any other parental lines and 100% of the resulting progeny will contain at least one of the dominant resistant alleles and show resistance. Multiplex breeding can be used in a similar way to develop parents in hybrid cross combinations.

5.2.7 The chi-square test

If plant breeders have an understanding of the inheritance of simply inherited characters, it is possible to predict the frequency of desired genes and genotypes in a breeding population. Breeders often ask questions relating to the nature of inheritance as well as the number of alleles or loci involved in the inheritance of a particular character of importance. For example, it is often valuable to determine whether a single gene is dominant, recessive or additive in inheritance. In the dominant/recessive case, the F_2 family will segregate in a 3:1 dominant : recessive ratio, while in the latter a 1:2:1 homozygous dominant : heterozygous : homozygous recessive ratio.

In segregating families such as those noted above, the ratios actually observed will not be exactly as predicted due to sampling error. For example, a coin is expected to fall as heads or tails in the ratio 1:1, but if tossed 10 times it does not always result in 5 heads and 5 tails. So the breeder is faced with interpreting the ratios that are observed in terms of what is expected, so it might be necessary to decode whether a particular ratio is really 1:2 or 1:3 or 3:4 or 9:7. How can this be done objectively? The answer is to use chi-square tests.

It should be clear that the significance of a given deviation is related to the size of the sample. If we expect a 1:1 ratio in a test involving six individuals, an observed ratio of 4:2 is not at all bad. But if the test involves 600 individuals, an observed ratio of 400:200 is clearly a long way off. Similarly, if we test 40 individuals and find a deviation of 10 in each class, this deviation seems serious:

observed	30	10
expected	20	20
obs−exp (difference)	10	−10

But if we test 200 individuals, the same numerical deviation seems reasonably enough explained as a purely chance effect:

observed	90	110
expected	100	100
obs−exp (difference)	−10	10

The statistical test most commonly used when such a problem arises is simple in design and application. Each deviation is *squared,* and the expected number in its class then divides each squared deviation. The resulting quotients are then all added together to give a single value, called the chi-square (χ^2). To substitute symbols for words, let d represent the respective deviations (observed minus expected), and e the corresponding expected values, then:

$$\chi^2 = \sum (d^2/e)$$

We can calculate chi-square for the two arbitrary examples above to show how this value relates the magnitude of the deviation to the size of the sample.

	Sample of 40 individuals		Sample of 200 individuals	
Observed (obs)	30	10	90	110
Expected (exp)	20	20	100	100
obs−exp (d)	10	10	10	10
d^2	100	100	100	100
d^2 / exp	5	5	1	1
χ^2		10		2

You might note that the value of chi-square is much larger for the smaller population, even though deviations in the two populations are numerically the same. In view of our earlier common-sense comparison of the two, this is a practical demonstration that the calculated value of chi-square is related to the significance of a deviation. It has the virtue of reducing many different samples, of different sizes and with different numerical deviations, to a common scale for comparison.

The chi-square test can also be applied to samples including more than two classes. For example, the table below shows the chi-square analyses of the tall 6-row × short 2-row barley example earlier. Suppose that a test cross was carried out where the heterozygous F_1 (*TtSs*) is crossed to a genotype with the recessive alleles at these loci. When this was actually done, and 400 test cross progeny were grown out, there were 112 tall and 6-row, 89 tall and 2-row, 93 short and 6-row, and 106 short and 2-row. Inspection of this would show that there is a higher proportion of each of the original parental types (tall/6-row and short/2-two) than the recombinant types. The question is, therefore: is this linkage, or simple random sampling variation? To determine this we would use the χ^2 test.

	Tall, 6-row	Tall, 2-row	Short, 6-row	Short, 2-row
Observed (obs)	112	89	93	106
Expected (exp)	100	100	100	100
obs−exp(d)	12	11	7	6
d^2	144	121	49	36
$d^2/\exp$	1.44	1.21	0.49	0.36

$$\chi^2_{3\,\mathrm{df}} = \sum (d^2/\exp) = 3.50, \text{ n.s.}$$

The number of degrees of freedom in tests of genetic ratios is almost always *one less than the number of genotypic classes*. To be more precise, it is the number of observable data that are independent. For example, in a two-gene test cross there are four possible phenotypes, expected in equal frequency. So, in our example, 400 plants were observed and there were 112 in the first group, 89 in the second group and 93 in the third group; then by definition to sum to 400 there must be 106 in the last group $(400 - 112 - 89 - 93 = 106)$. Therefore our chi-square test would have 3 degrees of freedom. In just the same way, if we were testing two groups in tests of 1:1 or 3:1 ratios, they will have one degree of freedom.

Do not confuse assigning degrees of freedom to genetic frequencies with degrees of freedom in χ^2 contingency tables. In a two-way contingency table, with pre-assigned row and column totals, one value can be filled arbitrarily, but the others are then fixed by the fact that the total must add up to the precise number of observations involved in that row or column. When there are four classes, any three are usually free, but the fourth is fixed. Thus, when there are four classes, there are usually three degrees of freedom.

Remembering the example given earlier with 40 individuals, the calculated chi-square = 10 for one degree of freedom. We can look up this value in the probability tables for chi-square, and in this case chance alone would be expected in considerably less than one in a hundred independent trials to produce as large a deviation as that obtained. We cannot reasonably accept chance alone as being responsible for this particular deviation; it represents an event that would occur, on a chance basis, much less often than the one-time-in-20 that we have agreed on as our point of rejection; this event would occur less often than even the one-time-in-a-100 that we decided to regard as highly significant.

In the case of the example we noted for barley, the expected frequency of each of the four phenotypes, if no linkage is present, would be 100 (total of 400, with four equal expectations), which would lead to deviations of 12, 11, 7 and 6; these squared and divided by the expected value $(144/100 + 121/100 + 49/100 + 36/100)$ gives a χ^2 value of 3.5. This value is compared in probability tables for χ^2 values and it falls just below the tabulated 50% probability value with 3 degrees of freedom, clearly not close to the accepted 5% probability we define as showing significance. We therefore say that there is no evidence that linkage exists, and that the observed deviations are likely to have occurred by random chance (sampling error).

Improper use of chi-square

The two most important reservations regarding the straightforward use of the chi-square method in genetics are:

- Chi-square can usually be applied only to numerical frequencies themselves, not to percentages or ratios derived from the frequencies. For example, if in an experiment one expects equal numbers in each of two classes, but observes 8 in one class and 12 in the other, we might express the observed numbers as 40% and 60%, and the expected as 50% in each class. A chi-square value computed from these percentages cannot be used directly for the determination of *the probability*. When the classes are large, a chi-square value computed from percentages can be used, if it is multiplied by $n/100$, where n is the total number of individuals observed.

- Chi-square cannot properly be applied to distributions in which the expected frequency of any class is less than 5. In fact, some statisticians suggest that a particular correction be applied if the frequency of any class is less than 50. However, the approximations involved in chi-squares are close enough for most practical purposes when there are more than 5 expected in each class.

5.2.8   Family size necessary in qualitative genetic studies

It is usual in genetic work for the scope of the experiments to be limited by such considerations as lack of available space, labour or money. It is therefore essential to make the best use of available resources, which will be a function of the number of plants/plots that need to be raised. Achieving this usually requires considerable care and planning of experiments. Often statistical considerations can be of great value.

In many experiments it is desirable to be able to pick out certain genotypes, usually (but not always) homozygous, with the aim of developing superior cultivars. It may also be necessary to detect some particular nonconformity. For example, to detect any linkage effects will involve test crossing F_1 lines onto a recessive parent. It would be advantageous to keep the population size down to a minimum while also assuring with high probability that the experiment will be sufficiently large so as to detect the required differences.

Let us consider now the question of detecting homozygotes in a segregating population. Any progeny that fails to segregate could have derived from a homozygous parent or could fail to show segregation because of sampling error. The greater the numbers of individuals that are examined, the lower the probability that sampling error will interfere with the interpretation of experimental results. The minimum size of progeny designed to test a particular ratio is then a statistical question involving consideration of the probability that any individual in a family derived from a heterozygote will be of the recessive type, and also the permissible maximum probability of obtaining a misleading result.

Consider the following example. Suppose you wish to test a series of individuals phenotypically dominant for a single gene, in order to identify the homozygous individuals, by using a test cross to a homozygous recessive. The progeny of the homozygotes will not show segregation whereas progeny from the heterozygotes will segregate in a 1:1 ratio (i.e. $\frac{1}{2}Aa : \frac{1}{2}aa$). The important error that can arise will be from our failure to detect any segregants in the progenies from crosses that actually do have a heterozygous parent. Let us also assume that we do not want the test to fail with greater probability than once in 100 experiments or tests (i.e. $p < 0.01$). In the progeny of a heterozygote, each individual has a chance of 1/2 of containing a dominant allele. Then a family of n will be expected to have $(1/2)^n$ individuals with one dominant allele. Therefore we can predict the possibility of having no such individuals, which represents the misleading (error) result which must be avoided and must not occur at a higher frequency than 1 in 100. Then the minimum value of n is given by the solution of the equation:

$$1 - (1/2)^n = 1/100$$

Taking natural logarithms this becomes:

$$n \times \ln(1 - 1/2) = \ln(1/100)$$

Therefore:

$$n = \ln(1/100)/\ln(1 - 1/2)$$

In our example, $n = 2/0.3010 = 6.6$. Therefore the smallest family size needed would be seven or more, in order to be at least 99% certain of detecting the difference.

This formula can be generalized to:

$$n = \ln(1 - p)/\ln(1 - x)$$

where n is the number that need to be evaluated, p is the probability of having at least one type of interest (i.e. 90% certain = $p = 0.9$) and x is the frequency of the desirable genotype/phenotype.

It is often necessary in plant breeding to estimate the number of individuals that need to be grown or screened in order to identify at least r individuals (where r is greater than one). It is never a good idea to simply multiply n (from above), the number that needs to be grown to ensure at least one by r the number required, as this will result in a gross overestimation.

An alternative is to use an extension to the above equation. The mathematics behind this equation is outside the scope of this book. However, the equation itself can be useful. It is:

$$n = \left(\frac{[2(r - 0.5) + z^2(1 - q)]}{+z[z^2(1 - q)^2 + 4(1 - q)(r - 0.5)]^{0.5}} \right)(2q)^{-1}$$

where n is the total number of plants that need to be grown; r is the number of plants with the required alleles that need to be recovered; q is the frequency of plants with the desired alleles; p is the probability of recovering the desired number of plants with the desired alleles; and z is a cumulative normal frequency distribution (area under standardized normal curve from 0 to z), a function of probability (p). For the sake of simplicity and to cover the most commonly used situations, $z = 1.645$ for $p = 0.95$ (i.e. 95% certainty) and $z = 2.326$ (for 99% certainty).

For example, consider the frequency of homozygous recessive genotypes (*aabb*) resulting from the cross $AAbb \times aaBB$ at the F_3 stage. How many F_3 lines would need to be evaluated to be 95% certain of obtaining ten homozygous recessive genotypes? Here the probability of *aabb* is 9/64 (i.e. $q = 0.141$), $p = 0.95$, therefore $z = 1.645$.

$$n = \left(\begin{array}{l} [2(10 - 0.5) + 2.706(1 - 0.141)] \\ +1.645[2.706(1 - 0.141)^2 \\ +4(1 - 0.141)(10 - 0.5)]^{0.5} \end{array} \right)(2 \times 0.141)^{-1}$$

$$= 110.69$$

Therefore a minimum of 111 F_3 lines would need to be grown.

5.3 Quantitative genetics

5.3.1 The basis of continuous variation

With qualitative inheritance, the segregating individual phenotypes are usually easily distinguished and fall into a few phenotypic classes (i.e. tall or short; white or red; resistant or susceptible). Characters that are controlled by multiple genes do not fall into such simple classifications. Let us consider a hectare field of potatoes, planted with one cultivar, and so every single plant in the field should be of identical genotype (ignoring mutations and errors). In that field there are likely to be 11,000 plants. At harvest, you have been given the task of harvesting all 11,000 plants and weighing the tubers that come from each plant separately. Would you expect all the potato plants to have exactly the same weight of potatoes? Probably not. Indeed, the weights can be presented in the form of a histogram (Figure 5.1) where 11,000 yields were divided into 17 weight classes. The variation in weight is obvious; some plants produce less than 0.5 kg of tubers, while others produce over 5 kg. Most, however, are grouped around the average of 3.2 kg per plant.

Yield in potato, as in other crops, is polygenically inherited. Yield is therefore not controlled by a single gene but by many genes, all acting collectively. Thus yield has more chance of some of the processes these many genes control being affected by differences in the 'environment'. As the example above involved harvesting plants that are all indeed clones of the same genotype, then the variation observed is due entirely to the environmental conditions to which each plant was subjected.

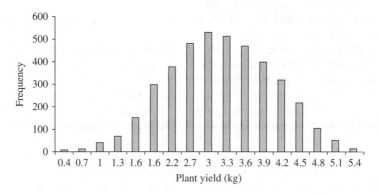

Figure 5.1 Yield of tubers from individual potato plants taken at random from a single clonally reproduced potato cultivar.

One of the major differences between single gene inheritance and multiple gene inheritance is that the former is relatively less affected or influenced by the environmental conditions compared with the later – at least in terms of the differences being expressed. When potato yields are being recorded we are recording the phenotype, but in reality as breeders we are interested in the genotype. The two are related by the equation:

$$P = G + E + GE + \sigma_e^2$$

where P is the phenotype, G is the genotypic effect, E is the effect of the environment in which the genotype is grown, GE is the interaction between the genotype and the environment, and σ_e^2 is a random error term. Unfortunately, the fact that agronomic practices are part of the environment plants face is often neglected. Therefore in order to reduce the overall environmental impact on the field expression of the genotype, it is essential to provide at all times the most uniform agronomic management to breeding trials.

As noted, relatively single gene characters are often less affected by environment, and so do not show much influence of genotype × environment interactions (i.e. the difference in expression of the two alleles is large in comparison with the variation caused by environmental changes). For example, a potato genotype with white flowers (a qualitatively inherited trait) will always have white flowers in any environment in which the plants produce flowers. Conversely, quantitatively inherited traits are greatly influenced by environmental conditions and genotype × environment interactions are common, and can be large. The greatest difficulty plant breeders face is dealing with quantitative traits, and in particular in detecting the better genotypes based on their phenotypic performance. This is why it is critically important to employ appropriate experimental design techniques to genetic experiments and also plant breeding programmes.

A major part of quantitative genetics research related to plant breeding has been directed towards partitioning the variation that is observed (i.e. phenotypic variation) into its genetic and non-genetic portions. Once achieved, this can be taken further to further divide the genetic portion into that which is additive in nature and that which is non-additive (quite often dominance variation). Obviously, in breeding self-pollinating crops, the additive genetic variance is of primary importance since it is that portion of the variation attributable to homozygous gene combinations in the population and is what the breeder is aiming for (i.e. homozygous lines). On the other hand, variance due to dominance is related to the degree of heterozygosity in the population and will be reduced (to zero) over time with inbreeding as breeding lines move towards homozygosity, but will be important when it comes to out-breeding species.

Let us now return to the potato weights as one of many possible examples of continuous variation. If the frequency distribution of potato yields is inspected, there are two points to note: (1) the distribution is symmetrical (i.e. there are as many high-yielding plants as really low-yielding ones); and (2) the majority of potato yields were clustered around a weight in the middle of the distribution. As we have taken a class interval of 0.3 kg to produce this distribution, the figure does not look particularly continuous. However, we know that potato yields do not go up in increments of 0.3 kg but show a more continuous and gradual range of variation. If we use more class intervals in this example we will produce a smoother histogram, and if we use infinitely small class intervals it will result in a truly continuous bell-shaped curve. The shape of this curve is highly indicative of many aspects of plant science and it is a distribution called a *normal distribution*, and occurs in a wide variety of aspects relating to plant growth, particularly quantitative genetics.

5.3.2 Describing continuous variation

The normal distribution

The 11,000 potato plant weights discussed above are a sample, albeit a large one, of possible potato weights from individual plants. It is possible to predict mathematically the frequency distribution for the population as a whole (i.e. every possible potato plant of that cultivar grown), provided it is assumed that the sample is representative of the population (i.e. that our sample is an unbiased sample of all that was possible).

It is not necessary to actually draw normal distributions (which, even with the aid of computer graphics, are difficult to do accurately). Most of the important properties of a normal distribution can be characterized by two statistics, the mean or average (μ) of the distribution and the standard deviation (σ), a measure of the *'spread'* of the distribution.

There are in fact two means, the mean of the sample and the mean of the population from which the sample was drawn. The latter is represented by the symbol μ, and is, in reality, seldom known precisely. The *sample mean* is represented by $\bar{x}$ (spoken *x* bar), and it can be known with complete accuracy. Generally, the best estimate of a *population mean* (μ) is the actual mean of an unbiased sample drawn from it ($\bar{x}$). The population mean is thus best estimated as:

$$\mu = \bar{x} = (x_1 + x_2 + x_3 + \ldots + x_n)/n = \sum_{i}^{n}(x_i)/n$$

where $\sum_{i}^{n}(x_i)$ is the sum of all *x* values from $i = 1$ to *n*.

The standard deviation is an ideal statistic to examine the variation that exists within a dataset. For any normal distribution, approximately 68% of the population sampled will be within one standard deviation from either side of the mean, approximately 95% will be within two standard deviations (Figure 5.2), and approximately 99% will be within three standard deviations of the mean.

Once again, it is necessary to distinguish between the actual standard deviation of a population, all of whose members have been measured, and the estimated standard deviation of a population based on measuring a sample of individuals from it. The standard deviation of a population is represented by the symbol σ (Greek letter sigma) and is defined thus:

$$\sigma = \sqrt{\left[\sum_{i=1}^{n}(x_i - \mu)^2/n\right]}$$

while the standard deviation of a sample is represented by the symbol $\hat{s}$, and is given by:

$$\hat{s} = \sqrt{\left[\sum_{i=1}^{n}(x_i - \bar{x})^2/(n-1)\right]}$$

Another measure of the spread of data around the **mean** is the *variance*, which is the square of the standard deviation. The *estimated variance* of a population (i.e. obtained from the sample) is given by:

$$\hat{s}^2 = \left[\sum_{i=1}^{n}(x_i - \bar{x})^2\right]\bigg/(n-1)$$

and the *actual variance* of an entire population (if every member of it has been measured) is given by:

$$\sigma^2 = \left(\sum_{i=1}^{n}(x_i - \mu)^2\right)\bigg/n$$

Calculators are often programmed to give means and either standard deviations or variances with a few key strokes once data have been entered. However, in case it is necessary to derive these descriptive statistics semi-manually, it is useful to know about alternative equations for $\hat{s}^2$ and σ^2:

$$\hat{s}^2 = \left\{\sum_{i=1}^{n}(x_i^2) - \left[\left[\sum_{i=1}^{n}(x_i)\right]^2/n\right]\right\}\bigg/(n-1)$$

$$\sigma^2 = \left\{\sum_{i=1}^{n}(x_i^2) - \left[\left[\sum_{i=1}^{n}(x_i)\right]^2/n\right]\right\}\bigg/n$$

Although these *look* more complicated than those given previously, they are *easier to use* because the mean does not have to be worked out first (which would entail entering all the data into the calculator twice). Note the difference between $\sum(x_i^2)$ (each value of *x* squared and then the squares totalled) and $\left[\sum(x_i)\right]^2$ (the values of *x* totalled and then the sum squared).

Standard deviations, as measures of spread around the mean, are probably intuitively more

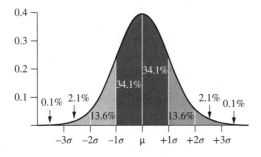

Figure 5.2 95% of a population that is normally distributed will lie within two standard deviations of the population mean.

understandable than variances; for example, 68% of the population fall within one standard deviation of the mean. Why introduce the complication of variances? Well, *variances are additive* in a way standard deviations are not. Thus, if the variances attributable to a variety of factors have been estimated, it is mathematically valid to sum them to estimate the variance due to all the factors acting together. Similarly, the reverse is also true, in that a total *variance* can be *partitioned* into the variances attributable to a variety of individual factors. These operations, which are used extensively in quantitative genetics, cannot be so readily performed with standard deviations.

Variation between datasets

Two basic procedures are frequently used in quantitative genetics to interpret the variation and relationship that exists between characters, or between one character evaluated in different environments. These are simple linear regression and correlation.

A straight-line regression can be adequately described by two estimates: the slope, or **gradient** of the line, (**b**) and the **intercept** on the y-axis (**a**). These are related by the equation:

$$y = a + bx$$

It can be seen that b is the gradient of the line, because a change of one unit on the x-axis results in a change of b units on the y-axis. If x and y both increase (or both decrease) together, the gradient is positive. If, however, x increases while y decreases or vice versa, then the gradient is negative. When $x = 0$, the equation for y reduces to:

$$y = a$$

and **a** is therefore the point at which the regression line crosses the y-axis. This intercept value may be equal to, greater than or less than zero.

The formulation and theory behind regression analysis will not be described here and is not within the scope of this book. However, the gradient of the best-fitting straight line (also known as the **regression coefficient**) for a collection of points whose coordinates are x and y is estimated as:

$$b = [SP(x, y)/SS(x)]$$

where $SP(x,y)$ is the **sum of products** of the deviations of x and y from their respective means ($\bar{x}$ and $\bar{y}$) and $SS(x)$ is the **sum of the squared deviations** of x from its mean. It will be useful to have an understanding of the regression analysis and to remember the basic regression equations.

Now, $SP(x,y)$ is given by the equation:

$$SP(x, y) = \sum_{i=1}^{n}(x_i - \bar{x})(y_i - \bar{y})$$

although in practice it is usually easier to calculate it using the equation:

$$SP(x, y) = \sum_{i=1}^{n}(x_i y_i) - \left[\sum_{i=1}^{n}(x_i)\sum_{i=1}^{n}(y_i)\right] \Big/ n$$

The comparable equations for $SS(x)$ are:

$$SS(x) = \sum_{i=1}^{n}(x_i - \bar{x})^2$$

$$SS(x) = \sum_{i=1}^{n}(x_i^2) - \left\{\left[\sum_{i=1}^{n}(x_i)\right]^2\right\} \Big/ n$$

Notice that a sum of squares is really a special case of a sum of products. You should also note that if every y value is exactly equal to every x value, then the equation used to estimate **b** becomes $SS(x)/SS(x) = 1$.

Having determined **b**, the intercept value is found by substituting the mean values of x and y into the rearranged equation:

$$a = \bar{y} - b\bar{x}$$

In regression analysis it is always assumed that one character is the dependant variable and the other is the independent variable. For example, it is common to compare parental expression with progeny expression (see Chapter 6), and in this case then progeny expression would be considered the dependant variable and parental expression independent. The expression of progeny is obviously dependent upon the expression of their parents, and not *vice versa*.

In addition, the degree of association between any two or a number of different characters can be examined statistically by the use of **correlation analysis**. Correlation analysis is similar in many ways to simple regression, but in correlations there is no need to assign one set of values to be the **dependant variable** while the other is said to be the **independent variable**. Correlation coefficients (r) are calculated from the equation:

$$r = \frac{\text{SP}(x, y)}{\sqrt{\text{SS}(x) \times \text{SS}(y)}}$$

where SP(x,y) is again the sum of products between the two variables, SS(x) is the sum of squares of one variable (x) and SS(y) is the sum of squares of the second variable (y), and:

$$\text{SP}(x, y) = \left[\sum_{i=1}^{n} (x_i - \bar{x})(y_i - \bar{y}) \right] \Big/ (n-1)$$

$$\text{SS}(x) = \left[\sum_{i=1}^{n} (x_i - \bar{x})^2 \right] \Big/ (n-1)$$

$$\text{SS}(y) = \left[\sum_{i=1}^{n} (y_i - \bar{y})^2 \right] \Big/ (n-1)$$

These can, of course, sometimes be calculated more easily by:

$$\text{SP}(x, y) = \left\{ \sum_{i=1}^{n}(x_i y_i) - \left[\sum_{i=1}^{n}(x_i) \sum_{i=1}^{n}(y_i) \right] \Big/ n \right\} \Big/ (n-1)$$

$$\text{SS}(x) = \left\{ \sum_{i=1}^{n} x_i^2 - \left[\sum_{i=1}^{n}(x_i) \right]^2 \Big/ n \right\} \Big/ (n-1)$$

$$\text{SS}(y) = \left\{ \sum_{i=1}^{n} y_i^2 - \left[\sum_{i=1}^{n}(y_i) \right]^2 \Big/ n \right\} \Big/ (n-1)$$

Correlation coefficients (r) range in value from -1 to $+1$. r values approaching $+1$ show very good positive association between two sets of data (i.e. high values for one variable are always associated with high values of the other). In this case, we say that the two variables are **positively correlated.** Values of r that are near to -1 show disassociation between two sets of data (i.e. a high value for one variable is always associated with a low value in another). In this case we say that the two variables are **negatively correlated.** Values of r that are near to zero indicate that there is no association between the variables. In this case a high value for one variable can be associated with a high, medium or low value of the other.

5.3.3 Relating quantitative genetics and the normal distribution

Consider two homozygous canola (*Brassica napus*) cultivars (P_1 and P_2). The yield potential of P_1 is 620 kg/plot, and is higher than P_2, which has a yield potential of 500 kg/plot. When these two cultivars were crossed and the F_1 produced, the yield of the F_1 progeny was exactly midway between both parents (560 kg/plot). This would suggest that additive genetic effects rather than dominance were present.

If yield in canola were controlled by a single locus and two alleles (which it is not), we would have:

$$P_1 \times P_2$$
$$AA \times aa$$

It should be noted here that upper and lower case letters denoting alleles do not signify dominance as in qualitative inheritance, but rather simply differentiate between alleles. It is common to assign uppercase letters to alleles from the parent with the greater expression of the trait, and designated as P_1.

When there is only one locus with two alleles involved, we would assume that the uppercase alleles each add 60 kg/plot to the base performance of a plant, and lowercase alleles add nothing. In this case the base performance is equal to P_2 = 500 kg/plot. Therefore, $P_1 = AA = 620$ kg/plant (500 + 60 + 60), $P_2 = aa = 500$ kg/plot (500 + 0 + 0), The $F_1 = Aa = 560$ kg/plot (500 + 60 + 0). In the F_2 we have a ratio of 1 AA : 2 Aa : 1 aa, and we would have three types of plants in the population: $AA = 620$ kg/plot; $Aa = 560$ kg/plot and $aa = 500$ kg/plot.

Obviously, yield in canola is not controlled by two alleles at a single locus. However, let us progress gradually and assume that two loci each with two alleles are involved. We now have:

$$P_1 \times P_2$$
$$AABB \times aabb$$

In this case (assuming alleles at different loci have equal effects), each of the two uppercase alleles would each add 30 kg/plot to the base weight. The $F_1 = Aa = 560$ kg/plot (the same as if only one gene was involved). However, in the F_2 we have 16 possible allele combinations that can be grouped according to the number of uppercase alleles (or yield potential).

		AAbb		
		AaBb		
	aabB	*aABa*	*AABb*	
	aaBb	*AabB*	*AAbB*	
	aAbb	*aAbB*	*AaBB*	
aabb	*Aabb*	*aaBB*	*aABB*	*AABB*
500	**530**	**560**	**590**	**620**

Extending in the same manner one more time, we see that the frequency distribution of the phenotypic classes in the F_2 generation when *three* genes having equal additive effects, and which segregate independently, are:

			aaBbCC			
			aAbBcC			
			aAbBCc			
			aAbbCC			
			AabbCC			
		aabbCC	*AabBcC*	*AABBcc*		
		aabBcC	*AabBCc*	*AABbCc*		
		aabBCc	*AAbbcC*	*AABbcC*		
		aaBbcC	*AabBCc*	*AAbBCc*		
		aaBbCc	*AaBBcc*	*AAbBcC*		
		aaBBcc	*aabBCC*	*AAbbCC*		
		aAbbcC	*aaBbCC*	*AaBBCc*		
		aAbbCc	*aaBBcC*	*AaBbCc*		
		aAbBcc	*aAbBcC*	*AaBbCC*		
	aabbcC	*aABbcc*	*aAbBCc*	*AabBCC*	*AABBcC*	
	aabbCc	*AAbbcC*	*aABBCc*	*aABBCc*	*AABBCc*	
	aabBcc	*AabBCc*	*AAbbcC*	*aABBcC*	*AAbBCC*	
	aaBbcc	*AabBcc*	*AAbBCc*	*aABBcC*	*AABbCC*	
	aAbbcc	*AaBbcc*	*AAbBcc*	*aAbBCC*	*aABBCC*	
aabbcc	*Aabbcc*	*AAbbcc*	*AABbcc*	*aaBBCC*	*AaBBCC*	*AABBCC*
500	**520**	**540**	**560**	**580**	**600**	**620**

In this case each single upper case allele adds only 20 kg to the base weight of 500 kg/plot. This is determined in the same way as for the one and two gene models, although it is considerably more involved. You should note once more that the F_1 would have had a yield potential of 560 kg/plot, exactly the same as in the single and two gene cases.

Even with only three loci and two alleles at each locus, it should be obvious that we are moving closer to a shape resembling a standard normal distribution. The frequency of different genotypes possible when four, five and six loci are considered has 9, 11 and 12 phenotypic classes, respectively.

It is fairly easy to visualize, therefore, that with only a modest number of loci, with segregating alleles acting in a more or less equal additive way, truly continuous variation in a character would be approximated quite closely. Quantitative inheritance deals with many loci and alleles, often too many to consider trying to estimate the number, and therefore explains the ubiquity of the normal distribution. Just as the mean and variance can describe the normal distribution, many of the important elements of the inheritance of a character can be described and explained using progeny means and genetic variances.

5.3.4 Quantitative genetics models

The relationship and importance of the normal distribution to quantitative genetics is clear; however, the closeness of the relationship between observed progeny performances and theoretical distributions will be related to **the model** upon which the relationship is based. For example, we assumed that all uppercase letter alleles were of equal additive value, which of course may not be true. It is important that an appropriate model of inheritance is applied; otherwise other derived statistics (i.e. heritabilities, see later in Chapter 6), which are potentially of great value in plant breeding, will be biased and are likely to be highly misleading.

Let us examine the basic model applied to quantitative situations and see how the model can be tested for its appropriateness to the situation or inheritance of specific characters.

Consider again the cross between two canola cultivars described above. The yield of the higher-yielding parent (P_1) is 620 kg/plot, the yield of the lower-yielding parent (P_2) is 500 kg/plot, and the yield of the F_1 is 560 kg/plot. Assume a model of additive genetic effects, where we have:

where the difference between the performance of the parents is divided in half (i.e. $120/2 = 60$ kg = [a]), and indicates the additive effect. Note that [a] carries no sign. *m* is called the mid-parent value and is midway in value between the performance of $\overline{P}_1$ and $\overline{P}_2$. Therefore:

$$[a] = (\overline{P}_1 - \overline{P}_2)/2$$
$$m = \overline{P}_2 + [a] \quad \text{or} \quad \overline{P}_1 - [a] \quad \text{or} \quad m = (\overline{P}_1 + \overline{P}_2)/2$$
$$\overline{P}_1 = m + [a] \quad \text{and} \quad \overline{P}_2 = m - [a]$$

The term $[a]$ is used to indicate the summation of the additive effects over all loci involved, however many this may be. In the example shown we assumed a completely additive model of inheritance and the $\overline{F}_1$ performance was indeed equal to m. Therefore in the absence of dominance, the mid-parent value will equal the performance of the $\overline{F}_1$. Dominance will be detected in cases where the performance of the $\overline{F}_1$ is not equal to m.

Let us return again to the canola cross, and continue to assume the relationship between uppercase alleles adding to a base yield and lowercase alleles adding nothing. Previously, we did not consider dominant alleles and their effect on the distribution.

Now let us assume a two loci and two alleles per locus model of inheritance for yield. Assume also that A is dominant to a, but B and b are additive. Therefore $Aa = AA$, so we have:

$$P_1 \times P_2$$
$$AABB \times aabb$$

One (or two) A alleles would add 60 kg/plot to the base weight. Therefore AA adds 60 kg/plot, $Aa = AA$ (dominance) = 60 kg/plot, and B adds 30 kg/plot. The $\overline{F}_1 = AaBb = 590$ kg/plot. Now $[a] = (\overline{P}_1 - \overline{P}_2)/2 = 60$ kg, so $m = \overline{P}_2 + [a] = 560$; clearly the $\overline{F}_1$ is not equal to m, and we have a case of dominance.

```
 P̄2                  F̄1                  P̄1
500                  590                  620
 ←───────────────── m ─────────────────→
 ←──────[a]──────→  ←──────[a]──────→
```

When the F_2 population is examined we see that the basic bell-shaped curve (below) has now been skewed to the right, as a greater frequency of progeny have higher yield due to the effect of the dominant A allele. The average (mean, shown by the arrow) performance of the F_2 is now 575 kg/plot.

```
                                    AABb
                        AAbb        AaBb
                        Aabb        aABb        aABB
            aabB        aABb        AabB        AaBB
aabb        aaBb        aaBB        aAbB        AABB
500         530         560         590         620
                          ↑
```

We can now expand this idea to a three-loci, two-allele example as before, and we have 64 possible genotypes with 7 possible phenotypes. Assume that A is dominant to a, but that B, b, C and c are all additive, and uppercase alleles add 20 kg/plot to the base yield of 500 kg/plot. We now have the $\overline{F}_1 = AaBbCc = 580$ kg/plot (as $Aa = AA = 20 + 20 = 40$ kg/plot). Again we see that the $\overline{F}_1$ performance is higher than the mid-parent value (m), reflecting the dominance that A exhibits.

As with the two loci case, the distribution of phenotypes with three genes is similarly skewed to the right. In this instance, the average (mean, shown by the arrow) performance of the F_2 generation would be 570 kg/plot.

```
                                 AABBcc
                                 AABbcC
                                 AABbCc
                       aabBCC    AAbBCc
                       aaBBcC    AAbBcC
                       aaBBCc    AAbbCC
                       aaBbCC    aAbBcC
             AAbbcC    aAbBCc    AABBcC
             AAbbCc    aAbbCC    AABBCc
             AABbCc    AabbCC    AAbBCC
   aabbCC    AABbcc    AabBcC    AABbCC
   aabBcC    aAbbcC    AabBCc    AaBBcC
   aabBCc    aAbbCc    AaBbcC    AaBBcC
   aaBbcC    aAbbcc    AaBbCc    AaBbCC
   aaBbCc    aABbcc    AaBBcc    AabbCC
aabbcC aaBBcc AabbcC   aAbbCc    aABBcC
aabbCc AAbbcc AabbCc   aAbbcC    aABBcC   AABBCC
aabBcc aAbbcc AabBcc   aABBcc    aAbbCC   aABBCC
aabbcc aaBbcc Aabcc    AaBbcc    aaBBCC   aAbBCC   AaBBCC
500    520    540      560       580      600      620
                         ↑
```

The keen observer will have noted two points:

- The F_1 performance was proportionally higher in the two gene case compared with the three gene case, because a higher proportion of alleles (2 out of 3) in the three gene case were not showing dominance. In Figure 5.3, the degree of skewness of a six-loci, two-allele system are shown for no dominance, one dominant loci, three dominant loci and five dominant loci.
- There was a relationship between the parent performance and the performance of the F_1 and F_2.

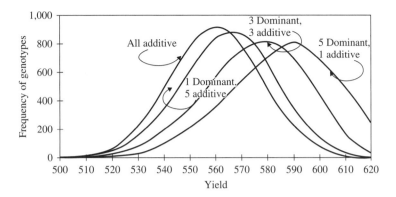

Figure 5.3 The degree of skewness of a six loci two-allele system is shown is shown for no dominance, one dominant locus, three dominant loci and five dominant loci. Note that increasing the number of dominant loci results in greater skewness.

The mean performance of the F_2 is midway between m, the mid-parent value (560 kg/plot) and the mean of the F_1 family (580 kg/plot). Not surprisingly the mean of the F_3 family would be 565 kg/plot, halfway between the mid-parent and the F_2 values. The mean of the F_∞, of recombinant inbred lines, would be equal to the mid-parental value, in the same way as if no dominance (or indeed linkage) existed. Therefore, if dominance effects are adversely effecting selection in plant breeding, then increased rounds of selfing will increasingly avoid these effects.

Consider again the inheritance model that we have for additive effects:

$$\overline{P}_2 \qquad\qquad \overline{F}_1 \qquad\qquad \overline{P}_1$$
$$500 \qquad\qquad 560 \qquad\qquad 620$$
$$\underset{\longleftarrow \qquad\qquad m \qquad\qquad \longrightarrow}{}$$
$$\underset{\longleftarrow \; [a] \; \longrightarrow \quad\cdot\quad \longleftarrow \; [a] \; \longrightarrow}{}$$

As we have seen, the $\overline{F}_1$ performance does not always coincide with the mid-parent value. In the two-loci case we had:

$$\overline{P}_2 \qquad\qquad \overline{F}_1 \qquad\qquad \overline{P}_1$$
$$500 \qquad 560 \;\; 580 \qquad 620$$
$$\qquad\quad m \quad\; [d]$$

So we now need to add a second parameter to the model which represents the amount of dominance $[d]$, and where:

$$[d] = \overline{F}_1 - m$$

and from this we have:

$$\overline{F}_1 = m + [d]$$

Unlike $[a]$ which is always positive, $[d]$ can be either positive or negative (i.e. the $\overline{F}_1$ has a higher or lower value than m, respectively).

Using these three parameters we can now proceed to determine the expected performance of any generation. We have:

$$\text{High parent } (\overline{P}_1) = m + a;$$
$$\text{Low parent } (\overline{P}_2) = m - a;$$
$$\overline{F}_1 = m + d$$

Assume that the $\overline{P}_1, \overline{P}_2$ and $\overline{F}_1$ generations are grown so that m, $[a]$ and $[d]$ can be calculated from their means; then it can be seen why m, a and d are referred to as the *components of the generation means*.

$$\overline{P}_2 \qquad\qquad \overline{F}_1 \qquad\qquad \overline{P}_1$$
$$m - [a] \qquad\qquad m + [d] \qquad\qquad m + [a]$$
$$\underset{\longleftarrow \qquad\qquad m \qquad\qquad \longrightarrow}{}$$
$$\underset{\longleftarrow \; [a] \; \longrightarrow \quad\cdot\quad \longleftarrow \; [a] \; \longrightarrow}{}$$

In addition to $\overline{P}_1, \overline{P}_2$ and $\overline{F}_1$, three further generations are commonly considered:

- The F_2 generation (i.e. F_1 selfed).
- The *back-cross generation* $F_1 \times P_1$ which is termed the $B_{1.1}$ generation.
- The *back-cross generation* $F_1 \times P_2$, which is termed the $B_{1.2}$ generation.

These three generations are typified by genetic segregation. It is therefore necessary to derive the proportions of the different genotypes, and their relative contributions to the means, in the various generations.

The *Aa* gene segregates with gamete formation in the F_1 thus giving, in the F_2 generation, the genotypes *AA*, *Aa* and *aa* in the ratio 1 *AA* : 2 *Aa*: 1 *aa* or, as proportions, $1/4 AA$, $1/2 Aa$ and $1/4 aa$. The mean expression of the *AA* plants is $m + a$. But since only a quarter of the F_2 generation is of the *AA* genotype, it contributes $1/4 (m + a)$ to the F_2 generation mean. Similarly, since the mean expression of the *aa* plants is $m - a$ and they make up a quarter of the F_2 generation, *aa* plants contribute $1/4 (m - a)$ to the F_2 generation mean. Finally, since half the F_2 generation has the genotype *Aa*, which has a mean expression of $m + d$, the heterozygotes contribute $1/2 (m + d)$ to the F_2 generation mean. The term [*a*] is a summation of additive effects over all loci, therefore, for simplicity, we assume (without proof):

$$\overline{F}_2 = 1/4\overline{P}_1 + 1/2\overline{F}_1 + 1/4\overline{P}_2$$
$$= 1/4(m + [a]) + 1/2(m + [d]) + 1/4(m - [a])$$
$$= 1/4m + 1/4[a] + 1/2m + 1/2[d] + 1/4m - 1/4[a]$$
$$= m + 1/2[d]$$

Considering a single gene model again, then the $B_{1.1}$ generation ($F_1 \times P_1$) is composed of $1/2 AA$ and $1/2 Aa$. Again assuming without proof that [*a*] is an accumulation of all additive effects and [*d*] an accumulation of all dominance effects, we have:

$$\overline{B}_1 = 1/2\overline{P}_1 + 1/2\overline{F}_1$$
$$= 1/2(m + [a]) + 1/2(m + [d])$$
$$= m + 1/2[a] + 1/2[d]$$

Similarly the $B_{1.2}$ generation ($F_1 \times P_2$) would be:

$$\overline{B}_2 = m - 1/2[a] + 1/2[d]$$

5.3.5   Testing the models

Earlier, a model which we shall now call the ***additive–dominance model*** was put forward that purports to explain the inheritance of quantitative (continuously varying) characters exclusively in terms of the additive and dominance properties of the single gene difference that underlies it. It is now necessary to consider whether a model based on only additive and dominance genetic differences is adequate. From a plant breeding standpoint it is important to know whether the inheritance of a character under selection is controlled by an additive–dominance model because many assumptions, notably the response to selection, are based on heritability, and in estimating heritability it is usually assumed that this model is appropriate. Testing the additive–dominance model in quantitative genetics should be regarded as directly comparable to testing for the absence of linkage or epistasis with qualitative inheritance. However, with genes showing qualitative differences the most common testing method to compare frequencies of genotypes or phenotypes is a χ^2 test. In quantitative genetics, genotypes (or more correctly the phenotypes) do not fall into distinct classes and hence frequency χ^2 tests are not appropriate. So we need an equivalent but appropriate test.

First, recall that the components m, [*a*] and [*d*] are derived from the generation means of the $\overline{P}_1$, $\overline{P}_2$ and $\overline{F}_1$ generations:

$$\overline{P}_1 = m + [a];$$
$$\overline{P}_2 = m - [a];$$
$$\overline{F}_1 = m + [d];$$

From these, similar equations can be formulated for the generation means of the $\overline{F}_2$, $\overline{B}_1$ and $\overline{B}_2$ generations:

$$\overline{F}_2 = m + 1/2[d]$$
$$\overline{B}_1 = m + 1/2[a] + 1/2[d]$$
$$\overline{B}_2 = m - 1/2[a] + 1/2[d]$$

These six equations lead to various *predictive* relationships between the means of different combinations of the generations. For example, if the additive–dominance model we have developed is correct, then it can be predicted that:

$$2\overline{B}_1 - \overline{F}_1 - \overline{P}_1 = 0$$

This relationship is easily be seen by the substitution of the appropriate combinations of m, [*a*] and [*d*] for the three generation means:

$$2\left(m + 1/2[a] + 1/2[d]\right) - (m + [d]) - (m + [a]) = 0$$
$$(2m + [a] + [d]) - (m + [d]) - (m + [a]) = 0$$
$$2m + [a] + [d] - m - [d] - m - [a] = 0$$

It can be seen that all the terms on the left-hand side of the last equation cancel out. This test is called the **A-scaling test**.

Another relationship is:

$$2\overline{B}_2 - \overline{F}_1 - \overline{P}_2 = 0$$

and:

$$2\left(m - \tfrac{1}{2}[a] + \tfrac{1}{2}[d]\right) - (m + [d]) - (m - [a]) = 0$$
$$(2m - [a] + [d]) - (m + [d]) - (m - [a]) = 0$$
$$2m - [a] + [d] - m - [d] - m + [a] = 0$$

Once again, all the terms on the left-hand side cancel out. This test is called the **B-scaling test**.

A final relationship, known as the **C-scaling test**, is:

$$4\overline{F}_2 - 2\overline{F}_1 - \overline{P}_1 - \overline{P}_2 = 0$$

This last test can be shown to hold true in the same way as for the two tests above.

These relationships are based on the model we have proposed for predicting the means of the various generations. Would you expect this relationship to hold if we substituted the means that we had actually measured? In other words would you expect the above equations to exactly equal zero? The answer is no, and in fact, it would be quite surprising if, for example, the sum of the means of the $\overline{P}_2$ and $\overline{F}_1$ generations exactly equalled twice the mean of the $\overline{B}_2$ generation. While $\overline{P}_2 + \overline{F}_1$ might be approximately equal to $2\overline{B}_2$, error variation would give rise to random variation in all three means resulting in some overall discrepancy.

Thus, from the above and thinking now in terms of measured generation means, we have the relationships:

$$2\overline{B}_1 - \overline{F}_1 - \overline{P}_1 = A$$
$$2\overline{B}_2 - \overline{F}_1 - \overline{P}_2 = B$$
$$4\overline{F}_2 - 2\overline{F}_1 - \overline{P}_1 - \overline{P}_2 = C$$

where A, B and C are all expected effectively to equal zero. If they do equal zero, or at least are not too far from it, then there is no reason to suspect that the additive–dominant model is inadequate as an explanation of the inheritance of the continuously varying character in question. On the other hand, if they do deviate markedly from zero, then there is reason to doubt the adequacy of the model as an explanation of the inheritance of the character in question. This is a classic instance of the need for

objective statistical tests to decide whether A, B or C differ significantly from zero, or whether any discrepancies observed could simply be fairly ascribed to chance. If the discrepancies could reasonably be attributed to chance, then the model can be provisionally accepted as an adequate description of reality. If this explanation is too unlikely, then the hypotheses (that A, B and C all equal zero within the bounds of sampling error) must be rejected and therefore also the additive–dominant model upon which they are based. The statistical tests are called the A-scaling test, B-scaling test and C-scaling test (and there are many others possible). However, the A-scaling test will be used here to represent the principles involved in all of them.

The A-scaling test

The basis of the A-scaling test is that the value of A is compared to the value predicted on the assumption that an additive–dominant model is adequate (i.e. $A = 0$). The question is then asked:

"If the hypothesis is true (i.e. $A = 0$), what is the probability that any difference between observation (A) and prediction (zero) could be due to chance?"

Conventionally, if the probability that the difference was due to chance is less than 0.05 (i.e. 5%, or 1 in 20) then the null hypothesis (in this case, that A *is equal to* 0) is rejected and the alternative hypothesis (that $A \neq 0$) is accepted. In accepting the alternative hypothesis, it is also accepted that $2\overline{B}_1 - \overline{F}_1 - \overline{P}_1 = 0$ is not equal to zero, and that an additive–dominant model is inadequate in this particular instance.

In comparing the actual value of A with its predicted value (zero), what factors must be taken into account? Clearly the magnitude of the discrepancy (i.e. the actual value of A itself, $2\overline{B}_1 - \overline{F}_1 - \overline{P}_1$ must be considered. Another important factor is the *variability in A* from one experiment to another. If A varied enormously from one experiment to the next, then the mean value of A would have to be relatively large for it to be significantly different from zero. On the other hand, if the value of A were relatively constant from experiment to experiment, then even quite a small value of A could be accepted as significantly different from zero.

Finally, values based on relatively few plants are likely to be less convincing than values based on the measurement of many plants. Thus sample size is also highly relevant.

All of this perhaps sounds like a pretty tall order. In fact, biometricians have provided us with a method of relating the difference between the actual and predicted values of A to the variability in A from experiment to experiment, and also a statistical table in which the probability of obtaining such a difference by chance, given the number of plants measured, can be looked up. The equation is:

$$t = \frac{\text{Actual value of A} - \text{predicted value of A}}{\text{standard error of A}}$$

$$t = \frac{A - 0}{\text{se(A)}}$$

$$t = \frac{A}{\text{se(A)}}$$

In fact, as you might have already noticed, the A-scaling test is just a particular application of Student's t test, which you may have come across elsewhere.

In order to calculate t, A has to be divided by its *standard error* (se). We have noted that $A = 2\overline{B}_1 - \overline{F}_1 - \overline{P}_1$, where $\overline{B}_1, \overline{F}_1$ and $\overline{P}_1$ are the **measured** (not the **predicted**) means of the $\overline{B}_1, \overline{F}_1$ and $\overline{P}_1$ generations. But what is the standard error of A? A standard error is, like a standard deviation, the square root of a variance, as shown earlier. In fact, the standard error of A is the square root of the variance of the mean of A (σ_A^2), and is represented by σ_A. Therefore:

$$\sigma_A^2 = 4\sigma_{\overline{B}_1}^2 + \sigma_{\overline{F}_1}^2 + \sigma_{\overline{P}_1}^2$$

where $\sigma_{\overline{B}_1}^2$ is the variance of the mean of the B_1 generation, $\sigma_{\overline{F}_1}^2$ is the variance of the mean of the F_1 generation and $\sigma_{\overline{P}_1}^2$ is the variance of the mean of the P_1 generation.

It is essential to realize that the variance of the mean of a generation (i.e. $\sigma_{\overline{B}_1}^2$) is not the same as the variance between plants in that generation. In principle, the variance of the mean is calculated by growing adequate numbers of plants representing the generation in several different plots or experiments, calculating a generation mean for each experiment, and then calculating the variance of these different means (effectively treating

them as the raw data). A variance of the mean so determined is less than the variance between all the individual plants grown in all the experiments. Fortunately, it is not necessary to perform several different experiments as described. Biometricians have demonstrated that a satisfactory estimate of the variance of the mean is obtained by dividing the variance derived from a single sample of plants by the number of plants measured that contribute to the estimate of the mean. That is:

$$\sigma_{B_1}^2 = \frac{\text{variance of } nB_1 \text{ plants}}{n}$$

As an example of this, consider that the height of 50 individual plants (i.e. $n = 50$) of a pure-line barley cultivar was recorded and that the average plant height of all plants measured was calculated to be 100 cm, with a variance of 40 cm². The variance and the standard error of the mean of this sample of plant heights would be:

$$\text{Variance of mean} = \frac{\text{variance of 50 plants}}{50}$$

$$= \frac{40}{50} = 0.80 \text{ cm}^2$$

and the standard error $= \sigma = \sqrt{0.80} = 0.89$ cm

therefore the mean is

$$100 \pm 0.89 \text{ cm}$$

So, given a set of individual measurements, you should be able to calculate the mean, the variance and the standard deviation of the population, of which the data you are given can be assumed to be an unbiased sample. Furthermore, you could calculate both the variance of the mean and therefore its standard error.

Let us consider now a simple example where two homozygous barley cultivars (P_1 and P_2) are cross-pollinated and a sample of F_1 seed is back-crossed to the higher yielding parent (P_1) to produce B_1 seed. Now if $11 P_1, 11 F_1$ and $11 B_1$ seeds were planted in a properly randomized experiment and the height of each plant recorded, then the following means and standard errors might have resulted:

$$\overline{P}_1 = 109.1 \pm 9.1$$
$$\overline{F}_1 = 105.5 \pm 8.6$$
$$\overline{B}_1 = 108.3 \pm 10.0$$

Now the variance of each family would be:

$$\sigma^2_{\overline{P}_1} = (9.1)^2$$
$$\sigma^2_{\overline{F}_1} = (8.6)^2$$
$$\sigma^2_{\overline{B}_1} = (10.0)^2$$

Therefore it follows that the variance of A ($\sigma_A{}^2$) would be:

$$\sigma^2_A = 4\sigma^2_{\overline{B}_1} + \sigma^2_{\overline{F}_1} + \sigma^2_{\overline{P}_1}$$
$$= 4(10.0)^2 + (8.6)^2 + (9.1)^2 = 556.77$$

and the standard error of A (σ_A) is given by:

$$\sigma_A = \sqrt{(\sigma^2_A)} = \sqrt{556.77} = 23.6$$

Now to consider the mean value of A, this is given by:

$$A = 2\overline{B}_1 - \overline{F}_1 - \overline{P}_1$$
$$= 2(108.3) - 105.5 - 109.1 = 2.0$$

Finally:

$$t = A/\sigma_A = 2.0/23.6 = 0.085$$

So, a value of t has been calculated for these data as 0.085. This is based upon both the deviation of A from its expected value of zero (i.e. 2.0) and the variability found in the P_1, F_1 and B_1 plants measured, all the variability being summarized in the standard error of A (i.e. 23.6). The following question now arises: Is the deviation we have observed statistically significant?

In order to decide this, it is necessary to account for the number of plants measured in each generation upon which the values of A and σ_A are based. In fact, it is not the number of plants as such that is used, but the relevant numbers of degrees of freedom, where the degrees of freedom of A = the degrees of freedom of $\sigma^2_{\overline{B}_1}$ + the degrees of freedom of $\sigma^2_{\overline{F}_1}$ + the degrees of freedom of $\sigma^2_{\overline{P}_1}$.

Degrees of freedom have previously been mentioned in connection with the χ^2 test. Generally, the number of degrees of freedom associated with a generation is one fewer than the number of plants representing that generation. Thus, if 11 plants of each of generations P_1, F_1 and B_1 were measured, then the degrees of freedom of A = (11 − 1) + (11 − 1) + (11 − 1) = 30 df.

It is necessary to look up the value of t (i.e. 0.085) for 30 degrees of freedom in a table of probabilities for t. As the t value we obtained is smaller in magnitude than the tabulated value with 30 degrees of freedom, there is no reason to reject the additive–dominance model in this instance, and so it is provisionally accepted as an adequate explanation of the inheritance of the character in question.

Joint scaling test

The procedure described above can be repeated in a similar way for the B- and the C-scaling tests. Indeed, sets of such scaling tests can be devised to cover any combination of types of family that may be available.

As an alternative, however, to testing the various expected relationships one at a time, a procedure was proposed by a researcher called L.L. Cavalli in 1952, which is known as the *joint scaling test*. This test effectively combines the whole set of scaling tests into one and thus offers a more general, more convenient, more adaptable and more informative approach.

The joint scaling test consists of estimating the model's parameters, m, [a] and [d] from the means of all the families available, followed by a comparison of these observed means with their expected values derived from the estimates of the three parameters. This makes it clear at once that at least three types of family are necessary if the parameters of the model are to be estimated. However, with only three types of family available, no test can be made of the goodness of fit of the model since in such a case a perfect fit *must* be obtained between the observed means and their expectations derived from the estimates of the three parameters. So to provide such a test, at least four types of family must be raised.

The procedure for the joint scaling test is illustrated by considering the example given by Mather and Jinks's seminal textbook *Introduction to Biometrical Genetics*. The data they presented have been truncated for simplicity, and so differences due to rounding errors may occur. Their example consists of a cross between two pure-breeding varieties of rough tobacco (*Nicotiana rustica*). The means and variances of the means for plant height of the parental, F_1, F_2 and first back-cross families

Table 5.7 Means and variances of the means for plant height of two parental lines ($\bar{P}_1$ and $\bar{P}_2$), the $\bar{F}_1, \bar{F}_2$ progeny, and the first backcross families ($\bar{B}_1$ and $\bar{B}_2$) derived from crossing $\bar{P}_1$ to $\bar{P}_2$.

Number of plants	V_x	Weight $1/V_x$	Model			Observed
			m	$[a]$	$[d]$	
$\bar{P}_1$ 20	1.033	0.968	1	1	0	116.30
$\bar{P}_2$ 20	1.452	0.669	1	−1	0	98.45
$\bar{F}_1$ 60	0.970	1.031	1	0	1	117.67
$\bar{F}_2$ 160	0.492	2.034	1	0	1/2	111.78
$\bar{B}_1$ 120	0.489	2.046	1	1/2	1/2	116.00
$\bar{B}_2$ 120	0.613	1.630	1	−1/2	1/2	109.16

Table 5.8 Coefficients of m, $[a]$ and $[d]$ in the parents ($\bar{P}_1$ and $\bar{P}_2$), the $\bar{F}_1, \bar{F}_2$ generation and both back-cross generations ($\bar{B}_1$ and $\bar{B}_1$) and the observed plant height of each family.

Generation	Model			Observed
	m	$[a]$	$[d]$	
$\bar{P}_1$	1	1	0	116.30
$\bar{P}_2$	1	−1	0	98.45
$\bar{F}_1$	1	0	1	117.67
$\bar{F}_2$	1	0	1/2	111.78
$\bar{B}_1$	1	1/2	1/2	116.00
$\bar{B}_1$	1	−1/2	1/2	109.16

(B_1 and B_2) derived from this cross are shown in Table 5.7.

Also shown in this table is the number of plants that were evaluated from each generation. Family size was deliberately varied with the kind of family. It was set as low as 20 for the genetically uniform parents and in excess of 100 for the F_2 and back-crosses, to compensate for the greater variation expected in these segregating families. All plants were individually randomized at the time of sowing so that the variation within families reflects all the non-heritable sources of variation to which the experiment is exposed. With this design the estimate of the variance of a family mean (V_x), valid for use in the joint scaling test, is obtained in the usual way, by dividing the variance within the family by the number of individuals in that family. Reference to these variances shows that the greater family size of the segregating generations has more than compensated for their greater expected variability, in that the variances of their family means are smaller than those of their non-segregating families.

Six equations are available for estimating m, $[a]$ and $[d]$, and these are obtained by equating the observed family means to their expectations as given above. The coefficients of m, $[a]$ and $[d]$ in the six equations are listed with the collected data in Table 5.8.

There are three more equations than there are parameters to be estimated (m, $[a]$ and $[d]$), therefore a least-square technique can be used. The six generation means to which we are fitting the m, $[a]$ and $[d]$ model are not known with equal precision;

for example, the variance of the mean (V_{P2}) of P_2 is almost three times that of the B_1. The best estimates will be obtained, therefore, if generation means are weighted in relation to the accuracy of the estimates. The appropriate weights in this instance are the reciprocals of the variances of the means. For the first entry in the data (Table 5.7), P_1, the weight is given by $1/1.0334 = 0.9677$ and so on for the other families.

The six equations and their weights may be combined to give three equations whose solution will lead to weighted least-squares estimates of m, $[a]$ and $[d]$ as follows. In order to obtain the first of these three equations each of the six equations is multiplied through by the coefficient of m that it contains, and by its weight, and the six are then summed. When we weight each line of the array by m (which is always equal to 1) we have:

	m	$[a]$	$[d]$		Observed
	+0.9677	+0.9677	0	=	112.541
	+0.6688	−0.6688	0	=	65.848
	+1.0310	0	+1.0310	=	121.327
	+2.0342	0	+1.0171	=	227.376
	+2.0458	+1.0229	+1.0229	=	237.316
	+1.6300	−0.8150	+0.8150	=	177.931
Σ	8.3775	+0.5067	+3.8860	=	942.340

The second and third equations are found in the same way using the coefficient of $[a]$ for the second and of $[d]$ for the third, along with the weights ($1/V_x$) as multipliers.

To illustrate, the next line is found in the same way by multiplying each of the lines by the coefficients of $[a]$ (i.e. $+1, -1, 0, 0, +1/2, -1/2$), and then summing columns thus:

	m	$[a]$	$[d]$		Observed
	+0.9677	+0.9677	0	=	112.541
	−0.6688	+0.6688	0	=	−65.848
	0	0	0	=	0
	0	0	0	=	0
	+1.0229	+0.5115	+0.5115	=	118.658
	−0.8150	+0.4075	−0.4075	=	−58.965
Σ	0.5067	+2.5555	+0.1040	=	76.385

Finally, the third line is obtained by multiplying through by the coefficients of $[d]$ (i.e. $0, 0, 1, 1/2, 1/2$), and then summing the columns thus:

	m	$[a]$	$[d]$		Observed
	0	0	0	=	0
	0	0	0	=	0
	+1.0310	0	+1.0310	=	121.327
	+1.0171	0	+0.5085	=	113.688
	+1.0229	+0.5114	+0.5114	=	118.658
	+0.8150	−0.4075	+0.4075	=	99.965
Σ	3.8860	+0.1040	+2.4585	=	442.639

We then have three simultaneous equations, known as normal equations, which may be solved in a variety of ways to yield estimates of m, $[a]$ and $[d]$. A general approach to the solution is by way of matrix inversion. The three equations are rewritten in the form:

$$\begin{bmatrix} 8.3775 & 0.5067 & 3.8860 \\ 0.5067 & 2.5555 & 0.1040 \\ 3.8860 & 0.1040 & 2.4585 \end{bmatrix} \times \begin{bmatrix} m \\ [a] \\ [d] \end{bmatrix} = \begin{bmatrix} 942.340 \\ 76.385 \\ 442.639 \end{bmatrix}$$
$$\mathbf{J} \qquad \times \quad \mathbf{M} \ = \ \mathbf{S}$$

where $\mathbf{J}$ is known as the information matrix, $\mathbf{M}$ is the estimate of the parameters and $\mathbf{S}$ is the matrix of the scores.

The solution then takes the general form $\mathbf{M} = \mathbf{J}^{-1} \times \mathbf{S}$ where $\mathbf{J}^{-1}$ is the inverse of the information matrix and is itself a variance–covariance matrix.

The inversion may be achieved by any one of a number of standard procedures; for our example,

inversion leads to the following solution:

$$\begin{bmatrix} m \\ [a] \\ [d] \end{bmatrix} = \begin{bmatrix} 0.4568 & -0.0613 & -0.7194 \\ -0.0613 & 0.4002 & 0.0800 \\ -0.7194 & 0.0800 & 1.5405 \end{bmatrix} \times \begin{bmatrix} 942.339 \\ 76.385 \\ 442.639 \end{bmatrix}$$
$$\mathbf{M} \ = \qquad \mathbf{J}^{-1} \qquad \qquad \times \quad \mathbf{S}$$

The estimate of m is then:

$$m = (0.4568 \times 942.339) - (0.0613 \times 76.385)$$
$$- (0.7194 \times 442.639)$$
$$= 107.244$$

The standard error (se) of m is $\sqrt{(0.4568)} = \pm 0.6759$.

In a similar way:

$$[a] = 8.1997 \pm 0.6326$$
$$\text{and } [d] = 10.0587 \pm 1.2412$$

All are highly significantly different from zero when looked up in a table of normal deviates.

The adequacy of the additive-dominance model may now be tested by predicting the six family means using these estimates of m, $[a]$ and $[d]$.

For example:

$$\overline{B}_2 = m - \tfrac{1}{2}[a] + \tfrac{1}{2}[d]$$

On the basis of this model and using the estimates obtained, it has the expected value:

$$107.3220 - \left[\tfrac{1}{2} \times 8.1997\right] + \left[\tfrac{1}{2} \times 10.0587\right]$$
$$= 108.2515$$

This expectation and those for the other five families are listed in Table 5.9.

The agreement with the observed values appears to be very close, and in no case is the deviation more than 0.83% of the observed value. The goodness of fit of this model can be tested statistically by a χ^2 test. Since the data comprise six observed means, and three parameters have been estimated (i.e. m, $[a]$ and $[d]$), then the χ^2 value has $6 - 3 = 3$ degrees of freedom.

The contribution made to the χ^2 by $\overline{P}_1$, for example, is the squared difference between Observed and Expected divided by the variance (or in our case we can multiply by one over the variance, that is, the weight. So, for example, $[116.300 - 115.522]^2 \times 0.968 = 0.5862$. Summing the six such contributions, one from each of the six

Table 5.9 Observed plant heights from both parents ($\bar{P}_1$ and $\bar{P}_2$), the $\bar{F}_1, \bar{F}_2$, and both backcross ($\bar{B}_1$ and $\bar{B}_2$) generations along with the expected plant height using the joint scaling test parameters, and the difference between the observed and expected plant height.

Family	Observed	Expected	Obs–Exp
$\bar{P}_1$	116.300	115.522	+0.778
$\bar{P}_2$	98.450	99.122	+0.672
$\bar{F}_1$	117.675	117.381	+0.294
$\bar{F}_2$	111.778	112.351	−0.573
$\bar{B}_1$	116.000	116.451	−0.451
$\bar{B}_2$	109.161	108.252	−0.090

types of family, gives a χ^2 of 3.411 for 3 degrees of freedom, which has a probability of between 0.40 and 0.30. The model must therefore be regarded as adequate (i.e. there is no evidence of anything beyond additive and dominance effects).

The individual scaling tests, A, B and C, referred to earlier can, of course, also be used to test the model. Thus with the present data:

$$A = 2\bar{B}_1 - \bar{F}_1 - \bar{P}_1$$
$$= 2(116.000) - 116.300 - 117.675$$
$$= -1.975$$
$$\sigma_A^2 = 4\sigma_{\bar{B}_1}^2 + \sigma_{\bar{P}_1}^2 + \sigma_{\bar{F}_1}^2$$
$$= [4 \times 0.4888] + 1.0334 + 0.9699$$
$$= 3.959$$

leading to $\sigma_A = \sqrt{(3.959)} = 1.990$.

Thus $A = -1.98 \pm 1.99$, which, when compared in Student's t test statistical tables, does not differ significantly from the expected value of 0.

The joint scaling test gives exactly the same answer as the A-, B- or C-scaling tests. However, the joint scaling test does more than test the adequacy of the additive–dominance model. It also provides the '*best*' estimates of all the parameters required (and their standard errors) to account for differences among family means when the model is adequate. If you try to estimate m, $[a]$ and $[d]$ with the procedure shown earlier you will find values of $m = 107.365$, $[a] = 8.925$ and $[d] = 10.310$. These estimates do not differ markedly from those estimated (107.322, 8.1997, 10.0587, respectively) from the joint scaling test. However, the difference may in some cases be of importance, and an additional factor of

relevance is that the joint scaling test can also be readily extended to more complex situations.

To conclude, in this example the best estimates show that the additive and dominance components are of the same order of magnitude, and since $[d]$ is significantly positive, on average alleles that increase final height must be dominant more often than alleles that decrease it.

What could be wrong with the model

In some instances the results from generation testing will lead to the conclusion that an additive–dominance model of inheritance does not adequately account for the data. There are many possible explanations for this, and here only three, in order of increasing genetic complexity, will be mentioned briefly.

- **Abnormal chromosomal behaviour**. In the case of single gene or multiple gene cases, the predicted means of the $\bar{F}_1$, $\bar{B}_1$ and $\bar{P}_1$ generations were derived in terms of m, $[a]$ and $[d]$. Throughout, the assumption was made that a heterozygote contributes equal proportions of its various gametes to the gene pool, following genetic segregation. If this assumption is unwarranted, then the frequencies of the different genotypes, and therefore their contributions to the generation means in terms of m, $[a]$ and $[d]$, would not be as predicted. This could result, for example, from the elimination of deleterious alleles through gametic selection.
- **Cytoplasmic inheritance**. If a character is determined, or affected, by non-nuclear genes (recall that plants have a small number of genes residing in organelles such as chloroplast and mitochondria), then the expression of the character will depend upon which of the genotypes was maternal. Under such circumstances, inheritance cannot be explained fully in terms of m, $[a]$ and $[d]$.
- **Non-allelic interactions or epistasis**. We have already mentioned the phenomenon of genetic dominance. Effectively, dominance-recessiveness are allelic interactions whereby the phenotypic expression of a character does not depend solely upon the additive effects of the different alleles at the same locus, and the mean expression of the heterozygote is not the same as the

mid-parent value. As we noted earlier, a similar phenomenon can occur between alleles at different loci, and this is known as **non-allelic interaction** or **epistasis**. An example of epistasis and how it might occur was presented in the qualitative genetics section. A further example might be:

$$AABB = 24; \quad AAbb = 12; \quad aaBB = 12; \quad aabb = 8$$

In the presence of *BB*, the difference between the *AA* and *aa* genotypes is $24 - 12 = 12$ units. However, in the presence of *bb*, the difference between *AA* and *aa* is $12 - 8 = 4$ units. Of course, another way of looking at the matter might be to say that the difference between *BB* and *bb* is $24 - 12 = 12$ units in the presence of *AA*, but $12 - 8 = 4$ units in the presence of *aa*. Either way, it can be seen that there is **interaction** between the alleles at different loci and that an additive–dominant model of inheritance cannot adequately account for the situation. In fact, it is possible to add epistasis to our model. This is usually done by adding symbols: *aa*, for interaction between loci that are homozygous, *ad* for those between loci where one is heterozygous and one homozygous and *dd* for loci that are heterozygous.

In general it is actually quite straightforward to take into account other genetic phenomena by inclusion of appropriate parameters in the basic additive–dominant model of inheritance, and thus increasingly account for more complex genetic inheritance.

Although you should be aware of the existence of these complications, they will not be taken into any further detail in this book. Moreover, it is often found that, for most characters of interest to plant breeders, the additive–dominant model is adequate – if it fails we are then aware that the situation is more complex and act accordingly. Also, since what is of primary practical interest is the ratio of the additive genetic variance in a generation to the variance attributable to all causes (environmental, additive, dominant and all other genetic phenomena), it is often unnecessary to itemize them individually.

5.3.6 Quantitative trait loci

The concept of linkage between different loci located on the same chromosome was introduced

in the qualitative genetics section. Quantitatively inherited characters are controlled by alleles at multiple loci. Yield, for example, is a highly complex character which is related to a multitude of other characters, like seedling germination and emergence, flowering times, partition, photosynthesis efficiency, nitrogen uptake efficiency, and so on, plus a susceptibility or resistance to a wide range of stresses including diseases and pests. Even if a single gene were to be responsible for all the individual factors that are involved in yield potential (which they are not), then it is easy to see that there will be hundreds or even thousands of genes that influence yield. Given that the number of chromosomes in crop species is small ($2n = 2x = 34$ in sunflower, $2n = 2x = 18$ in lettuce, $2n = 4x = 38$ in rapeseed, $2n = 2x = 20$ in maize, $2n = 6x = 42$ in wheat, $2n = 2x = 14$ in barley, $2n = 2x = 24$ in rice, $2n = 2x = 22$ in bean, and $2n = 4x = 48$ in potato), then linkage will always be a major factor in the inheritance of quantitatively inherited traits. So this, as with other quantitative effects, adds another level of complexity. In general, the complexity of the genetics has meant that many questions remain unanswered. Some questions that might be asked are:

- whether all loci have equal effects on the quantitative trial expression or whether there are some loci that have major effects while others have minor effects;
- whether the multiple loci are distributed evenly throughout the genome, or whether they are clustered on specific chromosomes, or in specific regions of the genome ('hot spots').

The concept of quantitative trait loci (QTL) was first raised by Karl Sax in 1923. Sax reported examining yield on a segregating F_2 progeny from a cross between two homozygous common bean (*Phaseolus vulgaris*) lines. One parent was homozygous for coloured seed while the other had white seed. A single gene at the P-locus determined seed colour, with *PP* alleles for coloured seed and *pp* for white seed. On inspection of seed weights, Sax found that *PP* lines produced seeds with an average weight of 30.7 g/100 seeds, heterozygotes (*Pp*) produced seed with 28.3 g/100 seeds, while *pp* lines had lowest seed weights (26.4 g/100 seeds). From this Sax introduced the concept that the quantitative loci determining seed weight were linked to the single gene locus for seed colour.

The potential of expanding this concept in plant breeding attracted the attention of many researchers after Sax's work was published. However, few advantages were achieved because plant breeders were forced to work with mainly morphologically visible single gene traits and major-gene mutants. These were not the most suitable for investigating QTLs because:

- they were relatively few in number;
- they were usually recessive, and their expression was masked in the phenotype by dominant alleles;
- they often had deleterious effects (or pleitropic effects) on the quantitative trait of interest (i.e. albinism, dwarfism, etc.)

These defects have been corrected by the introduction of molecular markers, which tend to be numerous, do not affect the plant phenotype, and are often co-dominant, allowing the heterozygotes to be differentiated from the homozygotes parental types.

In plant breeding, QTLs have greatest potential in marker-assisted selection for quantitatively inherited traits that have low heritability or that are difficult or expensive to screen or evaluate.

The process involved in QTLs will be illustrated using a simple simulated example where two homozygous parents are hybridized to produce F_1 plants. One parent was homozygous *AABBCC* at the A-, B- and C-bands, respectively, while the other parent was homozygous *aabbcc*. Traditionally, these bands were separated and observed on an autoradiograph and represent an allele. Currently though, these bands are more and more frequently resolved in sequencing machines and therefore appear as a peak. It should be noted that in this example, *A* is not dominant to *a*, etc.

Thirty-two homozygous lines were derived from the F_1 family using double-haploidy techniques (see Chapter 8). These lines were grown in a four replicate field trial to determine yield of each line. In addition, the lines were polymorphic for three loci that appeared to be located on the same chromosome. The molecular marker banding at the three molecular markers (identified simply as A-, B- and C-bands, *AA*, *BB*, and *CC*, respectively) along with the yield of each line, is shown in Table 5.10. We use doubled haploids in this example for simplicity as there will be no heterozygotes in the population. This makes some of the calculations simpler

as dominance effects can be ignored. However, the principle is the same and can be carried out using any segregating population resulting from a two-parent cross.

Mapping of the three qualitative loci is done according to the method described earlier, and the map is as follows:

A--------31.3 mμ--------B---18 mμ----C

which, when converted to cMs, is:

A--------49.5 cM--------B---22.5 cM---C

The first stage in QTL analysis is to determine whether there are indeed significant differences between the progeny lines. This is done by carrying out a simple analysis of variance. In our example, there were indeed significant differences between these lines (see Table 5.11).

Where there are significant differences in yield detected between the parental lines, can this difference in yield potential be explained by association between yield and the single marker bands?

Assume, for simplicity here, that genotypes with A-bands have genotype *AA*, and those without have genotype *aa*, and similarly for the B- and C-bands. Average yield of each single band genotype can be calculated by adding the yield of lines carrying the same bands at each locus and dividing by the number of individual lines in that class. For example, the average of all lines that have the *AA* bands is 109.52, while for those that have the *aa* bands it is 105.23. Similarly, yield of the *BB* band types is 114.31 compared with 100.44 for *bb*, and for *CC* types it is 112.05 compared with 102.70 for *cc* types. From this, there appears a pattern that lines carrying the *BB* band rather than the *bb* band have the largest yield advantage. Similarly, lines carrying the *CC* band over the *cc* band also have an advantage (albeit smaller than with the B-band). *AA* and *aa* lines differ only slightly. To apply significance to these differences requires partition of the sum of squares for differences between lines into:

- sum of squares due to the difference between the two genotypes at each band position, BG–SS (i.e. between *AA* types and *aa* types; *BB* and *bb* types; and *CC* and *cc* types);
- sum of squares due to the variation between lines within each genotype at each band, WG–SS.

Table 5.10 Yield of 32 double haploid canola lines, and genotype of each line at the A-, B-, and C-loci.

Line	A-loci	B-loci	C-loci	Yield	Line	A-loci	B-loci	C-loci	Yield
1	*AA*	*BB*	*CC*	107.80	17	*aa*	*BB*	*cc*	112.41
2	*AA*	*BB*	*CC*	113.57	18	*aa*	*bb*	*cc*	104.93
3	*AA*	*BB*	*cc*	111.68	19	*aa*	*bb*	*cc*	104.62
4	*aa*	*bb*	*CC*	101.09	20	*AA*	*BB*	*CC*	114.68
5	*aa*	*bb*	*cc*	91.29	21	*AA*	*BB*	*CC*	110.79
6	*aa*	*bb*	*cc*	112.24	22	*AA*	*bb*	*cc*	101.47
7	*aa*	*bb*	*cc*	97.17	23	*AA*	*BB*	*cc*	116.61
8	*aa*	*bb*	*cc*	95.75	24	*aa*	*bb*	*CC*	101.95
9	*aa*	*BB*	*CC*	113.52	25	*aa*	*bb*	*cc*	106.33
10	*aa*	*BB*	*CC*	119.27	26	*aa*	*bb*	*cc*	95.42
11	*AA*	*bb*	*cc*	98.40	27	*AA*	*BB*	*CC*	121.85
12	*AA*	*BB*	*CC*	106.82	28	*AA*	*BB*	*CC*	111.94
13	*AA*	*BB*	*CC*	117.61	29	*AA*	*bb*	*cc*	105.45
14	*AA*	*BB*	*CC*	112.88	30	*AA*	*bb*	*cc*	99.15
15	*AA*	*bb*	*CC*	101.58	31	*aa*	*BB*	*CC*	116.49
16	*aa*	*BB*	*CC*	119.27	32	*aa*	*bb*	*cc*	100.21

Table 5.11 Degrees of freedom and mean squares from the analysis of variance of seed yield on 32 doubled haploid lines grown in a three replicate randomized complete block design.

Source	df	Mean square
Between haploid lines	31	1241.8 ***
Replicate blocks	3	321.1 ns
Replicate error	93	401.4

$*** = p < 0.001$.

In this simple example, there are 16 lines that are *AA* and 16 that are *aa*. Similarly there are 16 lines that are *BB*, *bb*, *CC* and *cc*. Therefore it is completely balanced. In this instance the partition of the lines' sum of squares is by a simple orthogonal contrast. In actual experiments, the number of individuals in each class is likely to vary, and the BG–SS partition is completed by:

$$\text{BG} - \text{SS} = \left\{ \frac{(\bar{x}_{11} \cdot n_1)^2}{n_1} + \frac{(\bar{x}_{22} \cdot n_2)^2}{n_2} \right. $$
$$\left. - \frac{[(\bar{x}_{11} \cdot n_1) + (\bar{x}_{22} \cdot n_2)]^2}{n_1 + n_2} \right\} / \text{number of reps}$$

where $\bar{x}_{11}$ is the mean of lines with the 11 genotype, and n_1 is the number of lines with the 11 genotype.

In this example, for the *AA* and *aa* genotypes we would have:

$$\text{BG} - \text{SS} = \left\{ \frac{(\bar{x}_{AA} \cdot n_A)^2}{n_A} + \frac{(\bar{x}_{aa} \cdot n_a)^2}{n_a} \right. $$
$$\left. - \frac{[(\bar{x}_{AA} \cdot n_A) + (\bar{x}_{aa} \cdot n_a)]^2}{n_A + n_a} \right\} / 4$$
$$= \left\{ \frac{(109.52 \times 16)^2}{16} + \frac{(105.23 \times 16)^2}{16} \right. $$
$$\left. - \frac{[(109.52 \times 16) + (105.23 \times 16)]^2}{16 + 16} \right\} / 4$$
$$= 2,351$$

The sum of squares for variation within genotypes (WG–SS) is obtained by subtracting the variation between types (above) from the total sum of squares between lines:

$$\text{WG} - \text{SS} = \text{SS lines} - \text{BG} - \text{SS}$$

In the case of the *AA* and *aa* bands we have:

$$\text{WG} - \text{SS} = 38,498 - 2,351 = 36,147$$

The degree of freedom for the between-genotype sum of squares is 1, while the degrees of freedom for the within-genotype sum of squares is the total number of lines minus two (in our example, $32 - 2 = 30$).

Table 5.12 Degrees of freedom and mean squares from the analyses of variance of seed yield between and within progeny that are polymorphic at the *AA:aa*, *BB:bb* and *CC:cc* loci.

Source	df	*AA–aa* locus	*BB–bb* locus	*CC–cc* locus
Between genotypes	1	2,351 ns	24,618 ***	11,211 ***
Within genotypes	30	36,147 ***	463 ns	913 *
Replicate error	93	459	459	459

$* = 0.05 > p > 0.01$;
$** = 0.01 > p > 0.001$;
$*** = p < 0.001$.

Completing this operation for the other two bands, we have the mean squares from three analyses of variance (see Table 5.12).

The within genotypes effect is tested against the replicate error, while the between genotype effect is tested against the within genotype mean square.

Clearly, there is a significant relationship between seed yield and alleles at the B-bands. Similarly, some relationship exists between the C-band and yield, although the variability with genotypes *CC* and *cc* are highly significant, hence weakening the QTL relationship. There is no relationship between seed yield and bands at the A-band.

From the above analysis of variance, our best guess to the position of the QTL would be between the B- and C-bands, and nearer to the B than the C. Determination of the position of the QTL on the chromosome can be done using a number of statistical techniques. The simplest technique involves regression, and will be illustrated here.

Now the difference in yield between genotypes at each band (δ_i) is an indication of the linkage between the QTL and the single band position. In this example we have:

$$\delta_{A-a} = 2.145; \quad \delta_{B-b} = 6.935; \quad \delta_{C-c} = 4.475$$

Given a simple additive–dominance model of inheritance, we find that lines that have *BB* and QTL+ will have expectation of $m + a$, and this genotype will occur in the population with frequency $1/2R$, where R is the recombination frequency between the B-band and the QTL. The *BB* lines without the QTL (QTL−) will be $m - a$, and will occur in the population with frequency $1/2(1 - R)$, where R is the recombination frequency between the B-band and the QTL. Similarly,

for *bb* we have bb/QTL+ $= m + a$, frequency $= 1/2R$, bb/QTL− $= m - a$, frequency $= 1/2(1 - R)$. The difference between the *BB* and *bb* genotypes (δ_{B-b}) is therefore equal to $a(1 - 2R)$.

There is therefore a linear relationship between δ_is and $(1 - 2R)$:

$$\delta_i = a(1 - 2R)$$

where the regression slope is an estimate of a.

The value of a and the accuracy of fit of regression is dependent upon R, the recombination frequency between the three single band positions and the QTL. We know the map distance between A–B and B–C. Therefore all that is now required is to substitute in recombination frequencies to find the recombination frequency with the least departure from regression in a regression analysis of variance. It is usual to start with the assumption that the QTL is located at the A-band position and complete a regression analysis. Then assume that the QTL is 2 cM from A, towards B, and carry out another analysis. Repeat this operation until it is assumed that the QTL is located at the C-locus. Thereafter determine which of the regression analyses has the best regression fit (with least departure from regression term) and the QTL will be located at that map location. From this the recombination frequencies between the various single band positions and the QTL can be calculated to determine the usefulness of the linkage with the QTL and hence the usefulness in practice.

To avoid duplication, the speculative map distances from each single band position and the QTL in our example is shown from around the map location with minimum departure from regression:

	A-locus	B-locus	C-locus	Residual sum of squares
δ_i	2.145	6.935	4.475	
R_is	0.300	0.013	0.193	1932
	0.310	0.003	0.183	802
	0.3198	**0.01478**	**0.17148**	**0**
	0.330	0.017	0.163	339

From this the resulting map, including the QTL, would be:

A- - -30.5 mμ- - -B- -1.5 mμ- -QTL- -17.1 mμ- - -C

which, when converted to cMs, is:

A- - -47.1 cM- - -B- -1.5 cM- -QTL- - -21.0 cM- - -C

In this simple case the QTL and the B-bands are tightly linked and therefore selection based on the B-band would be highly effective in selecting for the QTL, and hence high seed yield. Actual examples in plant breeding, however, are rarely this close. The close proximity of the QTL to the B-band is also a reflection of the high recombination frequencies (low linkage) between the three bands in this simple example. If the recombination frequencies between A–B and B–C were halved (i.e. 15.7% and 9.0%), the QTL would have a recombination of 3% with the B-band position. Similarly, this example looked at only three bands on a single chromosome; in real situations, many chromosomes will be involved and more loci examined on each chromosome. However, the underlying theory is the same.

Think questions

(1) A cross is made between two homozygous barley plants. One parent is dominant for a tall stature gene (*TT*) and has a single dominant resistance to yellow rust (*YY*), but highly susceptible to powdery mildew (*mm*). So the genotype of the first parent can be represented by *TTYYmm*. The other parent has short stature (i.e. recessive dwarfing gene, *tt*) with single (completely dominant) resistance to powdery mildew (*MM*), but is susceptible

to yellow rust (*yy*). So the genotype of the second parent can be represented by *ttyyMM*. Consider if a sample of F_1 plants from this cross were self-pollinated and used to produce a population of 12,800 F_2 plants. At the harvest of these F_2 plants the short (dwarf) plants were selected and only seed from these plants was planted out in head-rows at F_3. How many of the F_3 head-rows would be resistant to both yellow rust and powdery mildew? How many will be resistant to powdery mildew but susceptible to yellow rust? And, how many will be susceptible to both diseases?

(2) In a breeding experiment similar to that in question 1, a cross is made between two homozygous barley plants. One parent is tall and with short leaf margin hairs (*TTss*), and the other is short with long leaf margin hairs (*ttSS*). Single genes control both plant height and leaf margin hair length, tall plants (*T_*) being completely dominant to short (*tt*), and long leaf margin hairs (*S_*) dominant to short (*ss*). A sample of F_1 plants were self-pollinated and 1,600 F_2 plants evaluated for plant height and leaf margin hair length. After evaluating the phenotypes it was found that there were:

893 tall plants with long leaf margin hairs;
313 tall plants with short margin hairs;
  0 short plants with long leaf margin hairs; and
394 short plants with short margin hairs.

Give a genetic explanation of what underlies the observed frequency of phenotypes and test your theory using a suitable statistical test.

(3) A spring wheat breeding programme aims to develop cultivars that are resistant to foot-rot (controlled at a single locus with a completely dominant allele (*FF*) conferring resistance), and that are also resistant to yellow stripe rust (controlled at a single locus with a single dominant (*YY*) allele conferring resistance). A cross is made between two parents where one parent is resistant to foot-rot and susceptible to yellow stripe rust (*FFyy*) while the other is resistant to yellow stripe rust but susceptible to foot-rot (*ffYY*). The heterozygous F_1 is crossed to a double susceptible homozygous line with

genotype *ffyy* and 2000 BC_1 progeny were evaluated for disease resistance. The following numbers of phenotypes were observed:

Resistant to both foot-rot and yellow
stripe rust = 90

Resistant to foot rot but susceptible to yellow
stripe rust = 893

Susceptible to foot-rot but resistant to yellow
stripe rust = 907

Susceptible to both foot-rot and yellow
stripe rust = 110

Determine the percentage recombination between the foot-rot and yellow stripe rust loci.

Given the frequency of phenotypes with the above percentage recombination, how many F_2 plants would need to be evaluated to be 99% certain of obtaining at least one plant that is resistant to both foot-rot and yellow stripe rust? In contrast, how many would need to be evaluated to be 99% certain of obtaining at least one plant that is resistant to both foot rot and yellow stripe rust, given independent assortment of the two loci involved?

(4) Potato cyst nematode resistance is controlled by a single locus with a completely dominant allele (R). A cross is made between two auto-tetraploid potato cultivars where one parent is resistant to potato cyst nematode and the other is completely susceptible. Progeny from the cross are examined and it is found that 83.3% of the progeny are resistant to potato cyst nematode. What can be determined about the genetic status of the resistant parent in the cross?

A breeding programme aims to produce potato parental lines that are either quadruplex or triplex for the potato cyst nematode resistant allele *R*. A cross is made between two parents, which are known to be duplex for the resistant allele *R*. What would be the expected genotype and phenotype of progeny from this duplex × duplex (i.e. *RRrr* × *RRrr*) cross?

(5) Why are plant breeders interested in conducting scaling tests for quantitatively inherited characters of importance in the breeding scheme?

A properly designed experiment was carried out in canola where parent lines (P_1 and P_2), were grown alongside F_1 and F_2 progeny plants from the cross $P_1 \times P_2$. The average yield of plants from each of the four families, the variance of each family and the number of plants evaluated from each family were:

Family	Mean yield	Variance of yield	Number of plants
P_1	1,901	74	21
P_2	1,502	68	21
F_1	1,429	69	21
F_2	1,888	102	31

Use an appropriate test to determine whether the additive–dominance model of inheritance is adequate to explain the genetic variation in yield in canola.

What can you conclude from the results obtained above? What could have caused this to occur?

(6) Assuming an additive–dominance mode of inheritance for a polygenic trait, set out the parameters to describe $P_1, P_2,$ and F_1 in terms of m, $[a]$ and $[d]$ and then similarly for F_2, B_1 and B_2.

From a properly designed field trial that included B_1, B_2 and F_1 families, the following yield estimates were obtained.

$$B_1 = 42.0 \, \text{kg/acre}; \quad B_2 = 26.0 \, \text{kg/acre};$$
$$F_1 = 38.5 \, \text{kg/acre}$$

From these family means, estimate the expected value of P_1, P_2 and F_2, based on the additive–dominance model of inheritance

(7) A cross is made between two homozygous barley plants. One parent has a two-row ear (controlled at a single locus by a recessive allele (*hh*) (as opposed to the dominant six-row (*HH*)) and is dominant for a tall stature at the 'T' locus (*TT*), is resistance to yellow rust (*YY* – completely dominant), but highly susceptible to powdery mildew (i.e. *TThhYYmm*). The other parent has a six-row ear (*HH*), short (dwarf, *tt*), resistant to powdery mildew, but is susceptible to yellow rust (i.e. *ttHHyyMM*).

All trait loci are located on different chromosomes. A sample of F_1 plants from this cross were self-pollinated, and a population of 92,160 F_2 plants grown. At harvest the short (dwarf) plants with two-row ears are selected and only seed from these plants are grown out in head-rows at F_3. How many of the F_3 head-rows would be resistant to both yellow rust and powdery mildew? How many will be resistant to powdery mildew but susceptible to yellow rust? And how many will be susceptible to both diseases?

(8) Two homozygous squash plants were hybridized and an F_1 family produced. One parent had long, green fruit ($LLGG$), and the other had round, yellow fruit ($llgg$). 1,600 F_2 progeny were examined from selfing the F_1s and the following numbers of phenotypes observed:

$LLGG$	L-gg	llG	$llgg$
891	312	0	397

Complete an appropriate analysis to explain and interpret this segregation pattern.

6
Predictions

6.1 Introduction

Plant breeders strive to make a wide range of predictions that allow them to act in the most effective way in creating genetic variation and selecting desirable genotypes. The ability to make sound predictions is critical, as this allows plant breeders to foresee situations which might not be realized or observed in field evaluations until several years later. Whether those predictions indeed translate into meaningful genetic progress and development of superior breeding lines will depend upon the questions being asked and the confidence in the predictions that are generated.

There are a number of questions that might be posed, but ones that a breeder might sensibly ask would include:

- What expression of what traits would be most successful when a cultivar is released?
- What will the cultivar need to display to meet agronomic requirements so as to fit ideally into the most effective management systems?
- How stable will be the expression of the important traits (e.g. yield) over a range of environments (especially locations and years)?
- Which parents will give the best progeny for further breeding or for commercial exploitation?
- Which traits will respond most significantly to the selection imposed?
- What type and level of selection will give the optimal response in the traits of interest?

- Based on the predictions made, what level of financial input and people's effort is needed to realize those expectations?

The first question is clearly difficult to answer and is one that faces breeders all the time. The second question is partly a matter of selection conditions, partly a matter of judging what is required, and partly luck! The third is one upon which a considerable amount of work has been carried out and is, of course, one of genotype by environment interaction. It is a fundamental aspect of breeding, and its formal treatment is beyond the scope of this book, therefore we intend to only spend a limited amount of time on it.

However, questions such as the last four (above) can be addressed through a combination of knowledge of genetics, experimental design and statistics – the better the knowledge we have and the more extensive it is, the more accurate our predictions will be.

Let us start by considering the stability of trait expression over environments and its genetic determination.

6.1.1 Genotype × environment interactions

The observed performance (in other words the phenotype) of a genotype will differ in different environments. Obviously one would expect that if we apply less fertilizer we generally get less yield, the further apart we space the plants of many

Plant Breeding, Second Edition. Jack Brown, Peter D.S. Caligari and Hugo A. Campos.
© 2014 John Wiley & Sons, Ltd. Published 2014 by John Wiley & Sons, Ltd.
Companion Website: www.wiley.com/go/brown/plantbreeding

species the more vegetative growth they make, and if the lifecycle of a newly developed cultivar does not fit the length of the growing season, it would run out of time and its yield potential would never be realized. We should never forget that even though plant breeders like to think in terms of genotypes and allelic effects, farmers only care about what they are able to see in their fields, i.e. a phenotype. This can clearly be handled in our selection programme by selecting under the conditions we think are most appropriate and, if necessary, in several different ones. But what is more complicated is that not all genotypes respond to the same extent or necessarily in the same way to differences in the numerous environmental variables.

So some genotypes are more drought-tolerant, some more disease-susceptible, some can withstand higher levels of salt in the soil than their counterparts, and so on. We cannot grow all the possible genotypes we are interested in under all the possible environmental conditions. We might note that some environmental variables are of a 'macro' nature and fairly obvious, but there are numerous possible differences in the environment that are experienced by individual plants – 'micro' ones.

So winter or spring sowings give a clear set of environmental differences that we might take specifically into account, just as different latitudes, temperatures, semi-arid and tropical climates might be important considerations in an international breeding programme. But differences in water availability at one end of a test area compared with the other, the row spacing produced by one piece of farm equipment compared with another, or even the source of the seed planted in breeding trials, are but a few of the possible subtle differences which might be very important to individuals or plots or fields or farms.

There are a number of ways in which plant breeders try to take genotype × environment interactions (G × E) into account in breeding, but it does mean that there is a need to carry out trials over a range of environments which might simply be different sites or over years/seasons, or running trials with defined differences such as water levels.

Genotype × environment interaction at its simplest can be examined by looking at the variance (or standard error obtained by taking the square root of the variance) of the phenotypes over the range of environments. It is then possible to

select for the lowest variance as being the most stable – remembering that we also need a good mean trait expression. A more sophisticated approach is to use the mean of all the material grown to provide a biological measure of the environment and compare (usually by using the slope of the regression line) the individual lines, families and clones against this.

It is not appropriate to go into greater details here about G × E or the various possibilities to take it into account. Needless to say, many breeding programmes effectively ignore G × E in an explicit way but take it into account, at least to a modest extent, by trialling the more advanced material at different sites, and by the fact that selection is carried out over a number of sequential years. A more detailed examination of analyses of multiple year and multiple location trials is presented in Chapter 7.

Let us now consider prediction protocols associated with answering which parents will give the best progeny, and which traits will respond most significantly to the selection we impose.

6.1.2 Genetically based predictions

Plant breeders use all the genetic information (qualitative and quantitative) in just the same way as we use the information from Mendelian genetics – in other words, to predict the properties of generations or families that have not actually been observed. So from an analysis of the observed variation, they firstly determine how much of the variation is due to environmental effects and how much due to genetic effects. Often it is desirable to go further and separate the genetic variation into additive, dominance, and other genetic effects, and to determine in which direction dominance is acting, and to what degree.

Let us consider one particular use of the information. How can we predict the response to selection? Before you start a selection programme you would obviously like to know what sort of response you might expect for any given input, which traits are worth targeting, and which populations or crosses are best to use. Is it a worthwhile venture? These questions involve many aspects of the biology of the crop, its handling in agriculture, the availability of other methods to affect the crop's performance, such as pesticides and herbicides, and so on. But one of the main components that will determine the outcome is the amount and type of variation that is present in the germplasm a breeder has at their disposal. For example, in the extreme case of no genetic variation, the breeder is wasting time and resources trying to select superior genotypes.

If genetic variation is present, but small compared with that due to the environment, then progress can be made but only very slowly, unless very large numbers are handled – in other words, much of the time the breeder will be selecting phenotypes that are 'superior', but as this is mostly due to the environment it will not give a reliable indication of a 'superior' genotype. If, on the other hand, the phenotype is a good reflection of the genotype, that is, most of the variation is genetic, then progress will be quick, since when the breeder selects a good phenotype and uses it as a parent it will pass on the superior attributes (via its genes) to its offspring.

One obvious question to ask is: can we estimate how much of the variation we observe, for example in the F_2 we are looking at, can be ascribed to genetic differences of segregating genes and how much to environmental causes? To do this we need to measure the proportion of the total observed variation that is genetic, and such a measure is called *heritability*.

6.2   Heritability

For a modern plant breeder (or indeed a farmer with no knowledge of genetics) to make economically meaningful progress in an organized programme of selective breeding, two conditions must be met:

- There must be some observable phenotypic variation within the crop. This would normally be expected, even if it were due entirely to the effects of a variable environment.
- At least some of this phenotypic variation must have a genetic basis.

This relates to the concept of heritability, the proportion of the phenotypic variance that is genetic in origin. This proportion is called the heritability and this section is concerned with the ways of estimating heritability.

Values of heritability (h^2) can range from zero to one. If h^2 is relatively high (e.g. close to 1), there is potential for a breeding programme to alter the mean expression of the character in future generations. On the other hand, if h^2 is close to zero, there will be little scope for advancement and there would probably be little point in trying to improve this character in a plant breeding programme, unless a new source of genetic variation became available (new germplasm).

There are three main ways of estimating heritability.

- Carrying out particular genetic crosses and observing the performance of their progeny so that the resulting data can be partitioned into genetic and environmental components.
- Based on the direct measurement of the degree of resemblance between offspring and one, or both, of their parents. This is achieved by regression of the former onto the latter in the absence of selection.
- Measuring the response of a population to given levels of selection (this will not be discussed until we cover selection later in Chapter 7).

The essential theoretical background of heritability was presented in the previous section on quantitative genetics.

Heritability is the ratio of genetic variance divided by total phenotypic variance. In a simple additive – dominance model of quantitative inheritance, the total genetic variance will contain *dominance genetic variance* (denoted by V_D) and *additive genetic variance* (denoted by V_A). Dominance genetic variance is variation caused by heterozygotes loci in the individuals in the population, whereas additive genetic variance is variation existing between homozygous loci in the segregating population.

Broad-sense heritability (h_b^2) is the total genetic variance divided by the total phenotypic variance. The total genetic variance in an additive – dominance model is simply $V_A + V_D$. The total phenotypic variance is obtained by adding to this genetic variance, the environmental variance.

The degree of heterozygosity within segregating populations will be related to the number of selfing generations to which it has been subjected. Maximum heterozygosity will be found in the F_1 family, and will be reduced, by half, in each subsequent selfed progeny. Similarly, the dominance genetic variance will be dependent upon the degree of heterozygosity in the population and will differ between filial generations. A more useful form of heritability for plant breeders, therefore, is ***narrow-sense heritability*** (h_n^2), *which is the ratio of additive genetic variance* (V_A) *to total phenotypic variance.*

Why should lack of resemblance between parents and their offspring be attributable to dominance, but not additive, components? Well, dominance effects are a feature of particular genotypes; but genotypes are 'made' and 'unmade' between generations as a result of genetic segregation during the production of gametes. Thus, the mean dominance effect in the offspring of a particular cross can be different from that of the parents, even when there is no selection. On the other hand, when selection is applied, there may be no change or even change in the 'wrong' direction. This is not true of additive genetic effects. The additive genetic component will remain more or less constant from one generation to the next in the absence of selection. If differential selection is applied, the change between generations must be in the direction corresponding to the favoured alleles. In addition, additive genetic variance is constant between filial generations, and so narrow-sense heritability of recombinant inbred lines can be estimated from early-generation segregating families.

In the first filial generation (F_1), after hybridization between two homozygous parents, there is no genetic variance between individuals of a progeny (they will all be genetically alike) and all variation observed between F_1 plants will be entirely environmental. The first generation for which there are both genetic and environmental components of phenotypic variance is the F_2. Partitioning of phenotypic variance and the calculation of the broad-sense (and ultimately narrow-sense) heritabilities will be confined in this section to this generation.

6.2.1 Broad-sense heritability

The first step is to derive an equation for the genetic variance of the F_2 generation. The genetic variance of the F_2 generation (without proof) is:

$$\sigma_{F_2}^2 = \frac{1}{2}V_A + \frac{1}{4}V_D$$

The total phenotypic variance of any generation is the sum of its genetic variance plus the environmental variance, E. So the total phenotypic variance of the F_2 generation can be written:

$$\sigma_{F_2}^2 = \frac{1}{2}V_A + \frac{1}{4}V_D + \sigma_E^2$$

In terms of V_A, V_D and σ_E^2, therefore, the broad-sense heritability of the F_2 generation is:

$$h_b^2 = \frac{\text{genetic variance}}{\text{total phenotypic variance}}$$
$$= \frac{\frac{1}{2}V_A + \frac{1}{4}V_D}{\frac{1}{2}V_A + \frac{1}{4}V_D + \sigma_E^2}$$

In order to estimate the broad-sense heritability of the F_2 family (or indeed any other segregating family), all that is required is an estimate of total phenotypic variation, and an estimate of environmental variation. The former is obtained by measurements on plants within F_2 families, while the latter is estimated from measurements on families or plants that have a uniform genotype (i.e. homozygous parental lines or F_1 families where plants are genetically identical and any variation between plants is due to environment).

To illustrate this, consider a simple numerical example. A field experiment with an inbreeding crop species such as wheat or barley was conducted which included 20 plants from Parent 1, 20 plants from Parent 2 and 100 plants from the F_2 family derived from selfing the F_1 generation, which was itself obtained by intercrossing the two parents (i.e. Parent 1 × Parent 2). These 140 plants were completely randomized within the experiment, and at harvest the weight of seeds from each plant was recorded. The variances in seed weight of the two parents were $\sigma_{P_1}^2 = 16.8 \text{ kg}^2$ and $\sigma_{P_2}^2 = 18.4 \text{ kg}^2$. The phenotypic variance (which included both genetic and environmental variation) of the F_2 was $\sigma_{F_2}^2 = 56.9 \text{ kg}^2$. Total phenotypic variance of the F_2 generation is represented by $\frac{1}{2}V_A + \frac{1}{4}V_D + \sigma_E^2$ and is estimated to be 56.9 kg^2 (the variance of the F_2). It therefore follows that the broad-sense heritability, h_b^2, for these data is:

$$h_b^2 = \frac{56.9 - \sigma_E^2}{56.9}$$

The problem reduces to: what is the value of the environmental component of the phenotypic variance σ_E^2? Since, by definition, both parents are completely homozygous inbreds, any variance displayed by either must be attributable exclusively to the environment. The best estimate of the value of σ_E^2 is therefore the mean phenotypic variance of these two generations. Thus:

$$\sigma_E^2 = (\sigma_{P1}^2 + \sigma_{P2}^2)/2$$
$$= (16.8 + 18.4)/2 = 17.6 \text{ kg}^2$$

and:

$$h_b^2 = \frac{56.9 - 17.6}{56.9} = 0.691$$

In other words, 69.1% of the phenotypic variance of the F_2 generation is estimated to be genetic in origin.

The other generation in which the phenotypic variance is also entirely attributable to environmental effects is the F_1. If the phenotypic variances of all three of these genotypically invariate generations were available, the environmental component of the phenotypic variance of the F_2 generation could be estimated as follows:

$$\sigma_E = \frac{\sigma_{P_1}^2 + 2\sigma_{F_1}^2 + \sigma_{P_2}^2}{4}$$

Quantitative geneticists often use more elaborate formulae, but this one will serve our purpose.

6.2.2 Narrow-sense heritability

Often it is of more interest, for reasons already noted, to know what proportion of the total phenotypic variation is traceable to additive genetic effects rather than total genetic effects. This ratio of additive genetic variance to total phenotypic variance is called the narrow-sense heritability (denoted by h_n^2) and is calculated as:

$$h_n^2 = \frac{\text{additive genetic variance}}{\text{total phenotypic variance}}$$

Therefore, in terms of V_A, V_D and σ_E^2, what is the narrow-sense heritability of the F_2 generation? Since:

$$h_b^2 = \frac{\frac{1}{2}V_A + \frac{1}{4}V_D}{\frac{1}{2}V_A + \frac{1}{4}V_D + \sigma_E^2}$$

it is logical that:

$$h_n^2 = \frac{\frac{1}{2}V_A}{\frac{1}{2}V_A + \frac{1}{4}V_D + \sigma_E^2}$$

In order to estimate the narrow-sense heritability it is therefore necessary to partition the genetic variance into its two components (V_A and V_D). This is done now by also considering the phenotypic variance of the two backcross families ($\sigma_{B_1}^2$ and $\sigma_{B_2}^2$). Without proof, the expected variances of $\sigma_{B_1}^2$ and $\sigma_{B_2}^2$ are:

$$\sigma_{B_1}^2 = \frac{1}{4}V_A + \frac{1}{4}V_D - \frac{1}{2}\left[\sum(a) \times \sum(d)\right] + \sigma_E^2$$
$$\sigma_{B_2}^2 = \frac{1}{4}V_A + \frac{1}{4}V_D + \frac{1}{2}\left[\sum(a) \times \sum(d)\right] + \sigma_E^2$$

The awkward expression $\frac{1}{2}[\Sigma(a) \times \Sigma(d)]$ disappears when the equations are added together. Therefore:

$$\sigma_{B_1}^2 + \sigma_{B_2}^2 = \frac{1}{2}V_A + \frac{1}{2}V_D + 2\sigma_E^2$$

As we noted earlier that:

$$\sigma_{F_2}^2 = \frac{1}{2}V_A + \frac{1}{4}V_D + \sigma_E^2$$

provided that numerical values for $\sigma_{F_2}^2$, $\sigma_{B_1}^2$, $\sigma_{B_2}^2$ and σ_E^2 can be estimated, there is sufficient information to calculate both V_A and V_D, and hence the narrow-sense heritability.

To illustrate this, consider the following example. A properly designed glasshouse experiment was carried out using the garden pea. Progeny from the F_1, F_2 and both backcross families (B_1 and B_2) were arranged as single plants in a completely randomized block design, and plant height recorded after flowering. The following variances were calculated from the recorded data:

$$\sigma_{F_2}^2 = 358 \text{ cm}^2; \quad \sigma_{B_1}^2 = 285 \text{ cm}^2$$
$$\sigma_{B_2}^2 = 251 \text{ cm}^2; \quad \sigma_E^2 = 155 \text{ cm}^2$$

now:

$$\sigma_{B_1}^2 + \sigma_{B_2}^2 - \sigma_{F_2}^2 = \left(\frac{1}{2}V_A + \frac{1}{2}V_D + 2\sigma_E^2\right)$$
$$- \left(\frac{1}{2}V_A + \frac{1}{4}V_D + \sigma_E^2\right)$$
$$= \frac{1}{4}V_D + \sigma_E^2$$

therefore

$$V_D = 4(\sigma_{B_1}^2 + \sigma_{B_2}^2 - \sigma_{F_2}^2 - \sigma_E^2)$$
$$= 4(285 + 251 - 358 - 155)$$
$$= 92 \text{ cm}^2$$

Rearranging the equation for $\sigma_{\overline{F}_2}^2$ (i.e. $\sigma_{\overline{F}_2}^2 = \frac{1}{2}V_A + \frac{1}{4}V_D + \sigma_E^2$) we have:

$$
\begin{aligned}
V_A &= 2\left(\sigma_{\overline{F}_2}^2 - \tfrac{1}{4}V_D - \sigma_E^2\right) \\
&= 2(358 - \left[\tfrac{1}{4} \times 92\right] - 155) \\
&= 360 \text{ cm}^2
\end{aligned}
$$

Therefore, the narrow-sense heritability for these data is:

$$
\begin{aligned}
h_n^2 &= \frac{\tfrac{1}{2}V_A}{\tfrac{1}{2}V_A + \tfrac{1}{4}V_D + \sigma_E^2} \\
&= \frac{0.5 \times 360}{(0.5 \times 360) + (0.25 \times 92) + 155} \\
&= 0.50
\end{aligned}
$$

However, this can be derived more simply by:

$$
\begin{aligned}
h_n^2 &= \frac{\tfrac{1}{2}V_A}{\text{total phenotypic variation}} \\
&= \frac{\tfrac{1}{2}V_A}{\sigma_{\overline{F}_2}^2} \\
&= \frac{0.5 \times 360}{358} = 0.50
\end{aligned}
$$

Thus, 50% of the phenotypic variation in this F_2 generation of peas is genetically additive in origin.

6.2.3 Heritability from offspring – parent regression

In this section we will consider another method of estimating the narrow-sense heritability. The option of predicting the response to selection using heritabilities will be discussed in the selection section (Chapter 7). However, the phenomenon does suggest another approach to measuring the heritability of a character, namely comparison of the phenotypes of offspring with those of one or both of their parents. Close correspondence, in the absence of selection, implies that the heritability must be relatively high. On the other hand, if the phenotypes of parents and offspring appear to vary independently of one another, it suggests that heritability must be low.

The foundations of this approach, which is termed *offspring – parent regression*, were laid in the nineteenth century by Charles Darwin's cousin Francis Galton in his study of the resemblance between fathers and sons. Therefore, the narrow-sense heritability of a metrical character can be estimated from the regression coefficient (slope) of the graph of offspring phenotypes on those of their parents.

In regression analysis, one variable is regarded as **independent**, while another is regarded as being potentially **dependent** upon it. Not surprisingly, in offspring – parent regression, the phenotype of the parent(s) corresponds to the independent variable and that of the offspring the dependent variable. So our aim is to predict the performance of progeny from crosses between chosen parents based on the phenotypic performance of the parent lines.

The narrow-sense heritability of a character in the F_2 generation is:

$$
h_n^2 = \frac{\tfrac{1}{2}V_A}{\tfrac{1}{2}V_A + \tfrac{1}{4}V_D + \sigma_E^2}
$$

and the regression coefficient from simple linear regression is:

$$
\mathbf{b} = \text{SP}(x, y)/\text{SS}(x)
$$

Therefore, there must be some relationship between h_n^2 and the slope (more formally termed regression coefficient) (**b**). The regression relationship when the offspring expression is regressed onto the expression of **one** of the parents (provided without proof) is:

$$
\mathbf{b} = \frac{\tfrac{1}{4}V_A}{\tfrac{1}{2}V_A + \tfrac{1}{4}V_D + \sigma_E^2}
$$

and since:

$$
h_n^2 = \frac{\tfrac{1}{2}V_A}{\tfrac{1}{2}V_A + \tfrac{1}{4}V_D + \sigma_E^2}
$$

it follows that, for the regression of offspring phenotypes on the phenotypes of one of their parents:

$$
h_n^2 = 2 \times \mathbf{b}
$$

In short, to estimate the narrow-sense heritability (h_n^2), it is necessary to perform a regression analysis of the mean phenotypic value of the offspring arising from individual parents onto the phenotypic value of those parents. The regression coefficient (**b**) is obtained in the usual way by dividing covariance (in this case of the offspring and parent) by the variance of the parental generation. The narrow-sense heritability is then double the value of this regression coefficient.

When the mean expressions of progenies are regressed onto the average performance of **both** parents (the mid-parental value) then the regression coefficient (or slope) is (given without proof):

$$b = \frac{\frac{1}{4}V_A}{\frac{1}{2}\left(\frac{1}{2}V_A + \frac{1}{4}V_D + \sigma_E^2\right)}$$

$$= \frac{\frac{1}{2}V_A}{\frac{1}{2}V_A + \frac{1}{4}V_D + \sigma_E^2}$$

and since:

$$h_n^2 = \frac{\frac{1}{2}V_A}{\frac{1}{2}V_A + \frac{1}{4}V_D + \sigma_E^2}$$

it follows that, for the regression of offspring phenotypes on the mean phenotypes of both their parents:

$$h_n^2 = b$$

In short, to estimate the narrow-sense heritability it is necessary to perform a regression analysis of the mean phenotypes of offspring (as before) but now onto the mean phenotypes of **both** their parents. The regression coefficient/slope (**b**), obtained by dividing the offspring – parent covariance by the variance of the parental generation, estimates the narrow-sense heritability directly.

Consider the following simple example. The data below are the phenotypes of parents and their off-spring from a number of crosses in a frost-tolerant winter canola breeding programme. The character is yield (kg/ha).

Female parent	Male parent	Mid-parent value	Offspring value
995	1,016	1,005.5	1,006
1,004	999	1,001.5	1,004
1,009	996	1,002.5	1,008
1,012	1,014	1,013.0	1,010
1,005	1,014	1,009.5	1,013
1,007	1,004	1,005.5	1,007
1,034	1,014	1,024.0	1,024
1,015	998	1,006.5	1,002
1,017	1,028	1,022.5	1,020
1,003	1,013	1,008.0	1,008

Then

Statistic	Female parent	Male parent	Mid-parent value
Regression slope (**b**)	0.476	0.468	0.813
se(**b**)	0.1632	0.1895	0.1269
t_{8df}	2.898	2.259	6.407

From the regression of offspring onto one parent:

$$\text{male } h_n^2 = 2 \times b = 2 \times 0.473 = 0.946$$
$$\text{female } h_n^2 = 2 \times b = 2 \times 0.468 = 0.936$$

From the regression of offspring onto the average phenotype of both parents (mid-parent) we have:

$$h_n^2 = 0.813$$

You will notice that the estimated heritability using either one of the parents is larger than that from both parents. However, it should be noted that the estimation based on both parents will be more accurate. Despite the differences it is obvious, however, that there is a high narrow heritability (80–90%) and therefore we can conclude there is a high degree of additive genetic variance for this character in this experiment.

Finally, always remember that any heritability estimate, no matter which method is used to obtain it, is only valid for that population, at that time, in that environment. Change the environment, carry out (or allow) selection to occur, add more genotypes, sample another population, and the heritability might be different. This should be clear from the descriptions and methods of calculating heritability, but you will find many examples in the literature where the basic limitations of the concept are forgotten.

6.3 Diallel crossing designs

It has been over 130 years now since the publication by Louis de Vilmorin that became known as Vilmorin's isolation principle or progeny test. He proposed that the only means to determine the value of an individual plant (or genotype) was to grow and evaluate its progeny. Ever since, of course, the progeny test has become well established and is frequently used by plant breeders to determine the genetic potential of parental lines. The diallel crossing scheme is simply a more sophisticated application of Vilmorin's progeny test.

The term ***diallel cross*** has been attributed to a Danish geneticist (J. Schmidt) who first used the design in animal breeding. The term and design came to plant breeding and began to be used by plant scientists in the mid-1950s.

The diallel cross was then described as ***all possible crosses amongst a group of parent lines***. With n parents there would be n^2 families. The n^2 families or progeny are called a ***complete diallel cross***. If the reciprocal crosses are not made, making $n[n-1]/2$ families, the result is called a ***half diallel***. A ***modified diallel*** is one in which all possible cross combinations are included but the parental selfs (diagonal elements) are excluded. This type of diallel will include $n^2 - n$ families. In a ***partial diallel*** fewer than the $n[n-1]/2$ cross combinations are completed. However, the crosses that are included are arranged in such a way that valid statistical analysis and interpretation can be carried out.

Initially, only inbred homozygous lines were used as parents in diallel crossing designs. Analytical approaches that allow for parents being non-inbred genotypes (i.e. heterozygous) are now available.

According to some critics of the designs, '*the diallel mating design has been used and abused more extensively than any other …*'. Whether this statement is true or otherwise, there is little doubt that if the theory of diallel analysis is adhered to, and if interpretation is carried out in a logical manner, then the use of diallel crossing designs can be of great benefit to plant breeders in aiding understanding of quantitative inheritance and providing invaluable information regarding the genetic potential of parental lines in cross combinations. The limitation of the design arises in terms of the sample of parental genotypes that can be handled, which is always somewhat restricted.

It will not be possible to cover the whole spectrum of information or even indeed the types of diallel crossing schemes that are available, or to investigate the interpretation of many examples within the space of this book. There are therefore several approaches to the analysis and interpretation of data derived from diallel cross, although only two will be covered briefly in this section:

- Analysis of general and specific combining ability. These methods are often referred to as **Griffing's analyses**, after B. Griffing who published his now famous paper *Concept of general and specific combining ability in relation to diallel crossing systems* in 1956.
- Analysis of array variances and covariances, often referred to as **Hayman and Jinks' analyses**, after B.I. Hayman and J.L. Jinks' paper of 1953, *The analysis of diallel crosses*.

6.3.1 Griffing's analysis

Griffing proposed a diallel analysis technique for determining general combining ability and specific combining ability of a number of parental lines in cross combination based on statistical concepts. Griffing's analyses have been used by many plant breeders and researchers over the past 40 years, and in many cases with good success. Much of the success found in applying Griffing analyses is the apparent ease of interpretation of results compared with other analyses available. Parents used in diallel crosses can be homozygous or heterozygous; for simplicity, diallel types are described here in terms of homozygous (inbred) parents. Four types of design analyses are available:

- **Method 1**. The full diallel where p parents are crossed in all possible cross combinations (including reciprocals). Therefore with p parents the design will consist of p^2 families ($p^2 - p$ segregating populations [i.e. say F_1s] and p inbred parents).
- **Method 2**. The half diallel where p parents are crossed in all possible combinations, parental selfs are included but no reciprocals. These types of design will contain $p[p+1]/2$ families ($p[p-1]/2$ segregating populations [i.e. say F_1s] and p inbred parents).
- **Method 3**. The full diallel without parental selfs, which consists of all cross combinations (including reciprocals) of p parents. Method 3 differs from Method 1 in that with Method 3 the inbred parents are not included in the diallel design.
- **Method 4**. The half diallel, without parental selfs, which consists of all p parents crossed in all possible combinations (but with no reciprocals). The Method 4 design differs from the Method 2 design as the inbred parents are not included in the Method 4 design.

Griffing's analysis allows the option to test for **fixed** (*Model 1*) or **random** (*Model 2*) effects. Fixed effect models are where inference is made only about the parents that are included in the diallel cross, while the random effects models are where inference is to be made regarding all possible parents from a crop species. Therefore, in fixed effect models the parents used in the diallel cross are specifically chosen (i.e. because a breeder wishes to have additional information regarding general or specific combining ability of chosen lines). In random effects models the parental lines should be chosen completely at random. If this is done then the analyses can be interpreted to cover the eventuality that *any parents are used*.

Obviously, in most cases where plant breeders are involved, it is often very difficult to decide whether the parental lines were *chosen* or identified at *random*. In many cases the parents in diallel crossing designs are a sample of already commercial cultivars. In this case, some would argue that being commercial cultivars they cannot be a *random sample of possible parents*, as by definition all commercial cultivars are a very narrow subset of all potential genotypes within a species. On the other hand, others have argued that plant breeders are only interested in genotypes of commercial or near-commercial standard, and they can therefore quite rightly term their choice as a random sample of commercially suitable cultivars.

There are no hard and fast rules regarding fixed or random models, and usually there is little to be lost or gained from either argument, provided that the analyses are not treated as one type and interpreted as another. For example, plant breeders and researchers often include diverse parental genotypes as parents in diallel crossing designs (and we believe this to be an excellent idea). However, they should not choose specific parental lines that show a range of expressions for (say) yielding ability, cross them in a diallel design, and try to infer from the results what would happen if other different lines were included.

It should be remembered that often the choice of either fixed or random parents is not so much *how* they were physically selected, but rather the overall goal that is to be achieved from the diallel analyses. Consider these two simple examples. In the first (say from a hybrid maize breeding program) the breeder wishes to intercross a selected subset of possible hybrid parents that may appear in hybrid cultivar combinations. In this case, the parents are obviously chosen to be fixed. The breeder's aim is to determine the general and specific combining ability of parents and specific parent cross combinations, with little interest in extrapolating beyond the parents actually used. In the second (say an inbreeding cultivar like wheat), now a wheat breeder may be interested in estimating the general combining ability (an estimate of additive genetic variance, see below for more details) to determine genetic gain in a selection programme. That wheat breeder will not wish to take a true random sample, as the chances are that those picked will have little or no adaptability or commercial worth in the target breeding region. So parents are *chosen* to be representative of the type of parents that would be used in a breeding programme, and from these the breeder does want to extrapolate to cover other parents not actually included in the design. In addition (as will be shown in the following chapter), genetic gain can only be achieved if there is suitable phenotypic variation within the progeny being selected. So it is common in breeding programmes to choose 'complementary' parents which have contrastingly different characteristics (i.e. high yield, poor disease resistance × low yield, good disease resistance). In which case, why would breeders not choose parents with known differences for diallel analyses? The arguments are likely to continue.

Griffing's analysis requires no genetic assumptions and has been shown by many researchers to provide reliable information on the combining potential of parents. Once identified, the 'best' parental lines (those with the highest general combining ability) can be crossed to identify optimum hybrid combinations or to produce segregating progeny from which superior cultivars would occur at a high frequency.

In simplest terms, the cross between two parents (i.e. parent i × parent j) in Griffing's analysis would be expressed as:

$$X_{ij} = \mu + g_i + g_j + s_{ij}$$

where μ is the overall mean of all entries in the diallel design, g_i is the general combining ability of the ith parent, g_j is the general combining ability of the jth parent, and s_{ij} is the specific combining ability between the ith parent and the jth parent.

General combining ability (GCA) measures the average performance of parental lines in cross combination. GCA is therefore related to (but not directly equal to) the proportion of variation that is genetically additive in nature.

Specific combining ability (SCA) is the remaining part of the observed phenotype that is not explained by the general combining ability of both parents that constituted the progeny. By definition, SCA is the portion of genetic variability which is not additive.

Griffing's analysis of a diallel is by analysis of variance, where the total variance of all entries is partitioned into: general combining ability; specific combining ability; and error variances. In cases where reciprocals are included, then reciprocals (or maternal effects) are also partitioned. Error variances are estimated by replication of families. To avoid excessive repetition, only Method 1 (complete diallel) and Method 2 (half diallel), both including parents, will be considered further.

Degrees of freedom (df), sum of squares (SS) and mean squares (MSq) from the analysis of variance for Method 1 for the assumption of model 1 (fixed effects) are shown in Table 6.1. Also shown are the expectations for the mean squares (EMS).

Similar expected mean squares for Method 1, model 2 (random effects) are shown in Table 6.2.

Considering now Method 2 (the half diallel), the degrees of freedom (df), sum of squares (SS), mean squares (MSq) and expected mean squares (EMS) for model 1 are shown in Table 6.3 and Method 2, model 2 in Table 6.4. For Method 1, where r is the number of replicates; p is the number of parents; S_g is $\frac{1}{2}p\Sigma_i(X_{i.}+X_{.i})^2-2/p^2X_{..}^2$; S_s is $\frac{1}{2}\Sigma_{ij}x_{ij}(x_{ij}+x_{ji})-\frac{1}{2}p\Sigma_i(X_{i.}+X_{.i})^2+1/p^2X_{..}^2$; S_r is $\frac{1}{2}\Sigma_{i<j}(x_{ij}-x_{ji})^2$ and $X_{i.}$ is $\Sigma_jx_{ij}=x_{i1}+x_{i2}+x_{i3}+\ldots,$

Table 6.2 Degrees of freedom, sum of squares and mean squares from the analysis of variance of a full diallel including parent selfs (Method 1) assuming random effects. Also shown are the expectations for the mean squares.

Source	df	SS	MSq	EMS
GCA	$p-1$	S_g	M_g	$\sigma^2+2p(1/(1-p))\sigma_s^2+2p\sigma_g^2$
SCA	$p(p-1)/2$	S_s	M_s	$\sigma^2+2((p^2-p+1))/p^2\sigma_s^2$
Reciprocal	$p(p-1)/2$	S_r	M_r	$\sigma^2+2\sigma_r^2$
Error	$(r-1)p^2$	S_e	M_e	σ^2

Table 6.3 Degrees of freedom, sum of squares and mean squares from the analysis of variance of a half diallel including parent selfs (Method 2), assuming fixed effects. Also shown are the expectations for the mean squares.

Source	df	SS	MSq	EMS
GCA	$p-1$	S_g	M_g	$\sigma^2+(p+2)(1/(1-p))\Sigma g_i^2$
SCA	$p(p-1)/2$	S_s	M_s	$\sigma^2+2(p/(p-1))\Sigma_js_{ij}^2$
Error	$(r-1)\{p(p+1)/2\}$	S_e	M_e	σ^2

Table 6.4 Degrees of freedom, sum of squares and mean squares from the analysis of variance of a half diallel including parent selfs (Method 2) assuming random effects. Also shown are the expectations for the mean squares.

Source	df	SS	MSq	EMS
GCA	$p-1$	S_g	M_g	$\sigma^2+\sigma_s^2+(p+2)\sigma_g^2$
SCA	$p(p-1)/2$	S_s	M_s	$\sigma^2+\sigma_s^2$
Error	$(r-1)[p(p+1)/2]$	S_e	M_e	σ^2

that is, sum over rows; $X_{.j}$ is $\Sigma_ix_{ij}=x_{1j}+x_{2j}+x_{3j}+\ldots$, that is, sum over columns; and $X_{..}$ is $\Sigma_{ij}x_{ij}$, the sum of all observations. Where r is the number of replicates; p is the number of parents; S_g is $1/(p+2)\{\Sigma_i(X_{i.}+x_{ii})^2-4/pX_{..}^2\}$; S_s is $\Sigma_{i<j}x_{ij}^2-1/(p+2)\Sigma_i(X_{i.}+x_{ii})^2+2/((p+1)(P+2)X_{..}^2)$ and $X_{i.}$ is $\Sigma_jx_{ij}=x_{i1}+x_{i2}+x_{i3}+\ldots$, that is, the sum over rows; $X_{..}$ is $\Sigma_{ij}x_{ij}$, the sum of all observations.

When SCA is relatively small in comparison with GCA, it should be possible to predict the performance of particular cross combinations based only on the values obtained for GCA of the parents.

Table 6.1 Degrees of freedom, sum of squares and mean squares from the analysis of variance of a full diallel including parent selfs (Method 1) assuming fixed effects. Also shown are the expectations for the mean squares.

Source	df	SS	MSq	EMS
GCA	$p-1$	S_g	M_g	$\sigma^2+2p(1/(1-p))\Sigma g_i^2$
SCA	$p(p-1)/2$	S_s	M_s	$\sigma^2+2/(p(p-1))\Sigma_{ij}s_{ij}^2$
Reciprocal	$p(p-1)/2$	S_r	M_r	$\sigma^2+2(2/(p(p-1)))\Sigma_{i<j}r_{ij}^2$
Error	$(r-1)p^2$	S_e	M_e	σ^2

A relatively large SCA/GCA ratio implies the presence of dominance and/or epistatic gene effects. It should also be noted that if dominance × additive effects are present, the GCA component will also contain some of these effects in addition to pure additive effects.

For inbred lines, the closer that the following equations are equal to one (i.e. as SCA becomes small or very small compared with GCA), then greater predictability based on GCA will be possible. The ratio equations for each model are:

$$\text{Model 1} : 2g_i^2/[2g_i^2 + s_{ij}^2]$$
$$\text{Model 2} : 2\sigma_g^2/[2\sigma_g^2 + \sigma_s^2]$$

where g_i^2, σ_g^2 are the general combining ability mean square and variance, respectively and s_{ij} and σ_s^2 are specific combining ability mean square and variance, respectively.

The choice of Griffing method will depend on the plant breeder or researcher's preference and on the characters of the crop and trial under investigation. If, for example, there is a suspicion that the particular inheritance has a maternal or cytoplasmic effect, then Method 1 or Method 3 may be the desired choice. If, however, there is no evidence of reciprocal differences, then Method 2 or Method 4 would be chosen. When the variance components are of major importance, then it has been suggested that Method 1 will result in a more accurate and consistent variance estimation compared with the other methods available. Conversely, it has been reported that the inclusion of the parental genotypes in the diallel design can cause an upward bias in the estimation of the GCA and SCA variances.

Normally the F_1 generation is considered in Griffing's analysis. However, as no genetic assumptions are involved then there are no reasons why F_2 or indeed other segregating generations cannot be analysed. This is a tremendous advantage in crop species where seeds per hybridization event are few. For example in garbanzo bean (chick pea) a single emasculation and pollination will result in 1 or 2 seeds, so multiple hybridizations are needed to obtain quantities of seed suitable for a proper F_1 diallel analyses. In this case it is easier to increase limited quantities of F_1 seed to obtain larger quantities of F_2 seed (which can simply be obtained by bagging flowers to prevent cross-pollination) for the diallel analysis.

Despite the attraction and simplicity of Griffing's analysis, several researchers have criticized the diallel technique. In open-pollinated species such as corn, where GCA is the only parameter of interest, then it has been suggested that other designs such as topcross or polycross would yield equally reliable results with less effort, and that these alternative methods provide the opportunity to test many more parental lines. Similarly it has been argued that in many instances North Carolina I designs (where a set of p parents to be tested are each intercrossed with a set number of other parents, and where each parent under test is not necessarily crossed to the same tester) or North Carolina II designs (where a set of p parents are crossed to a common set of n different parents and where each parent under test is crossed to the same set of non-test parental (or tester) lines) would offer a better alternative to diallel designs and Griffing's analysis.

Many studies have shown that the GCA values of parents from diallel analyses are similar to actual phenotypic performance of the parents. It has, therefore, been argued that it is not necessary to progeny test potential parents in a plant breeding programme, but simply to 'cross the best with the best'. Many practical plant breeders often add to this statement, however, 'cross the best with the best, and hope for the best', but perhaps that is what we would be doing anyhow.

Example of Griffing's analysis of half diallel

Let us consider now an example of a half diallel. A half diallel crossing design between ten homozygous lines of spring canola (*Brassica napus*) was carried out at the University of Idaho, Moscow, Idaho, US, in the spring of 1992. The parental lines were: Global, Helios, Jaguar, Starr, 93.C.3.1, Westar, DNK.89.213, Cyclone, Hero and Reston. Hero and Reston are both industrial rapeseed cultivars, while the others are canola (edible) types. Crossing resulted in $n[n-1]/2 = 45$ different F_1 families. Over the following winter, each of the 45 F_1 families was grown in a two-replicate randomized complete block design which also included the 10 parent selfs, making a design with 55 entries ($n[n+1]/2$) and two replicates (i.e. 110 plots).

Throughout the growth of this experiment a number of different traits were recorded on each of the 110 plots. To avoid excessive repetition we will only

Table 6.5 Average plant height (cm) of each of the 45 F_1s and the 10 parents in a half diallel with selfs.

	Global	Helios	Jaguar	Starr	93.C.3	Westar	DNK.89.213	Cyclone	Hero	Reston
Global	164									
Helios	171	176								
Jaguar	168	155	132							
Starr	165	136	146	149						
93.C.3	135	154	136	156	156					
Westar	128	175	139	149	162	137				
DNK.89.213	142	157	145	133	150	135	156			
Cyclone	160	132	140	120	143	121	137	101		
Hero	123	130	158	131	120	128	125	122	116	
Reston	153	163	147	143	182	148	137	135	124	139
GCA	+7.5	+11.5	+3.2	−0.6	+6.0	−1.2	−1.7	−12.2	−15.7	+3.7

consider one of these characters, plant height at the end of flowering.

The average plant height of each of the 45 F_1s and the 10 parents are shown in Table 6.5. The data used were the average of two plant heights (cms) as two representative plants were measured in each of the replicate plots.

From the data, the total variance (sum of squares) is partitioned into differences between the two replicate blocks (Reps), general combining ability, specific combining ability and an error term (based on interactions between replicates and other factors). Sum of squares (SS) and mean squares (MS) obtained are shown in Table 6.6.

The basic assumption of this experiment was that the ten parental lines were specifically **chosen**, although they were taken to be representative of a wide range of spring canola cultivar types. We are therefore analysing a *fixed effect model* and all the mean squares in the analysis are tested for significance (using the 'F' test) against the error mean square (i.e. 153).

From the analysis, the overall replicate block effect (i.e. difference between replicate one and replicate two) was not significant. An F-value is obtained for specific combining ability as $162/153 = 1.05$. This 'F' value is compared to F-values found in statistical tables at differing probability levels with 45 and 54 degrees of freedom. When this is done, it is found that our observed F-value is smaller than the value from the statistical tables at the 5% level. Therefore in this example, specific combining ability is not significant at the 5% level, and hence there should be good opportunity to predict progeny performance based on parent GCA values.

Table 6.6 Degrees of freedom, sum of squares and mean squares from the analysis of variance of plant height of a half diallel including parent selfs. In the analysis, the total variance is partitioned into differences between the two replicate blocks (Reps), general combining ability, specific combining ability and an error term (based based on the replicate differences).

Source	df	Sum of squares	Mean squares
General combining ability	9	8,283	920
Specific combining ability	45	7,301	162
Replicate blocks	1	194	194
Replicate error	54	8,262	153
Total	109	25,120	

Consider now the variance ratio for general combining ability. The appropriate F-value is $920/153 = 6.01$. When this value is compared with the appropriate F-values in statistical tables with 9 and 54 degrees of freedom, we find that it exceeds the appropriate expectation based on 99.9% confidence (i.e. approximately 3.54), and so we say that general combining ability is highly significant. This, in combination with the marginal significance of specific combining ability, suggests a genetic model with highly additive effects.

Now the expected mean square for specific combining ability of a half diallel with fixed effects is:

$$\sigma^2 + 2(p/(p-1))\Sigma_i s_i^2$$

Therefore

$$162 - 153 = 2(10/(10-1))\sum_i s_i^2$$
$$9 = 2.2\sum_i s_i^2$$

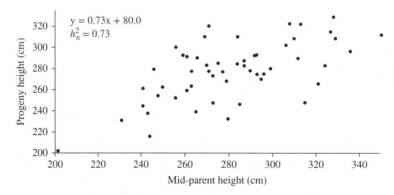

Figure 6.1 Scatter diagram of mid-parent phenotype height against average offspring progeny phenotype height from a 10×10 half diallel in *Brassica napus*.

so

$$\sum_i s_i^2 = 9/2.2$$

$$= 4.1$$

Similarly for general combining ability, the expected mean square is:

$$\sigma^2 + (p+2)(1/(1-p))\Sigma g_i^2$$

Therefore

$$920 - 153 = (10+2)(1/(1-10))\sum g_i^2$$
$$767 = 1.33 \sum g_i^2$$

so

$$\sum g_i^2 = 767/1.33$$
$$= 576.7$$

Now, from the equation above we can compare GCA and SCA effects, so as noted earlier we have:

$$2g_i^2/[2g_i^2 + s_{ij}] = 2 \times 576.7/[(2 \times 576.7)+4]$$
$$= 0.996$$

As this value is very close to one, it indicates that, as expected, s_{ij}^2 is relatively small compared to g_i^2. Therefore additive genetic effects predominate. This means there is a good chance that plant height at the F_1 stage in this *B. napus* breeding programme can be predicted with good accuracy, depending upon the general combining ability of the chosen parental lines.

In many instances there is good agreement between the general combining ability of a genotype and the phenotypic performance of the line. If this is the case, then all that is necessary

is to determine the expression of the parents and from these the expected expression of the offspring can be estimated (compare with h_n^2). Therefore in this example, consider the regression of offspring against the average parental performance (Figure 6.1). It can be clearly seen that there is relatively good agreement between offspring and parents. The regression equation gives the offspring mean = (0.7265 × parent) + 80.0. Therefore the narrow sense heritability of these data is approximately 0.73, which is very high in that 73% of the total variation can be considered as additive genetic variance.

6.3.2 Hayman and Jinks' analysis

Hayman and Jinks developed an analysis for diallels that has been widely used by many plant researchers to evaluate the mode of inheritance. This analysis is based on a model that, for any one locus, i, with two alleles, the difference between the two homozygotes is $2a$. The difference between the heterozygous (F_1) and the mid-parent value (m) is d.

To simply interpret a Hayman and Jinks' analysis, the following assumptions are made:

- Diploid segregation
- Homozygous parents
- No difference between reciprocal crosses
- No epistasis
- No multiple alleles
- Genes are distributed independently between the two parents.

But these assumptions are tested in the approach.

The parents and all possible F_1 progenies are evaluated for the trait of interest. All the offspring of one parent used in crosses is called an **array** – that is, all the crosses in which the particular parent was used. Seven kinds of variances and covariances are calculated, including:

V_p = variance among the parent lines

V_r = variance among family (F_1 and reciprocal) means within an array

V_{xr} = variance among the means of the arrays

$\overline{V}_r$ = mean value of all V_r over all arrays

W_r = the covariance between familieswithin the ith array and their non-recurrent parent

$\overline{W}_r$ = mean value of all W_r over all arrays

σ_E^2 = error variance

From these, a number of parameters can be estimated, including:

$$V_A = 4/7 \left[V_p + \overline{W}_r + V_{xr} \right] - \sigma_E^2$$
$$V_D = 4\overline{V}_r - V_A$$

The estimates of V_A and V_D indicate the amounts of additive variance and dominance variance among the crosses. This estimate of V_D assumes that F_1 progeny are being evaluated (although other generations can be accommodated). Obviously the frequency of heterozygous alleles in a population will determine the degree of dominance variation, and this will vary with successive rounds of selfing.

The most useful aspect of Hayman and Jinks' analysis for plant breeders involves examination of variance and covariance relationships and estimation of V_A or V_D. Therefore we will only cover the within-array variances and between-array covariances, how they can help in determining the inheritance of the character of interest, what the relationship of these two parameters means in comparing different parental lines, and estimation of h_n^2.

Based on the assumptions (listed above) of Hayman and Jinks analyses, we have:

$$\overline{V}_r = \frac{1}{4}(V_A + V_D)$$
$$\overline{W}_r = \frac{1}{2}V_A$$

Consider now the relationship between W_r and V_r. If we plot W_r against V_r, the regression line must have a slope that will pass through the point $(\overline{V}_r, \overline{W}_r)$ and will have an expected value of 1 only if the additive – dominance model is adequate to explain the variation observed. It should therefore be noted that the relationship between W_r and V_r provides a test of the additive – dominance model of gene action. If the contributions of many genes are not independent, that is, if there are genes interacting in their effects (epistasis), we would not expect the relationship between W_r and V_r to be as described. Hence if the additive – dominance model is not adequate, then the regression of W_r against V_r will not result in a regression slope of 1.

In addition, regression of W_r against V_r will result in a gradient that will pass through the point $\frac{1}{4}(V_A + V_D)$, $\frac{1}{2}V_A$ and which will cut the y-axis at $\overline{W}_r - \overline{V}_r = \frac{1}{2}V_A - \frac{1}{4}(V_A + V_D) = \frac{1}{4}(V_A - V_D)$. So we can learn something about the average dominance relationships of the quantitative inheritance system. If additive genetic variance (V_A) is greater than dominance genetic variance (V_D), then the regression line will cut the y-axis (the W_r-axis) above zero. Similarly, the reverse will be true of V_D is greater than V_A.

The relative position of each array (V_r and W_r) will indicate the relative frequency of the dominant to recessive alleles that the array parent has. Therefore, the relative position of the array points on the line will reflect the direction of dominance. If an array has a scatter of points close to the origin (i.e. low V_r and W_r values), this indicates that the parent common to that array has a high frequency of dominant alleles for that character of interest. If an array has a scatter of points at a distance from the origin (i.e. high V_r and W_r values), then the common parent in that progeny array will have a relatively high frequency of recessive alleles.

This graph (W_r/V_r) can therefore provide a great deal of information about the genetic situation between the parents in the diallel. In plant breeding terms, the frequency of dominant (or recessive) alleles, combined with the average progeny performance, can be useful indicators for selection – for example, given two possible parents, if one has a high frequency of recessive alleles and the other a high frequency of dominant alleles for, say, yield. If both parents have similar general combining ability, then a plant breeder should choose the recessive parent, as it will be easiest to select and fix for high yield. Similarly, selection for high yield, which is related to a high frequency of dominant alleles, will probably have a lower narrow-sense heritability compared with the case of high recessive allele frequency.

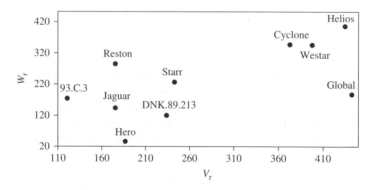

Figure 6.2   Scatter diagram of W_r and V_r values from arrays in a 10 × 10 half diallel in spring canola (*Brassica napus*).

To illustrate further, consider the example from a half diallel design involving 10 homozygous parents of spring canola (*B. napus*). Although many traits have been recorded from this trial, we will again consider plant height, which was explained as an example of the Griffing's analysis earlier. The means, over replicates, have therefore been shown earlier. From the array means, values of within array variances (V_r) and covariances (W_r) were obtained. Similar V_r and W_r values were calculated from each of the two replicates.

Regression analysis of V_r against W_r resulted in a regression equation:

$$W_r = 0.60 \times V_r + 59.07$$

The analysis of regression resulted in a mean square for linearity of 47,473 and a mean square for departures from linearity to be 9,286. From this we calculate an 'F' value of 5.11, which is significantly ($p < 0.05$) larger than would have been expected if the relationship between W_r and V_r had no linear component.

The standard error (se_b) of the regression coefficient (b) was 0.265. From this we can calculate the Student's t as:

$$t = \frac{b-1}{se_b} = \frac{0.400}{0.265} = 1.51$$

This did not exceed the value from t-tables with 8 degrees of freedom ($n - 2$) and $p < 0.05$. Therefore **b** is not significantly different from a regression slope of 1 and so we have a good indication that the additive – dominance mode is adequate to explain the inheritance of plant height in spring canola.

The regression line cuts the y-axis at ($V_r = 0$) above the origin (+59.07), so we can say that additive effects are greater than dominance effects.

A scatter diagram of the W_r and V_r values from the 10 arrays (parents) is shown in Figure 6.2. From the diagram we see that the cultivars Hero, DNK.89.213, Jaguar and 93.C.3 are relatively close to the origin, while Helios, Cyclone and Westar are further from the origin. From this we can deduce that those closest to the origin have a higher frequency of dominant alleles for plant height and those further from the origin have highest frequency of recessive alleles for plant height. In the extremes, the cultivar Hero has highest relative frequency of dominant alleles and the cultivar Helios has highest frequency of recessive alleles for plant height.

Moving on to the $W_r + V_r$ and $W_r - V_r$ from each array, $W_r + V_r$ and $W_r - V_r$ will contain all the information that W_r and V_r contain. Now if dominance is present, then $W_r + V_r$ will vary from array to array. If there is non-allelic interaction, then $W_r - V_r$ will vary from array to array. If only dominance is present, then $W_r - V_r$ will not vary between arrays more than would be expected by sampling variation. We can calculate the values of $W_r + V_r$ and $W_r - V_r$ from the V_r and W_r values obtained from each replicate, and carry out a one-way analysis of variance on the resulting data. When this is done we have the two analyses of variance tables:

$$W_r + V_r$$

Source	df	MSq	F-value	Significance
Between arrays	9	17,217	4.64	(0.01 < p < 0.05)
Within arrays	10	3,708		

$$W_r - V_r$$

Source	df	MSq	F-value	Significance
Between arrays	9	5,396	2.49	n.s.
Within arrays	10	2,170		

Therefore values of $W_r + V_r$ vary significantly between arrays ($p < 0.05$), so we can say that dominance is present. Values of $W_r - V_r$ between different arrays are not significantly different and therefore we can say that there is no evidence of non-allelic interaction. From these two analyses we can conclude that the only significant non-additive effects present are ascribable to dominance.

These data are from homozygous parents and F_1 progenies. In many cases it is difficult to obtain large quantities of F_1 seed, and the actual diallel analysis needs to be carried out on the F_2 (or higher) generations. When this is done the same six assumptions listed at the beginning of this section still apply. The regression of W_r/V_r will still have an expected slope of unity if the additive – dominance model is adequate to explain the inheritance, and $W_r - V_r$ should be constant across arrays if no epistasis is present. Therefore, as far as the example above is concerned, then it would make no difference if F_2 data were used. However, if a more detailed analysis is to be carried out and the components V_A and V_D are to be estimated, then some modifications are needed. The modifications are not within the scope of this book.

Estimating h_n^2 from Hayman and Jinks' analysis

If the crop under investigation in a diallel crossing design complies with all the restraints of the Hayman and Jinks design, then it is possible to obtain accurate estimates of additive genetic variance (V_A) and dominant genetic variance (V_D) straightaway and hence determine the narrow-sense heritability (h_n^2). The average W_r value ($\overline{W}_r$) is an estimate of $\frac{1}{2}V_A$. The V_p value is a direct estimate of V_A, and the V_{xr} value is an estimate of $\frac{1}{4}V_A$. These relationships hold true irrespective of the generation (i.e. F_1, F_2, F_3, etc.) that is analysed. From these three estimates of V_A, we can produce a weighted

mean where:

$$V_A = 4/7[V_p + \overline{W}_r + V_{xr}]$$

The dominance genetic variance (V_D) will vary from generation to generation. Greatest V_D will be observed in the F_1 generation, as there is the greatest frequency of heterozygotes compared with other generations. In F_1 the average V_r value ($\overline{V}_r$) is an estimate of $\frac{1}{4}[V_A + V_D]$, and V_D can easily be estimated by substituting the already calculated V_A value into this equation. Therefore, when analysing data from F_1 family diallels:

$$D = 4\overline{V}_r - V_A$$

From estimation of V_A and V_D we can now calculate h_n^2:

$$h_n^2 = \frac{1}{2}V_A / \left[\frac{1}{2}V_A + \frac{1}{4}V_D + \sigma_E^2\right]$$

where σ_E^2 is the replicate error term obtained from the analysis of variance in the *B. napus* example shown earlier in the Griffing's analysis.

In F_2 families, $V_D = \frac{1}{4}[V_A + \frac{1}{4}V_D]$, and so $V_{D-F_2} = 16\overline{V}_r - 4V_A$, and in F_3 families, $V_{D-F_3} = \frac{1}{4}[A + 1/16V_D]$. It should be noted that the proportion of V_D in each family is decreased each generation by $[1/2]^n$, where n is the generation number (i.e. $[1/2]^1 = 1/2$ at F_1; $[1/2]^2 = 1/4$ at F_2; $[1/2]^3 = 1/8$ at F_3, etc.).

6.4 Cross prediction

There is one further way that it is possible to predict the response to selection in the long term, although not necessarily the rate of response. This approach, based on the genetics underlying the traits, was proposed by Jinks and Pooni, and is currently attracting considerable attention in terms of experimental investigations and in applying it to practical breeding. This will be covered in more detail in the next chapter, but needs mentioning here to keep in view the options available to the breeder in terms of making predictions.

If, to start with, we assume that we have an inbreeding species and wish to produce a final variety that is true-breeding. What we want to know of any population or cross is what is the distribution of inbred lines that we predict can be derived from it, and what is the probability of one of these lines having a phenotype equal to, or exceeding, any target level that we set, in other words, that we would be aiming for with selection.

If we assume that the distribution of the final inbred lines that are derivable have a normal distribution, as is generally the case in practice, then this distribution can be described by the mean and standard deviation. Since they are inbred lines they will have a mean of m and a standard deviation of $\sqrt{V_A}$, we can predict the properties of the distribution of all inbred lines possible and hence we can obtain the frequency (= probability) of inbreds falling into a particular category. In other words, we can simply use the properties of the normal probability integral given in tables to say what is the probability of obtaining an inbred line with expression falling in a particular category. If the probability is low, it will obviously be difficult to actually obtain such a line. If the probability is high, it will be easy to produce.

How do we put this into practice? If we have a set of genotypes for use as parents, which ones do we cross to produce our desired new inbred lines? Do we take $A \times B$ and $C \times D$ or $A \times D$ and $C \times Z$, etc.? We will need to decide between the crosses before we invest too much time and effort, otherwise we may well be spreading our efforts over crosses that will not easily produce the phenotypes we want. If we take the crosses and estimate m and V_A for each, then we can estimate the probability of obtaining our desired target values. From this we can rank the crosses on their probabilities and then only use the ones with the highest probabilities of producing lines with the required expression of character deemed to be important.

In fact, the approach is even more general in that it can be used to predict the properties of the F_1 hybrids derived from the inbred lines. It can also be used to predict the probability of combination of characters that is the probability of obtaining desirable levels of expression in a series of characters.

What are the drawbacks to the approach? Firstly, we need to estimate m and $\sqrt{V_A}$, and this involves a certain amount of work in itself, but it is fairly modest.

Secondly, it also assumes that the estimates we use are appropriate to the final environment in which the material is to be grown. In other words, as in the case of heritabilities, if we carry out the experiments in one environment at one site in one year, we are

assuming that this is representative of other years and sites. We can, of course, carry out suitable experiments to obtain estimates in more years and sites, but this involves extra time and effort.

The use of cross prediction techniques in selection will be discussed in greater detail in Chapter 7.

It is unfortunate that, in reality, few breeding programmes, both of an academic or commercial nature, seek to gain insight into the topics addressed in this chapter. To some extent this is a consequence of the time and resource constraints under which breeders operate. For instance, many breeders would rather spend resources developing more or larger breeding populations, or running more field trials, than preparing specific populations and analysing the information they provide. Nevertheless, we argue that most, if not all, breeding programmes would benefit, achieve greater genetic progress and therefore develop superior varieties, if just a small fraction of their efforts was devoted to gaining insight as noted above.

Think questions

(1) Given values for the variance of the mean of the F_2, and variance of the F_1, estimate h_b^2 and explain what this tells us about the genetic determination of the trait.

$$V_{\bar{F}_2} = 436.72$$
$$V_{\bar{F}_1} = 111.72$$

Given below are the variances of the mean from two parents (P_1 and P_2), the F_1, F_2, and both backcross families (B_1 and B_2). Estimate h_n^2 and explain what the value means in genetic terms.

$$V_{\bar{P}_1} = 14.1 \quad V_{\bar{P}_2} = 12.2$$
$$V_{\bar{F}_1} = 13.3 \quad V_{\bar{F}_2} = 40.2$$
$$V_{\bar{B}_1} = 35.2 \quad V_{\bar{B}_2} = 34.6$$

(2) List the six assumptions necessary for a straightforward interpretation of a Hayman and Jinks' analysis of diallels.

Below are shown values of array means, within array variances (V_r) and covariances between array values and non-recurrent parents (W_r) from a Hayman and Jinks' analysis of a 7×7 complete diallel in dry pea.

The trait of interest is pea grain yield. Regression of W_r against V_r resulted in the equation: $W_r = 0.837 \times V_r - 4.817$, with standard error of the regression slope equal to $se_b = 0.0878$. An analysis of variance of $W_r + V_r$ showed significant differences between arrays, while a similar analysis of $W_r - V_r$ showed no significant differences between arrays. What can be deduced regarding the inheritance of pea yield from the information provided? If you were a plant breeder interested in developing high-yielding dry pea cultivars, on which two parental lines would you concentrate your breeding efforts? Briefly explain why.

Parent name	V_r	W_r	Array mean
'Souper'	34.1	19.3	456
'Dleiyon'	99.9	79.3	305
'Yielder'	21.0	11.2	502
'Shatter'	99.4	68.4	314
'Creamy'	49.6	39.4	372
'SweetP'	59.1	48.8	361
'Limer'	61.8	49.2	393

(3) Below is shown an analysis of variance of plant yield from a 6×6 half diallel (including parents). The analysis of variance is from a Griffing's analysis (Model 2). GCA = general combining ability, SCA = specific combining ability, Error = error obtained by replication, df = degrees of freedom, and SS = sum of square.

Source	df	SS
GCA	5	4,988
SCA	15	6,789
Error	21	5,412

Discuss the results from the analysis, given that the six parents were: (a) specifically chosen, and (b) chosen completely at random.

(4) It is desired to determine the narrow-sense heritability for flowering date in spring canola (*B. napus*). Both parents and their offspring from ten cross combinations were grown in a properly designed field experiment. At harvest, yield was recorded for each entry,

and using these data the average phenotype of two parents (i.e. $[P_1 + P_2]/2$) was considered to be x, the independent variable, while the performance of their offspring was considered as y, the dependent variable. A regression analysis is to be carried out by regression of the offspring (y) onto the average parent (x). The following data are derived: $SP(x, y) = 345.32$; $SS(x) = 491.41$; $SS(y) = 321.45$. Estimate the slope of the regression line (**b**), test whether this slope is greater than zero, and estimate the narrow-sense heritability from the regression equation.

How would the relationship between the regression and the narrow-sense heritability differ if the regression were carried out between only the offspring and the male parent?

(5) Four types of diallel can be analysed using a Griffing's analysis. Describe these types.

Families from a 5×5 half diallel (including selfs) were planted in a two-replicate yield trial at a single location. The parents used in the diallel design were chosen to be the highest yielding lines grown in the Pacific-Northwest region. Data for yield were analysed using a Griffing's analysis of variance. Family means, averaged over two replicates, degrees of freedom and sum of squares (SS) from that analysis are shown below. Interpret the results from the Griffing's analysis. What differences would there be in your analytical methods if the parents used had been chosen at random?

	Parent 1		Parent 2	Parent 3	Parent 4	Parent 5
Parent 1	62.0	Parent 2				
Parent 2	71.0		69.5	Parent 3		
Parent 3	55.5		52.5	50.5	Parent 4	
Parent 4	72.5		80.5	56.5	76.5	Parent 5
Parent 5	70.5		66.5	36.5	71.0	64.5

Source	df	SS
GCA	4	6,694.058
SCA	10	825.676
Replicates	1	8.533
Error	14	317.467
Total	29	7,845.733

From the same diallel data (above), within-array variances (V_r) and between-array and non-recurrent parent covariances (W_r) were calculated. The values of V_r and W_r for each parent along with the mean of V_r, mean of W_r, sum of squares of $V_r (\Sigma V_r^2)$, sum of squares of $W_r (\Sigma W_r^2)$ and sum of products $\Sigma V_r W_r$ are shown below.

	V_r	W_r
1	98.0	102.0
2	66.3	53.2
3	161.8	207.4
4	50.3	65.2
5	71.4	83.1

Mean of $V_r = 89.56$; mean of $W_r = 102.19$; $\Sigma V_r^2 = 7,702.01$; $\Sigma W_r^2 = 15,192.05$; $\Sigma V_r W_r = 10,535.29$. From these data, test whether the additive – dominance model is adequate to describe variation between the progenies. What can be determined about the importance of additive versus dominance genetic variation in this study? From all the results (Griffing, and Hayman and Jinks, above), which two parents would you use in your breeding programme and why?

(6) Given the variance from two parents (P_1 and P_2), the F_1, F_2, B_1 and B_2 families, estimate the narrow-sense heritability (h_n^2) and explain what the value means in genetic terms.

$$\sigma_{P_1}^2 = 9.5; \quad \sigma_{P_2}^2 = 7.4$$
$$\sigma_{F_1}^2 = 8.6; \quad \sigma_{F_2}^2 = 17.7$$
$$\sigma_{B_1}^2 = 14.3; \quad \sigma_{B_2}^2 = 15.2$$

(7) A new oil crop (*Brassica gasolinous*) has been discovered that may have potential as a biologically renewable fuel oil substitute. This diploid species is tolerant to inbreeding and is self-compatible. A preliminary genetic experiment was designed to examine the inheritance of seed yield (Yield) and percentage oil content (%Oil). This experiment involved a 4×4 half diallel (including selfs). The four homozygous parental lines are represented by the codes AAA, BBB, CCC and DDD. The half diallel array values (averaged over two replicates), array means, general combining ability (GCA) values, mean squares from the analyses of variance (Griffing style), V_r and W_r values (Hayman and Jinks' analysis), variance of array means (V_{xr}) and parental variances (V_p), and the one-way analyses of variance for $V_r + W_r$ and $V_r - W_r$ are shown below for each character.

		Yield		
AAA	40.5			
BBB	38.5	29.5		
CCC	37.0	28.0	19.5	
DDD	32.5	20.5	18.5	10.0
	AAA	BBB	CCC	DDD
Array means	37.1	29.1	25.8	20.4

		%Oil		
AAA	20.5			
BBB	20.5	25.0		
CCC	23.0	26.0	30.5	
DDD	24.5	27.5	31.0	36.0
	AAA	BBB	CCC	DDD
Array means	22.1	24.7	27.6	29.8

Source	df	Yield	%Oil
GCA	3	796.5	180.6
SCA	6	90.5	30.1
Replicate blocks	1	0.4	0.1
Replicate error	9	51.5	5.7

V_r and W_r values

	Yield		%Oil	
	V_r	W_r	V_r	W_r
AAA	46.2	171.3	15.6	50.7
BBB	218.2	376.7	36.3	75.0
CCC	297.7	436.7	58.3	98.0
DDD	344.3	472.3	97.3	132.3

V_{xr} and V_p values

	Yield	%Oil
V_{xr}	196.9	44.4
V_p	687.6	180.7

Source	df	Yield $V_r + W_r$	$V_r - W_r$	%Oil $V_r + W_r$	$V_r - W_r$
Between	3	8,760	28.0	628	7.68
Within	4	133	17.0	115	4.77

Source	df	MS
GCA	5	30,769
SCA	10	10,934
Reciprocal	10	9,638
Error	49	5,136

Without using regression, estimate the narrow-sense heritability (h_n^2) for seed yield, and explain this value in terms of genetic variance. Explain the analyses for Yield and outline any conclusions that can be drawn from these data. Explain the analysis for %Oil (percentage of seed weight that is oil) and outline any conclusions that can be drawn from these data. Which **one** of these four genotypes would you choose as a parent in your breeding programme? Explain your choice. Describe any difficulties suggested from these analyses in a breeding programme designed for selecting lines with high yield and high percentage of oil.

(8) F_1, F_2, B_1 and B_2 families were evaluated for plant yield (kg/plot) from a cross between two homozygous spring wheat parents. The following variances from each of the four families were found:

$$\sigma_{F_1}^2 = 123.7; \quad \sigma_{F_2}^2 = 496.2$$
$$\sigma_{B_1}^2 = 357.2; \quad \sigma_{B_2}^2 = 324.7$$

Calculate the broad-sense (h_b^2) and narrow-sense (h_n^2) heritabilities for plant yield. Given the heritability estimates you have obtained, would you recommend selection for yield at the F_3 stage in this wheat breeding programme, and why?

(9) Griffing has described four types of diallel crossing designs. Briefly outline the features of each of Methods 1, 2, 3 and 4. Why would you choose Method 3 over Method 1? Why would you choose Method 2 over Method 1? A full diallel, including selfs, was carried out involving five chickpea parents (**assumed to be chosen as fixed parents**), and all families resulting were evaluated at the F_1 stage for seed yield. The following analysis of variance for general combining ability (GCA), specific combining ability (SCA) and reciprocal effects (Griffing's analysis) was obtained:

Complete the analysis of variance and draw conclusions from the analysis. If it was actually the case that the parents were chosen at random, how would this change the results and your conclusions?

Plant height was also recorded on the same diallel families and an additive–dominance model found to be adequate to explain the genetic variation in plant height. Array variances (V_rs) and non-recurrent parent covariances (W_rs) were calculated and are shown alongside the general combining ability (GCA) of each of the five parents, below:

Parent	V_r	W_r	GCA
1	491.4	436.8	−0.76
2	610.3	664.2	+12.92
3	302.4	234.8	−14.32
4	310.2	226.9	−15.77
5	832.7	769.4	+17.93

Without further calculations, what can be deduced about the inheritance of plant height in chickpea?

(10) A 4×4 half diallel design (with selfs) was carried out in cherry and the following seed yields of each possible F_1 family were observed:

	Small reds			
Small reds	12	Big yields		
Big yields	27	36	Jim's delight	
Jim's delight	21	35	27	Jacks' best
Jack's best	28	27	26	21

From the above data, determine the narrow-sense heritability for yield in cherry.

(11) A crossing design involving two homozygous pea cultivars was carried out, and both

parents were grown in a properly designed field experiment with the F_2, B_1 and B_2 families. Given the following **standard deviations** for parents (P_1 and P_2), the F_2, and both backcross progeny (B_1 and B_2), determine the broad-sense heritability and narrow-sense heritability for seed size in these dry peas. What can be concluded from the values of heritability you obtain?

Family	Standard deviation	Variance
P_1	3.521	12.39
P_2	3.317	11.00
F_2	6.008	36.10
B_1	5.450	29.70
B_2	5.157	26.59

(12) It is common in plant breeding programmes to carry out diallel analyses to determine the genetics of specific parents and their progeny. With particular reference to diallel interpretation, explain the terms **fixed** and **random** parents, and how choosing one or other would influence the analyses of a Griffing ANOVA. Griffing's analysis of diallels is a statistical analysis based on the following model:

$$Y_{ijk} = \mu + g_i + g_j + s_{ij} + e_{ijk}$$

Describe the meaning of the terms in the g_j and s_{ij} in the Griffing model, above.
Below are the mean seed yields (kg/plot), averaged over four replicates, of the parents and offspring from a 3×3 half diallel, with selfs.

	Sunrise			
Sunrise	16.4	Premier		
Premier	14.3	11.1	Clearwater	
Clearwater	18.3	15.2	17.6	Hyola.401
Hyola.401	13.2	10.7	16.2	15.3

Using these data, determine the g_i value of each parent and estimate the s_{ij} value in the cross Hyola.401 × Premier.

(13) A newly appointed barley breeder was interested in the inheritance and heritability of yield in spring barley. He chose five cultivars from a range that were grown throughout his target region and intercrossed them in all possible cross combinations (excluding reciprocals) including selfs, and he completed a Hayman and Jinks' analysis. Some of the results are shown below.

Cultivar	Mean yield	V_i	W_i
Premier	62	1.8	−8.1
Sunrise	44	64.1	51.5
Reaper	39	43.8	38.0
Lifeline	66	7.1	9.4
Star	39	71.0	51.1
Mean		37.56	28.38

The regression equation of W_i onto V_i is: $W_i = 0.8134 \times V_i - 2.15$, and the standard error of the slope is 0.103.
The variance of parents (σ_p^2) is 42.37, and the variance or array means (σ_{xr}^2) is 16.77.
Complete the Hayman and Jinks' analysis of variance, determine whether the additive – dominance model is appropriate, calculate the narrow-sense heritability, and describe what can be determined about the inheritance of yield in spring barley.

7
Selection

7.1 Introduction

Selection amongst all living organisms has been going on since life first existed. Natural selection (i.e. evolution) has resulted in the diversity of plant and animal life that exists on Earth today. All selection results in a change of gene frequencies. Throughout evolution, species have been changing, '*more fit*' genotypes have predominated, while those that are less fit with regard to survival or reproduction have become extinct. The aim of plant breeding is to mimic this process, but to direct selection towards increasing the frequency of desirable gene combinations that best suit agricultural systems and increasing consumer needs.

In order to be successful in a selection programme, two criteria need to be satisfied:

- There is variation between plants within the unselected population and the breeders must be able to distinguish between different phenotypes.
- At least some of that variation must be genetic in nature.

Obviously, if a plant breeder cannot distinguish any differences between plants within a population (or different populations), then it will be impossible to select those individuals that appear superior. Secondly, if the variation observed between plants within a population is the result purely of the environmental response of lines, with no genetic component, then there will be no progress made in a selection scheme as any observed variation will not be inherited.

7.2 What to select and when to select

Having decided that the two criteria above are indeed satisfied, among the first tasks to be addressed by a plant breeder are to decide *what characters are to be selected for*, and *at what stage in the breeding scheme will selection be applied*.

Consider the first question of *what to select for?* To address this, a plant breeder must refer to the ***breeding objectives***. These will have been set according to criteria such as:

- the potential market size of the crop;
- the region targeted for propagation;
- the major deficiencies existing within cultivars that are presently available;
- the economic implications of addressing deficiencies such as disease and pest resistance;
- needs of the farmer, such as rapid establishment, early maturity, plant height, resistance to lodging, and harvesting ability;
- the needs of the end-user, including: appearance; uniformity; storability; processing quality; and so on.

Many more factors may need to be included in setting the breeding objectives, and the above list mentions but a few of the more important questions that need to be addressed.

Plant Breeding, Second Edition. Jack Brown, Peter D.S. Caligari and Hugo A. Campos.
© 2014 John Wiley & Sons, Ltd. Published 2014 by John Wiley & Sons, Ltd.
Companion Website: www.wiley.com/go/brown/plantbreeding

It is not usually possible to select for all the wide range of characters needed for a successful new cultivar in a single season. Plant breeders therefore screen plant populations over several years, sometimes addressing a number of different traits at each evaluation stage. Having decided what characters are to have greatest priority, it is necessary to follow an organized scheme of selection to determine which characters will be addressed at the various stages.

The inheritance of traits will be of great importance in determining not only whether selection is to be carried out, but also the complexity of field and laboratory experimentation needed in order to identify the desirable types.

7.2.1 Qualitative trait selection

Characters, or variation within a character, whose expression show a qualitative form of inheritance can be easily selected for, provided that a suitable screening method is available to determine the presence or absence of the single gene in plants of seed-propagated crops. If the expression of the qualitative character is determined by a recessive allele, then a single round of selection should ensure that all selected plants are fixed for the particular trait. If the desirable allele is completely dominant, several rounds of recurrent selection will be necessary to ensure that the character is genetically fixed in selected plants.

Qualitative characters can often be selected relatively quickly and using very small plots (sometimes even single plant plots) compared with quantitative inherited traits. The ease of selecting for single gene traits has resulted in these characters having high selection intensity in the early generation stages where most genotypes are evaluated, and where it may only be possible to grow small plots because of seed availability.

Selection for such qualitative expression can indeed be a powerful tool in reducing the number of genotypes to be processed, selected and advanced in a plant breeding scheme, although it should never be forgotten that it is often the quantitatively inherited characters that add greatest value to a new cultivar (i.e. yield, quality and durable plant resistance). If early generation selection is to be carried out for single gene traits, then the breeder must be sure that this selection is not having an adverse effect on the selected populations (i.e. no linkage between advantageous qualitative traits and adverse quantitative traits, or any unwanted non-allelic interactions, or pleiotropic effects).

7.2.2 Quantitative trait selection

Quantitatively inherited characters are usually more difficult to evaluate due to the higher potential for modification of expression by the environment and the larger number of genes involved in their control. Greater experimentation (replication or plot size) is necessary to maximize selection response. As a result, many of the quantitative traits are not positively selected for in the early generation selection stages. Selection for these characters is often delayed until the numbers of different genotypes that require testing are reduced, and where greater amounts of planting material are available for more sophisticated tests. For example, it is common practice in most plant breeding schemes not to select the early generation lines for quality traits that involve either large quantities of produce, or which provide only crude estimates of worth with small samples, or that are expensive.

Obviously any character that is considered of high importance should be selected for at the earliest stages of a plant breeding scheme, where greatest variation will exist among families or populations, but where the trial designs and amount of material make selection effective. Despite the simplicity of this statement, in practice it is often completely ignored.

The characters that are evaluated at different stages of a plant breeding scheme will be discussed in later sections.

7.2.3 Positive and negative selection

Two forms of selection are said to be available to plant breeders: positive and negative selection. It is sometimes difficult to clearly define the difference between the two types (and indeed, some wise and seasoned breeders do not distinguish between them). In simple terms, negative selection is where the very worst plants or families are discarded, while positive selection is where the very best plants or families are selected. In either case, the focus is on the extreme end of the distribution of variation in the trait. Perhaps the simplest description would be related to the proportion of plants that are selected from a population. If more than 50% of the original population is selected, then this can be considered negative selection. If less than 50% of the population is retained, then this would be positive selection.

7.3 Response to selection

It has already been stated that selection will only be successful if there is sufficient phenotypic variation and if at least some of this variation is genetic in origin. It should be of no surprise, therefore, that the response to selection is related to heritability. Indeed, consider the equation:

$$X_1 - X_{n-1} = R = i\sigma h^2$$

where X_1 is the mean phenotype of the selected genotypes, X_{n-1} is the mean phenotype of the whole population, R is the advance as a result of one round of selection, h^2 is the appropriate heritability (narrow-sense heritability for sexually reproducing species, or broad-sense heritability for clonally or apomictic breeding crops), σ is the phenotypic standard deviation of the whole population, and i is the **intensity of selection**, which is a statistical factor dependent upon the proportion of the population selected. The above equation is probably the most fundamental equation in plant breeding and should be kept in mind.

The intensity of selection (i) is related to the proportion (or percentage) of the population that is selected (k), and takes the values:

Proportion selected (k)	i
0.01	2.665
0.05	2.063
0.10	1.755
0.20	1.400

Although the intensity of selection (i) has been tabulated for a range of different selection rates, in cases where the initial population is large (i.e. greater than 50 genotypes) and the proportion of genotypes selected less than 20%, then the following equation can provide an estimate of i:

$$i = 0.77 + (0.96 \times \log(1/k))$$

From the tabulated values of selection intensities and the estimation equation it can be seen that there is not a linear relationship between higher selection rates (k) and greater response to the selection applied. Retaining 10% of the selected population results in an intensity of selection value of 1.755, while retaining only 1% (i.e. a 10-fold reduction in selections) results in an intensity of selection value of only 1.52 times larger (i.e. $i = 2.665$).

Consider a simple example which is represented diagrammatically in Figure 7.1. Selection is to be carried out on a base population with an average, or mean, of 560 kg yield and with phenotypic standard deviation (σ_p) of 19.0 kg. We assume that from past research it is known that the heritability (h_2) is equal to 0.6 and selection is to be carried out at the 10% level (i.e. $k = 0.1$, $i = 1.755$).

From this we have that:

$$\sigma i = 19.0 \times 1.755 = 33.34$$

From this we can estimate the performance of the selected fraction in the following year, as the response to selection, would be $\sigma i h^2$ and equal to $33.34 \times 0.6 = 20.0$ kg. The mean of the selected plants would therefore be $560 kg + 20 kg = 580 kg =$ the average performance of the top 10% selected lines in the next year.

It should be noted that the phenotypic standard deviation in the selected population must be less than the whole (unselected) population. As it can be assumed that the error variance remains constant, then this must mean that the genetic variance is smaller and the error variance is the same. From this, the heritability between the selected population and further selection years must be less than from the base population in the first selection year.

Therefore, if selection continues, then there would be decreasing response with increasing rounds of selection (Figure 7.2).

Return now to the response equation given above, and recall that the formulae for the broad-sense and narrow-sense heritabilities are σ_g/σ_p and σ_a/σ_p, respectively, where σ_g is the genetic variance component, σ_a is the additive genetic variance and σ_p is the phenotypic variance. From this we can write the average performance of a selected population after selection is:

$$P = X + \sigma i h^2$$

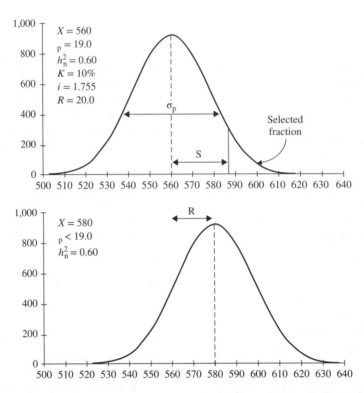

Figure 7.1   Illustration of the response from selection given population parameters from the unselected population (top) to predict the selected population (bottom).

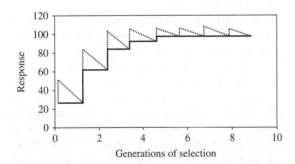

Figure 7.2 Response to selection from successive rounds of selection. The dashed line indicates the phenotypic expression and the solid line represents the genetic gain. Note that greatest gains are from the initial rounds of selection and that after several rounds of selection there is little or no gain.

where X is the average performance of the initial population (i.e. the unselected family mean), i is the selection intensity, h^2 is the heritability and σ is the phenotypic standard deviation between plants in the population.

This means that the very best responses from selection are based on high family means, high selection intensity (although limited increase in return for very high selection), heritability, and the phenotypic variance. From this, breeders should be aiming to:

- Identify highly productive families with high average performance (i.e. high means).
- Maximize heritabilities by minimizing non-genetic errors. This can best be achieved by good experimentation (i.e. selected appropriate approaches for statistical designs and data analysis), improved agronomy, increasing plot sizes and replication levels, and standardized, automated procedures to collect trait data in the field.
- Select as intensely as considered feasible, although remember the efficiency will only increase as a reciprocal beyond 20% selected.
- Choose parents that are genetically diverse for characters that require improvement or change, and hence attempt to increase the phenotypic

variance. Sometimes this could become a hard decision, as the pressures breeders operate under and the limited resources available often encourage them to only cross "best" by "best". Some breeders develop "pre-breeding" populations as a mean to enhance genetic diversity without affecting their probabilities of commercial success, and select and advance lines from these pre-breeding efforts into their mainstream populations.

On the other hand, if a plant breeding programme is not producing the expected response, the same equation can be used to identify possible reasons for the failure.

The close correspondence between heritability and the proportional change in a selected character from one generation to the next, when selection is applied, has already been pointed out. Having considered estimation of narrow-sense heritability, h_n^2, in some detail earlier, it is now appropriate to return to the issue of estimating heritability.

A third definition of narrow-sense heritability, usually termed the *realized heritability* (h_r^2), is:

$$h_n^2 = R/S$$

where R is the **response to selection** (the same as described above) and S is the **selection differential**. The response to selection is the difference between the mean of the selected genotypes for a particular character and the mean of the population before selection was applied. The selection differential is the average phenotypic superiority for the character in question of the selected genotypes over the whole of the population from which they were selected.

Consider the following example: the average seed yield of an F_3 family is 15 kg. Suppose plants that produced the highest seed yields (with a mean = 20 kg) were selected to be grown in the next generation. What would be the selection differential? Since the selection differential is the average performance of the selected plants over the base population (i.e. all original unselected plants in the family) as a whole, $S = 20 - 15$ or 5 kg seed yield.

Now, if the mean seed yield of the selected progeny in the following year (F_4) was found to be 17.5 kg, what would be the response to selection? Since response to selection is the difference between the mean of the progeny and the mean of

the parental generation before the application of selection, $R = 17.5 - 15$ or 2.5 kg of seed yield.

Finally, the realized heritability (h_r^2) would be given by:

$$h_r^2 = R/S = 2.5/5.0 = 0.5$$

It should be noted that in the above example it is assumed that there are no dramatic year effects. In a practical situation, actual performance from year to year is highly variable and largely unpredictable. This can be taken into account in part by growing a **random sample** of progeny and/or control cultivars the next year. Assuming that the random sample is indeed representative of the whole sample, it will be possible to use this in order to adjust the values and to obtain a direct indication of response to selection.

Similarly, a plant breeder is quite likely to want to know what response might be expected from a given selection differential when the narrow-sense heritability has already been estimated (from the partitioning of phenotypic variances or from offspring–parent regression).

Therefore, if the selection differential (S) applied was 5 kg of seed yield and the narrow-sense heritability had been estimated to be 0.5, the response to selection expected would be:

$$R = h_n^2 \times S = 0.5 \times 5.0 = 2.5 \text{ kg}$$

Thus the average seed yield of the selected progeny might be expected to be $15 + 2.5$ or 17.5 kg, and the gain from the selection operation would be 2.5 kg.

7.3.1 Association between traits or years

The degree of association between any two, or a number, of different characters can be examined statistically by the use of **correlation analysis**. As noted earlier, correlation analysis is similar in many ways to simple regression, but in correlations both variables are expected to be subject to error variance, and there is no need to assign one set of values to be the **dependent variable,** while the other is said to be the **independent variable.** Correlation coefficients (r) are calculated from the equation (see Chapter 5, page 86 for more details):

$$r = \frac{SP(x, y)}{\sqrt{[SS(x) \times SS(y)]}}$$

Positive correlation

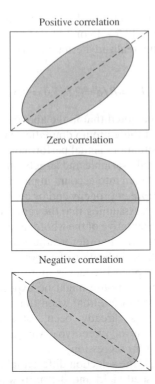

Zero correlation

Negative correlation

Figure 7.3 Diagrammatic representation of positive correlation (top), zero correlation (middle) and negative correlation (bottom).

Diagrammatically, the association between two variables is shown in Figure 7.3 with positive correlation (top), no correlation (middle) and negative correlation (bottom) indicated. Recall that a correlation coefficient does not in any way establish a cause–effect relationship, only the existence of an association between two variables which may well reflect their association with a third unidentified variable.

As can be seen, if there is strong positive correlation between two traits it will be possible to select individual genotypes that have high expression in both traits. Conversely, if there is strong negative correlation between traits of interest, it may be very difficult to select genotypes with high expression in both characters. The magnitude of the correlation value, in absolute terms, can be associated with underlying physiological processes or even pleiotropy (i.e. the same genes directly control expression in the two characters), or may be a reflection of genetic linkage.

It would seem obvious that there must be some relationship between r, the correlation coefficient, and h^2, the heritability. If, therefore, characters are recorded on the same set of genotypes grown in two different environments (say locations or years), then the magnitude of the correlation coefficient indicates the relationship between performances in the different environments. Squaring the correlation coefficient (r^2 – "the coefficient of determination") provides an estimate of the proportion of the total variation between the environments that is explained by the correlation. The total variation between sites can be considered the total phenotypic variation, and as the proportion accounted for by regression must have a genetic base, then a simple relationship exists whereby r^2 is a direct estimate of h^2.

7.3.2 Heritability and its limitations

In this short but important section, a critical look is taken at the concept of heritability, its uses and misuses.

Four distinct methods of estimating narrow-sense heritabilities have been outlined:

- partitioning of phenotypic variances;
- offspring–parent regression;
- response to selection;
- correlation analyses.

How response to selection can be predicted from a given selection differential, when the narrow-sense heritability is already known from other experiments, has also been covered above. The concept of heritability, and estimates of it, have been of great value to plant breeders and to population geneticists interested in continuously varying characters in natural populations. However, it is very important that the limitations of heritability estimates are realized. These limitations occur on at least three levels:

- There are many technical assumptions inherent in the theory as presented (e.g. that genes assort independently, that alleles segregate independently, and that there is no epistasis). It is possible to allow for many of these complications, but only at the expense of making the theory more complicated.
- An estimate of narrow-sense heritability strictly applies to a *particular* character, in a *particular* population, at a *particular* moment and in a

particular environment. Thus, even for a single character, heritability is not constant. It is obvious that h^2 is particularly vulnerable to changes in environmental variance, changes that can occur at the same place at different times, different places at the same time, or both. Indeed, widely different estimates of h_n^2 for the same character can be found in different populations investigated at the same place and time. Also, it has been seen that additive genetic variance, and hence narrow-sense heritability, generally declines over generations of selection, even in a constant environment. Caution must therefore be exercised in interpreting estimates of h_n^2 if it is not known that every precaution has been taken to expose different populations and/or characters to the same range of environments, and one is interested only in the response in the same, or very similar, environments.

- While means are so-called first degree statistics, variances are second degree. Second degree statistics are usually 'less precise' than first degree statistics. As h_n^2 and h_b^2 (with the exception of h_n^2 from offspring onto mid-parent regression) are based on the ratios of variances, they thus share the weakness of the same lack of precision as all second degree statistics.

7.3.3 Methods of selection

When a plant breeder is selecting a particular population for only a single trait, the operation is usually relatively simple. The population is evaluated for the character in question and those phenotypes with desirable (whether this is high, low or intermediate) expression are *selected* while the phenotypes with less desirable expression are *rejected*. Therefore the only variable decision is the *selection intensity*, or the proportion of the total population that will be selected for further evaluation in relation to the proportion that are to be rejected or discarded. This form of selection is called *cull selection*. In such a scheme a target value is set and all phenotypes that meet the target are said to fulfil the selection criteria and are retained, while those that do not reach the target value are rejected as failing to meet the selection criteria.

A new cultivar is rarely successful due to desirability in only a single character, but it is rather a reflection of increase of several different traits. Therefore

deciding which individuals in a population are to be retained and which are to be discarded usually involves simultaneous evaluation of more than a single character. Successfully achieving these trade-offs between multiple, and sometimes antagonistic, traits represents a real example of the 'art' component of a breeder's job, rather than based upon hard theory, molecular biology or quantitative genetics. These trade-offs are based instead upon experience, intuition and the overall insight that a breeder has about his breeding populations and what the market and customer expects in terms of improved cultivars.

When more than a single character is to be considered in a selection scheme, a plant breeder can make selection by either *independent culling* of a number of characters, or by using some defined *selection index*.

Independent cull selection

To examine this, consider a simple case where there are only two traits to be included in the selection decision. If independent culling is used, then the breeder will choose **target values** for each of the two characters independently. In order for a genotype to be selected, then the phenotype must exceed (or be less than, depending upon the trait of interest) the target values of both of the characters simultaneously. Therefore each of the genotypes from the initial base population will fall into one of four possible categories, which, for example if we are selecting for greater expression of both characters, will be:

- Greater than the target value set for both trait 1 and trait 2.
- Greater than the target value set for trait 1, but less than the target value set for trait 2.
- Less than the target value set for trait 1, but greater than the target value set for trait 2.
- Less than the target value set for both trait 1 and trait 2.

With this form of selection, only the genotypes that fall into category 1 (i.e. greater than the target values for both traits) would be retained, while all other categories would be discarded.

Index selection

Index selection involves creating an equation that includes values recorded for both traits. Selection indices can be either additive or multiplicative. For example, an additive selection index for the *i*th genotype with only two traits would be represented by:

$$I_i = (w_1 \times x_{i1}) + (w_2 \times x_{i2})$$

where I_i is the index value, w_1 and w_2 are the **weights** for each trait and x_1 and x_2 are the actual recorded values for each trait of the *i*th genotype. Obviously if **n** traits were included in the index value, then the index equation would be represented by:

$$I_i = (w_1 \times x_{i1}) + (w_2 \times x_{i2}) + \cdots + (w_n \times x_{1n})$$

A similar multiplicative index with only two traits would be:

$$I_i = (w_1 \times x_{i1}) \times (w_2 \times x_{i2})$$

where I_i, w_1, w_2, x_{i1} and x_{i2} are as above. Finally, if **n** traits were included in a multiplicative selection index we would have:

$$I_i = (w_1 \times x_{i1}) \times (w_2 \times x_{i2}) \times \cdots \times (w_n \times x_{in})$$

The difference in results between index selection and independent culling are primarily related to the association between the two (or more) traits and the differences in the relative weighting of them. If there is good association between the traits (i.e. high expression in one trait is related to high expression in the other, and *vice versa*) and both are nearly equally valued, then there may be little difference between the genotypes selected by either method (Figure 7.4). If, however, there is poor association between traits (i.e. high expression in one trait is not related to a similar high expression in the other trait), or one character is of vital importance while the other is valued much less, then there could be a large difference in the genotypes that would be selected by index selection over independent culling (Figure 7.5).

In almost all studies carried out it has been shown that index selection is more effective in identifying genotypes that are 'superior' for many different traits. The difficulty in all selection index schemes is how to determine the *index weights* (i.e. the w_is).

It will not be possible within the scope of this book to fully explore the possibilities available with selection indices. In simple terms, however, trait or character values in a selection index can be weighted either by:

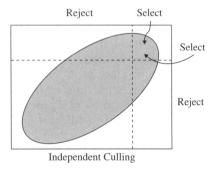

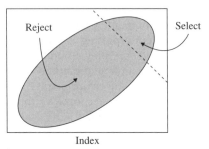

Figure 7.4 Association between independent culling and index selection when there is high correlation between two selectable traits.

- **Economics:** where the potential economic impact of each trait is estimated and the datum recorded of each trait expression is weighted by that value. For example, the average price paid per unit weight can usually be predicted from past seasons, and an increase in productivity could be related in money terms by an appropriate weight. Similarly if a particular insecticide costs a unit more per acre than if biological resistance is incorporated, then that resistance will accrue monetary value. The problem with economic weights is that they change from year to year. If there is overproduction of a product in any year then there is a tendency for the unit weight price to drop.

- **Statistical features:** where the weight values are derived according to some statistical procedure. The most commonly used routines have involved *multivariate transformations* such as principal component analysis, canonical analysis or discriminant function analysis. Each of these statistical techniques, although all called analysis, are in fact statistical transformations that produce various equations of multivariate data,

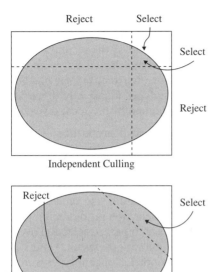

Independent Culling

Index

Figure 7.5 Association between independent culling and index selection when there is low correlation between two selectable traits.

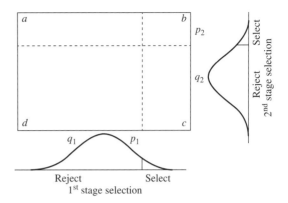

Figure 7.6 Classification of phenotypes based on independent culling of a single trait over two stages of selection.

usually with minimum correlation between traits or maximum discrimination between genotypes. The problem with statistical weights is again that they will be different from dataset to dataset. In some cases the weights do indeed show some biological meaning, but in other cases there appears to be no coherent association.

Selection indices can be extremely useful in plant breeding and their true value is perhaps yet to be realized. If index selection is carried out in a meaningful manner, then it should be more effective than independent culling.

7.3.4 Errors in selection

Each time selection is applied there is a chance that an error will occur. Errors in selection happen because the true genotype value is masked by environmental effects or because of administrative or clerical error.

Consider the illustration in Figure 7.6, which shows a two-stage selection process. This could refer to selection over two different years, or at two different locations, or indeed for two different

traits. The distributions at the side and bottom of the figure show the frequency distribution of each selection stage. For simplicity let us consider an experiment where we score the plants as if we were going to select or reject them, but in fact we retain all the plants and grow them again in a second year, that is, we simulate selection being carried out over a two-year cycle. In the first year, a proportion of the total genotypes (p_1) will be selected while the remainder (q_1) are theoretically discarded, where ($p_1 + q_1 = 1$). Similarly, in year 2 a proportion of genotypes would be selected (p_2) and the remainder discarded (q_2) from this bivariate normal distribution, so each genotype can be classified as:

a. rejected in the first year and selected in the second year;
b. selected in both years;
c. selected in the first year but rejected in the second year;
d. rejected in both years.

From this, there are two areas of misclassification that can be considered as errors in selection. (It should be noted that at this point we are not concerned with those genotypes that were discarded in both years, i.e. those designated d in Figure 7.6, as they are of little interest in the context of practical plant breeding.) These selection errors can be termed:

Type I errors: which are when genotypes are selected in the first year (i.e. $b + c$) but rejected in the second year (i.e. c) and so their frequency is $c/(b + c)$.

Type II errors: which are when genotypes that would have been selected in the second year (i.e. $a + b$) had already been rejected in the first year (i.e. a) and are estimated as $a/(a + b)$.

In terms of the practicalities of plant breeding, Type II errors are potentially far more important than Type I errors. Type II errors result from wrongfully rejecting a genotype, based on phenotypic performance in the early selection stage, which really should have been selected, as evidenced by the second year trials (which are usually more extensive or more precise). So this effectively results in discarding potentially valuable genotypes.

Conversely, Type I errors result from selecting genotypes in the first year which really should have been discarded and therefore result in a waste of resources which should have been better used in other areas. However, such a situation is normal in a breeding programme where genotypes are selected to continue at one stage and then are subsequently rejected upon more detailed examination in the latter ones.

A second aspect of interest in the data from two-stage selection is to estimate selection ratios. This ratio basically gives an estimate of how much more likely you are to select in year 2 from those genotypes selected in year 1, as compared with those genotypes that you discarded in year 1.

The possible selection pattern over two stages of selection is again as shown in Figure 7.6. Assume once again that the x-axis relates to selection in the first year of a breeding programme, and the y-axis relates to selection in the following year, albeit that all of the first-year genotypes here are planted and evaluated in the second year. Again, in the first year selection is carried out such that a proportion of the population (p_1) is selected (i.e. $b + c$) while the remainder (q_1) is discarded (i.e. $a + d$). Assume that the selection conducted in year 1 is ignored and all the breeding lines are re-evaluated in the second year in more elaborate trials. Now, the breeding lines are re-evaluated in year 2 and again selection is carried out such that a proportion (p_2) is selected (i.e. $a + b$) and the remainder (q_2) rejected (i.e. $c + d$). From this, of the genotypes selected in year 2, we can ask what proportion of these was selected in the first year and what proportion was discarded in the first year. In other words, we have the two categories as:

- Correctly selected in the first year ($b/(b + c)$); this represents the genotypes correctly selected in the first year as defined by the second year results.
- Should have been selected but were incorrectly discarded in the first year ($a/(a + d)$). This represents the genotypes incorrectly discarded in the first year as defined by the second year.

The ratio of these two proportions is termed the selection ratio and is given by:

$$[b/(b + c)]/[a/(a + d)] = b(a + d)/a(b + c)$$

Values of selection ratios can range from 0 (zero) to infinity. If the selection ratio is equal to 0, then it means there was no repeated selection over the two stages. A selection ratio between 0 and 1 indicates that a higher proportion of genotypes were selected in year 2 from those discarded in year 1 than were selected from year 1 (i.e. negative correlation). A selection ratio less than 1 therefore suggests that selection has had a negative effect (i.e. all the good genotypes have been discarded). A selection ratio equal to 1 indicates that selection has occurred at random (i.e. a zero correlation coefficient between the two stages). If the selection ratio is greater than 1, then selection has been better than random (i.e. a positive correlation between the stages). A larger value of the selection ratio means an increased efficiency of selection. For example, if a selection ratio of 2.0 is obtained, then the genotypes selected in year 1 would be twice as likely to be re-selected in year 2 as genotypes that were discarded in year 1.

It should be obvious that the selection ratio is related to the heritability of the character being selected. It also should be noted that the selection ratio will be influenced by the selection intensity. Obviously, regardless of the heritability for a character, the selection ratio will be zero, if the selection intensity is set so high that no genotypes survive repeat selection. Similarly, the selection ratio will always be 1.0 where the selection intensity is so low that no genotypes are discarded. So in summary, the relationship between selection ratio values and heritability is linear and related to selection intensity (Figure 7.7).

Similarly, it is possible to estimate selection ratio values for different selection intensities if the heritability is known. Where the heritability is zero, then there is no response to selection and hence the selection ratio is zero. The selection ratios with different

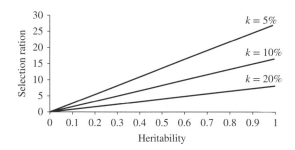

Figure 7.7 Selection ratio values with increasing heritability and different selection intensities.

Table 7.1 Selection ratios values with different selection intensities and heritability values.

Selection intensity		Heritability			
Year 1 (%)	Year 2 (%)	0.2	0.4	0.6	0.8
5	5	6.98	13.95	20.93	27.91
5	10	3.04	6.07	9.11	12.14
5	15	1.82	3.63	5.45	7.28
5	20	1.22	2.45	3.68	4.91
10	10	4.02	8.03	12.05	16.07
10	15	2.27	4.55	6.83	9.11
10	20	1.50	3.01	4.51	6.02
15	15	2.88	5.76	8.65	11.53
15	20	1.87	3.74	5.60	7.47
20	20	2.33	4.67	7.00	9.34

Table 7.2 Number of genotypes (per 1000) that would be selected in: (i) both stages of a two-stage selection scheme (**b** values in Figure 7.6), or (ii) would be selected when two variables are being selected for, based on inverse tetrachoric correlations from different selection intensities (0.05, 0.10, 0.15 and 0.20) in each of two years and with correlation coefficients between years of r = 0.1, 0.2, 0.3, 0.4, 0.5, 0.6, 0.7, 0.8 and 0.9.

Year 1

	0.05	0.10	0.15	0.20
0.05	004 005 007 009 012 016 020 025 032			
0.10	007 009 012 016 019 024 029 035 043	013 017 022 027 032 039 047 056 069		
0.15	010 013 017 020 025 030 035 041 047	019 024 030 036 043 051 059 070 083	028 035 042 049 058 067 078 091 108	
0.20	013 017 020 025 029 034 039 044 049	025 031 037 044 052 060 069 079 091	037 044 052 061 070 081 093 107 125	048 057 066 076 087 099 113 129 150
	0.05	0.10	0.15	0.20

Year 2

selection intensities and heritabilities are shown in Table 7.1.

Type I and II errors and selection ratios can be useful in setting the selection intensity levels at each stage in a breeding scheme. To estimate any of these it is necessary to determine the frequency of genotypes that fall into the a, b, c and d classes (Figure 7.6). In order to achieve this, it is necessary to 'virtually' select and reject genotypes in one stage and to re-evaluate all selected and rejected lines in a second stage.

It has been mentioned above that correlation analysis can be useful in determining selection efficiency. This can be done by the use of ***inverse tetrachoric correlation***. Tetrachoric correlations were first described by Peter Digby. He showed that it was possible to determine the correlation coefficient between two stages of evaluation (i.e. two years of testing) from frequency tables like the one shown in Table 7.1.

Inverse tetrachoric correlations are where, given the correlation coefficient between two selection stages (or indeed between two traits selected at the same time), it is possible to estimate the values of a, b, c and d (from Figure 7.6) and hence estimate Type I and Type II errors and selection ratios at different selection intensities.

The theory of tetrachoric correlations is beyond this book; however, values of **b** (the frequency of genotypes that would be selected at both stages of a two-stage selection) for varying selection intensities used at the different stages and with correlation values between the stages ranging from 0.2 to 0.9) are shown in Table 7.2.

To illustrate the use of this table, consider the following example. It is known that the correlation coefficient for seed yield between two assessment years (year 1 and year 2) is equal to 0.7. What would be the level of Type I error and the level of

Type II error given that selection was carried out at the 20% level in year 1 and at the 15% level in year 2. From Table 7.2, with p_1 at 0.20 (20%), p_2 at 0.15 (15%) and with a correlation coefficient of 0.7, we have a **b** value of 093 (or 93 genotypes out of 1,000).

From the fact that we selected 20% in year 1 out of 1000 genotypes, this would have resulted in $0.2 \times 1000 = 200$ total selected in year 1, then the number of genotypes selected in year 1 and discarded in year 2 (**c**, in Figure 7.6) would be $200 - 93 = 107$. Therefore the Type II error would be $107/200 = 0.535$, or 53.5% of genotypes selected in year 1 will be discarded in year 2.

Similarly, 15% ($p_2 = 0.15$) were selected in year 2. So given that 1,000 genotypes were screened, this would result in $0.15 \times 1000 = 150$ genotypes being selected in year 2. From this we have that $150 - 93 = 57$ genotypes would have been selected in year 2 which would have been discarded in year 1. Therefore the Type I error is $57/150 = 0.380$, or 38% of all selections that would have been made in year 2 were in fact discarded in year 1.

It can be seen, therefore, that even with relatively high correlations between different years of evaluation (i.e. $r = 0.7$) there will still be a high potential selection error.

By subtraction we have that $1,000 - (93 + 107 + 57) = $ **d** $= 743$, the number that would be discarded in both stages. From this the selection ratio would be $[93/107]/[57/743] = 11.19$. Therefore a genotype selected in the first year would be more than 11 times more likely to be reselected in a second year than a genotype discarded in the first year. Thus, despite the relatively large Type I and II errors, with a correlation between selection stages of $r = 0.7$, then selection is more than effective at these selection intensities.

It is interesting to try the same operation with:

- Uniform selection intensity and varying correlation coefficients.
- Uniform correlation coefficient and varying selection intensity.

Inverse tetrachoric correlations can also be used in a similar way to determine the association between selections for two different characters, if the correlation coefficient between the traits is known.

7.4 Applied selection

Selection in a plant breeding programme is an operation which is carried out over several years. After genetic variation is created, then a population of genotypes will be evaluated under different environmental conditions. At each stage the '*most desirable*' lines are selected, while the lines with defects or that are less desirable are discarded.

For simplicity, the various stages of selection can be divided into three types:

- **Early generation selection.** This is the first stage where the initial unselected population is screened. In most programmes this critical stage can involve many thousands of individuals (Figure 7.8).
- **Intermediate generation selection.** After the least adapted lines have been discarded, more detailed evaluation of remaining lines is carried out. Intermediate selection usually involves hundreds rather than thousands of lines.
- **Advanced generation selection.** At this stage the initial population has been reduced such that only tens of lines survive. Trials in the advanced selection stage are more accurate and detailed, including multiple location evaluation trials.

7.4.1 Number of genotypes in initial populations

To many, plant breeding is a '*numbers game*' where the more genotypes that are screened, the greater the chance of identifying desirable recombinants that will eventually become new cultivars.

Why therefore is it necessary to evaluate so many different genetic lines in a plant breeding programme? Consider a simple wheat breeding scheme with the objective of developing new cultivars that have high yield, good breadmaking quality, qualitative resistance (dominant) to stripe rust, quantitative resistance to mildew, good establishment, cold tolerance, short in stature, early maturity and resistance to lodging. Consider that the selection intensities in Table 7.3 are needed to ensure that at least some individuals will meet the required standards. Also shown is the accumulative frequency of desirable individuals, given that all characters simultaneously selected for are independently inherited.

Figure 7.8 Selecting single potato plants based on visual appearance.

Table 7.3 Possible selection intensities used in a wheat breeding scheme to develop cultivars that have high yield, good breadmaking quality, qualitative resistance to stripe rust, quantitative resistance to mildew, good establishment, cold tolerance, short in stature, early maturity and resistance to lodging. Also shown is the cumulative frequency of desirable genotypes, given that all characters are independently inherited.

Character	Selection intensity	Cumulative frequency
High yield	5	1 : 20
Good quality	5	1 : 400
Stripe rust resistance	50	1 : 800
Mildew resistance	20	1 : 4,000
Crop establishment	50	1 : 8,000
Cold tolerance	20	1 : 40,000
Short stature	10	1 : 400,000
Early maturity	10	1 : 4,000,000
Lodging resistance	10	1 : 40,000,000

Obviously, even when this very limited set of criteria is used as the breeding target, and with relatively modest selection intensities, the number of plants that need to be screened can be very high. If a breeder wishes to have some chance of success in having ***at least one*** individual that meets the criteria that are set, then the numbers that need to be screened will be large.

Plant breeding, therefore, does require evaluation of many thousands (or millions) of plants to have any chance of producing a successful new cultivar. So given that the initial population needs to be large and that selection of the better lines should be carried out as efficiently as possible, how can the best genotype be identified?

To further examine the different stages of selection, consider two examples. First is the selection scheme used to develop new potato cultivars at the Scottish Crop Research Institute, as shown in Table 7.4.

In this scheme, years 1, 2 and 3 could be considered to be early generation selection, years 4, 5 and 6 would be intermediate generation selection, and years 7 and 8 would be advanced selection.

Secondly, the wheat breeding programme at the University of Idaho, USA, has the selection scheme as shown in Table 7.5.

In this scheme, years 1 and 2 could be considered as early generation selections, years 3 and 4 would be intermediate selections, and years 5 and 6 would be advanced generation selection. In year 7, and subsequent years, remaining selections (2 to 3 lines) would be entered for regional testing where they would be evaluated at many western US locations.

Table 7.4 Selection scheme used to develop new potato cultivars at the Scottish Crop Research Institute.

Year	Number of genotypes	Number of replicates	Plot size	Characters assessed
1	140,000	1	1	Visual assessment of commercial worth
2	40,000	1	1	Visual assessment of commercial worth
3	4,000	1	3	Visual assessment of tuber size, shape, tuber number, yield and defects
4	1,000	2	5	Actual assessment of yield, initial quality tests, visual assessment of appearance and defects
5	500	2	10	Actual assessment of yield, fry quality, boil quality, initial disease testing for late blight and common scab, visual assessment of appearance and defects
6	100	2	40	Yield, fry and boil quality assessment from early and late harvest, initial taste testing, multiple disease testing, initial virus testing, visual assessment of appearance and defects
7	50	4	40	Yield, quality and disease testing at seven locations throughout the target region
8	10	4	40	Repeat multiple locations testing, initial on-farm testing (large field-scale trials)

Table 7.5 Selection scheme used by Dr Robert Zemetra in the soft white wheat breeding programme at the University of Idaho.

Year	Number of genotypes	Plot type	Characters assessed
$1-F_3$	1,200,000	Single plants	Visual selection of plant types, plant height and stripe rust resistance
$2-F_4$	20,000	Head rows	Visual selection of plant types, uniformity, yield, height, stripe rust resistance, lodging resistance. Actual assessment of protein content and kernel hardness
$3-F_5$	500	Preliminary yield trials, at one location	Actual yield performance, stripe rust yield resistance, lodging, stand establishment, heading date, test weight, dough viscosity, milling quality, baking quality, protein content and kernel hardness
$4-F_6$	80	Preliminary yield trials at two locations	Actual yield performance, stripe rust yield resistance, lodging stand establishment, heading date, test weight, dough viscosity, milling quality, baking quality, protein content and kernel hardness, yield stability, Russian wheat aphid resistance, dwarf bunt, cephalosporium stripe
$5-F_7$	10	Advanced yield trials at eight locations	Actual yield performance, stripe rust yield trials, resistance, lodging stand, establishment, heading date, test locations weight, dough viscosity, milling quality, baking quality, protein content and kernel hardness, yield stability, Russian wheat aphid resistance, dwarf bunt, cephalosporium stripe, foot rot, Hessian fly resistance and winter-hardiness
$6-F_8$	6	Advanced yield trials at eight locations	Actual yield performance, stripe rust resistance, lodging, stand establishment, heading date, test Weight, dough viscosity, milling quality, baking quality, protein content and kernel hardness, yield stability, Russian wheat aphid resistance, dwarf bunt, cephalosporium stripe, foot rot, Hessian fly resistance and winter-hardiness

Therefore early generation selection should eliminate the very worst genotypes, intermediate selection would identify the very best genotypes, and advanced selection would confirm genotypic performance over differing locations, assess environmental stability and identify the superior (cultivar quality) genotypes from those which may just fail to become cultivars.

After each stage of selection, fewer genotypes will remain for further testing. Increased rounds of selection will also be associated with decreased genetic variation between selected lines. It will therefore require more detailed evaluation studies to differentiate between remaining selections.

7.4.2 Early generation selection

Selection in the early generation stages differs from later selection because:

- Many thousands of genotypes/lines are to be screened.
- Only small amounts of planting material are available from each genotype and so sophisticated experimental designs with large numbers of plots and high replication are not possible.
- Selection is often carried out on populations with highly heterozygous genotypes, where dominance effects can be large and can mask the true genotype being selected.

The first two points can be considered as a single problem because, even in cases where large quantities of planting material are available, it may be impractical to have very large plots and high replication of so many different lines. Staff and land are not usually available to carry out such large screens.

Early generation selection is therefore carried out on small plots and most often on unreplicated plots or even on individual plants. Even when the test entries are not replicated, it is possible to increase the efficiency of testing by including a wide range of control entries interspaced within the test plots. Direct comparison can be made between the control lines (often existing cultivars) and those under test. The control plots can be included more than once over the whole trial area, and from this an estimate of plot-to-plot error variance can be obtained.

Several forms of analysis are available by which test entries can be compared with, or adjusted to, adjacent control lines. It should be noted that it is unlikely that all controls within a trial will have been included as showing equal performance of important characters. Some controls will have higher yield, others would have better disease resistance or high expression in a single character, and this needs to be accounted for if comparisons are to be made with the test lines.

Visual assessment

The large number of genotypes that need to be assessed in the early generations usually dictates that most selection is by visual assessment. Visual assessment of genotype performance is based upon a mental image of the desirable attributes (*ideotype*) that will constitute a successful variety. Such assessments are therefore similar to an informal selection index. The efficiency of visual assessment can be influenced by breeders' experience and also the time taken over assessment. Visual assessment has been shown to be more effective when more than a single breeder is involved in assessment and selection is carried out based on the average assessment rating.

Throughout the season different traits can be visually assessed and then genotypes culled based upon this information. At harvest, only the lines that have met all selection criteria are retained and others are discarded.

Visual assessment is often very subjective, and different people have been shown to give different emphasis in screening, depending on individual preference. Overall, however, different evaluations based on visual ratings can be remarkably similar in the lines that are chosen. Despite the problems with visual selection, it can be carried out relatively quickly so that many lines can be evaluated and at low cost. The biggest failing of visual selection alone is that characters can only be assessed if they can be seen. Therefore it is not usually possible to use visual evaluation to screen, for example, for food quality characters. If more objective selection is to be achieved then it may be necessary to actually record information on yield, disease rating or quality, or carry out specific tests.

Even when there is no replication it is possible to obtain some indication of error in visual assessment if the assessment operation is repeated. This will not provide environmental error estimation, but can often be useful in determining the repeatability of visually assessed characters.

Mass selection

It is often possible to use mass selection in the early generations. Examples of this would include selection for short plants by cutting tall ears from populations in wheat. Mass selection can also be used to select larger seeds, higher specific gravity in tubers and morphological traits such as fruit colour.

Mass selection in the early generations can be achieved by growing the early population under specific environmental conditions. The more adapted lines will be more productive, and the frequency of less adapted genotypes will reduce. Bulk selection has been shown to be effective in increasing the frequency of drought, heat, salt and other stress tolerances.

Efficiency of early generation selection

The efficiency of early generation selection has been examined in a number of different crops. When breeding a pure-breeding crop species, for example wheat or barley, selection will be influenced by the highly heterozygous nature of the breeding lines in the early generations. Segregation effects can be avoided by advancing towards homozygosity prior to selection, but this has not been common in the past because of time restraints or cost factors.

Visual assessments of yield and yield components have been examined and visual evaluations of yield from single rows or small plots have proved unreliable in predicting actual yield in subsequent generations. The highest yielding progeny bulks, derived from F_2 and F_3 single plants, do not necessarily produce the highest yielding segregants. Visual selection for yield on individual plants in cereals results in only a random reduction in population size with little or no effect in increasing yield. Even when the actual yield of an early generation of a cereal pedigree bulk breeding scheme (say F_2 or F_3) was measured and it was found to be significantly correlated with yield in later generations (say F_5 or F_6), the association found between segregating populations was usually so poor that it was questionable whether selection at the early stages (along with the expense that this would incur) would be justified.

Selection for yield *per se* in the early segregating generations of other pure-breeding species has also been shown to produce an effect that is no better

than random. Examples from past research include chickpea, cotton, soybean and rice. In addition, selection for yield components such as seed size in chickpea and grains per ear in spring barley were shown to be slightly more effective in the early generations than selection for yield itself.

The large numbers and small plots used in the early generations dictates that selection is only carried out for characters that are highly heritable. Often these mainly include single gene traits. As might be expected, selection for qualitative disease resistance in the early generations has been found to be more effective than selection for quantitatively inherited resistance or other polygenic characters.

The efficiency of selection in a pedigree bulk breeding scheme has been related to the heterozygosity of the bulks under selection. As homozygosity increases, selection becomes more effective. Homozygosity can be accelerated by single seed descent. However, care is needed to ensure that in single seed descent there is not a non-random loss of genetic material. Homozygosity can also be accelerated by various doubled haploid techniques. Again, however, care must be taken in the use of these procedures, as there is evidence of non-random success and a strong genotypic response to *in vitro* regeneration.

Most research into the efficiency of selection on clonally reproduced crops has been on potatoes and sugarcane. Early generation selection in potatoes (see Figure 7.8) has been shown to be, at best, a random reduction in the number of genotypes in the breeding scheme. However, care must be exercised as there has been some evidence that selection in the early generations was producing an undesirable response, and that selections were not always the genotypic lines most suited to agricultural conditions.

Significant correlations have been found between sugar cane seedlings and later clonal generations for stalk and stalk diameter. These associations, although statistically significant, would result in large selection Type I errors and selection ratios of less than 2.0. With such results it may be difficult to justify the expense and effort that such selection involves.

Early generation selection of grasses using small plots resulted in identifying lines thst did not perform well under sward conditions where inter-plant competition was greater. A similar response has been noted in potato where selected lines were less competitive under 'field stand' conditions due to selection being carried out under wide plant spacing.

Despite the relative inefficiency of selection in the early generation stages, most plant breeding schemes usually discard by far the greatest proportion of genetic variation in the first and second rounds of selection. It is certainly not uncommon to have cases where 99.9% of genotypes are discarded in the first or second selection stage, and where this has been achieved using small plots and without replication.

In summary, selection in the early generations is usually affected by:

- **limited amounts of planting material** so it is not possible to have sophisticated experiments involving large plots, high replication and multiple sites;
- **large numbers of genotypes** need to be evaluated which also usually results in small plots (often single plants), low levels of replication and single location trials.

As a result, many of the initial evaluations are carried out by visual inspection rather than, say, actually recording yield. Similarly, many of the more 'difficult to assess' traits, including polygenic disease or pest resistance, cannot easily be taken into account.

It is difficult to determine exactly what characters should have priority in early generation selection. Unfortunately there is no simple equation that allows breeders to say that it is best to select for this character now and at this intensity. However, there are some simple questions a breeder can ask him/herself which can help in making these decisions. These include:

- What are my breeding objectives and what characters are to be included throughout the **whole selection process**?
- What characters can be most easily and most economically assessed on small plots with minimal replication?
- Which characters are most heritable (i.e. high h^2)? Which have low heritability? Selection in the early generation should be based on the most heritable traits.

- Which characters have highest priority? For example, which characters **must** a new cultivar have? Either by being important (i.e. high yield or specific quality) or by legislation (i.e. low glucosinolates and erucic acid content in canola).

Never forget the golden rule of any selection stage: that a breeding scheme should never carry more individual genotypes than can be **efficiently screened**. It is almost always more effective to evaluate fewer lines with greater accuracy than to use an ineffective selection scheme.

7.4.3 Intermediate generation selection

It is assumed that at the intermediate generation selection stage, the large initial population (usually thousands) has been reduced to a practical number which will allow more detailed assessment (usually hundreds of lines). It is also assumed that by this stage there has been a simultaneous increase in the availability of planting material. As a result of fewer lines and more planting material, it is possible to organize evaluation trials with reasonable plot sizes (may differ according to the crop), and replication of all test entries.

The number of lines that require testing at the intermediate stage will still be large enough to dictate that evaluation is restricted to very few (or one) locations.

Field trials

Field testing is a major part of all selection stages, and intermediate selection is no exception. Test entries should always be evaluated in comparison with control entries in replicated trials. Randomized complete block (RCB) designs are commonly used for the first rounds of intermediate generation selection. These designs provide reasonable error estimates and are fairly robust. One major advantage of RCB designs is that they can be used for any number of entries. As the number of surviving test entries is reduced, then more detailed incomplete block designs such as lattice squares, rectangular lattices and partially balanced incomplete block designs may be more attractive.

Lattice squares are amongst the most efficient designs that can be used for field testing in a plant breeding programme. A lattice design is similar to a RCB in that each entry appears once in all replicate blocks. However, within each replicate block, plots are arranged into sub-blocks. Analysis of data from lattice designs allows the actual mean performance of each test entry to be adjusted due to two-dimensional (row and column) environmental variation. The major problem with lattice squares is that the number of test entries must be an exact square (i.e. 9, 16, 25, 36, 49, 64, 81, 100, etc.). A second restriction is that the most efficient use of the designs requires high replication. For example, a 16 entry design (4×4 lattice square) requires $4 + 1 = 5$ replicates. Larger lattice squares can be used with $n - 1$ replicates, where n^2 is the number of entries in the whole trial.

Rectangular lattice designs allow greater flexibility in the number of entries and replicates used, although each replicate must be a rectangle (i.e. 10 plots $\times$ 5 plots, where sub-blocks would be either 5 or 10 plots). Rectangular lattice designs are, however, not as efficient as lattice squares in reducing error variance, as sub-block adjustments are made in only one direction.

One advantage of all lattice designs is that they are resolvable (i.e. data collected from them can be analysed as an RCB design, if desired).

Split-plot designs are often used in the latter stages of intermediate selection. The main use is often to evaluate a number of lines under differing environmental conditions at a single location. For example, several test entries may be assessed under differing nitrogen levels where genotypes would be main-plots and varying nitrogen levels are sub-plots. Similarly, split-plot designs can be used for differential chemical treatments or harvesting dates.

A more detailed overview of experimental designs available to plant breeders is provided in Chapter 10.

Traits recorded

In the early generation stages of selection there are many thousands of test lines for which only a few traits are recorded. In the intermediate stage of selection the number of traits upon which selection is based is increased, often considerably.

Data are likely to be collected prior to planting, throughout the growing season, at harvest and post-harvest. Therefore, a major part of plant breeding is managing the vast datasets that can arise and interpreting this information to best advantage in selection.

Data analysis and interpretation

It is useful to analyse data as they are collected throughout the year, so as not to have a backlog of analyses needed for decision-making at the end of the season. It is common in plant breeding to have a relatively quick turnover. For example, in winter wheat breeding, evaluation plots are harvested, yields recorded, samples taken for quality assessment, assessment carried out and decisions made within a few months, so that selection procedures can effectively use all possible information while still being able to plant selected lines at the appropriate seasonal time.

If selection is to be successfully applied for any character, it is important that there are indeed significant differences between test entries. Obviously if an analysis of variance shows no significant difference between test lines for yield, then there is little or no genetic variation for the character, and hence there will be no response to selection.

It can often be useful to estimate narrow or broad-sense heritabilities from yield trials. Broad-sense heritabilities can easily be obtained by simply estimating the genetic component of variance. In cases where test lines are highly heterozygous, then environmental variances can be estimated from homozygous control entries in trials. Narrow-sense heritabilities can be estimated by regression, if a sufficient number of the parental lines are included in the evaluation trial.

Heritabilities can also be estimated in relation to response to selection. To achieve this with any accuracy it is necessary to retain a certain proportion of the unselected population and to include these *random* samples along with the deliberate selections in the following season's trial. The practice of retaining a random sample is highly recommended, as it allows a continual check on what advances selection is making in terms of producing more desirable lines.

Bar charts or histograms of data can be helpful in visualizing and so understanding the variation and distribution of data for individual traits. Inspection of distributions along with trait means and variances can help to determine possible culling levels (i.e. target values that must be met for entries to be retained). This also helps to identify transgressive segregants – segregants showing a phenotypic value higher (or lower, depending upon whether the breeder is selecting towards or against a given trait value) than either parent.

After each character has been analysed individually, it is very important to consider the relationships that exist between different traits. This can be achieved by simple correlation analysis.

If two characters of interest are positively correlated, then there may not be any difficulty in selection (except that there will tend to be greater emphasis in either independent culling or index selection with positively correlated traits). However, if the expression of two characters is negatively correlated, it may be impossible to select for high expression in both traits simultaneously. Lower culling levels or index selection will then be necessary to identify lines that may be intermediate in performance for both characters, as the most desirable recombinants are unlikely to be present in the sample of materials evaluated.

Correlation analysis can easily be carried out using a variety of different computer software packages. The use of computers in all aspects of plant breeding will be discussed later. It is sufficient to say at this time that selection is one area where statistical analyses are particularly helpful in understanding the vast datasets which are likely to arise, and can also act as useful tools to select the better lines based on the data collected.

Selection

The most effective selection will result from the most accurate data collection and highest heritability. These in turn are related to good experimentation.

A good understanding of the relationship between different characters and genetic variation within characters can be of tremendous help in deciding whether to apply independent culling (along with the cull levels) or whether a selection index is more appropriate.

7.4.4 Advanced generation selection

At the advanced selection stage it is assumed that all remaining genotypes in the selection scheme have previously been assessed for all (or the majority of) characters of interest to the breeding objectives. At this stage there are relatively few (under 100, and usually no more than between 10 and 50 lines) selections that have survived the previous selection stages. Selections would therefore be expected to have shown acceptable levels of yield, quality, disease resistance and pest resistance, and be relatively free from obvious defects.

At this stage it is also assumed that there are relatively large amounts of planting material to allow evaluation at a number of different locations throughout the target environments for a breeder.

The major aims of advanced selection are:

- To confirm the past performance of selected lines over a wide range of different environments (locations and years).
- To identify the superior lines based on either specific or general adaptability.
- To identify breeding lines to be considered as parents in the next round of crosses, to allow a new breeding cycle to start.

There is usually little response to selection at the advanced stages because most of the genetic variation existing at the early and intermediate selection stages has been reduced; the few remaining genotypes represent a highly selected (see response to selection in Figure 7.2) group amongst which it is hoped simply to select the very best.

Choice of advanced trial locations

The choice of land suitable for trials is discussed in the field plot techniques section. It is sufficient to emphasize some points and state a few additional factors here.

Locations for advanced testing must be representative of the environments in which the potential new cultivars will be grown. If the target region has several diverse locations (i.e. large differences in rainfall, temperature, soil type), or if different agronomic practices are applied in different regions (i.e. irrigation versus rain-fed), then attempts should be made to ensure that at least one location is chosen to represent each main environment type.

If trial sites are a long way from the central research offices, then it may be necessary to find good collaborators (farmers, local agricultural agents, extension personnel) who will take care of the research plots.

In some instances trials may be grown in other states or different countries. In this case it may be necessary to arrange appropriate phytosanitary inspections of the previous seed crop, the seed or planting material that will be used, as well as appropriate regulatory approval if the lines to be included have been 'genetically modified'.

Number of locations

The number of locations used for testing will be dependent upon:

- Availability of planting material. This is not usually a major constraint at the advanced selection stage but may need to be considered.
- The diversity of environmental conditions throughout the target region. Clearly if, for example, the target region is the whole of the US corn belt or the Cerrados Region in Brazil, then many sites will be needed, while if a small region where the extent of genotype × environmental interactions is small or predictable is the only target area of interest, then perhaps one or two local trials will be sufficient.
- The magnitude of error variances and genetic variances applicable to specific trials in any one year or location. If, from past experience, the environments for which a new cultivar are targeted are all very similar (i.e. small variance between sites) then fewer locations need be considered. If different regions are markedly different, then more sites would be required.
- The cost of individual trials and the availability of sufficient funds to pay for off-station trials. In the real world, most breeding programmes have restricted budgets, and multi-location trials can often be expensive in shipping, land rental, staff time and travel.

There have been many debates regarding the substitution of more locations in advanced trialling at the expense of reducing the number of testing years. Obviously if more locations are evaluated in each year, then fewer years may be necessary to fully evaluate environmental response of test lines. However, it should be noted that year-to-year environmental variation is almost always unpredictable (i.e. climatic), while with

between-location variation there will be many predictable environmental effects such as soil type. Therefore it will always be necessary to assess advanced breeding lines over more than a single season, and several years testing may be required before a satisfactory decision of commercial worth can be made. Also, in practice, when appropriate data have been analysed, the interactions with years are often larger than those with site or location.

It should be noted that selection is still being carried out amongst lines in the advanced breeder trials, and that this can affect the average performance over years. For example, consider that in the first advanced trial 50 breeding lines are tested. The *best* 25 lines may be re-evaluated in year 2 (this is therefore based on their *phenotypic* performance in the year 1 trial). Say, the *best* 5 lines, now based on phenotypic year 2 performances, are retained from the year 2 trial and tested in the third year. After the third year trial, the *best* breeding line will be considered for cultivar release. At this stage, and before, it is common practice to examine the performance averaged over 3–4 years of advanced testing and to compare this performance to standard control cultivars that were included in the trials. The performance of the breeding lines is likely to be somewhat biased as they had specifically been chosen in the previous years' testing because they *had better than average* performance. In order to get a true representation of the new cultivar's worth, it is common to evaluate the newly released cultivar for several years in breeder trials after selection is complete. These post-selection trials are commonly conducted on a large scale, on large (on-farm) plots, and utilizing farm-scale equipment (planters, harvesters, etc.). Often it is the produce from these on-farm tests that offers breeders the first opportunity to have sufficient volume of material for realistic quality evaluations (Figure 7.9).

Experimental design

A limited number of test entries combined with large amounts of planting material allows the use of the most sophisticated experimental design at the advanced stages. Therefore it is common to use lattice squares or rectangular lattice designs for location trials. It should be noted that such trials are often managed by collaborators and/or farmers

Figure 7.9 Breeder's trial of advanced selections planted within a farmer's field crop.

who may be inexperienced in handling trials, and so the more highly sophisticated the designs, the more easily it can be planted or harvested incorrectly. With this in mind, randomized complete block designs may still offer the most practical design for advanced trials, despite the availability of more sophisticated experimental designs.

Genotype by environment interactions

A major goal of advanced selection is to determine the response of selected lines over differing environments. It is therefore difficult to consider this aim without specifically considering genotype by environment interactions.

As noted earlier, genotype × environment (G × E) interactions occur because some genotypes perform to a high degree under some environmental conditions, while others perform poorly in that same environment; conversely the lower yielding lines may exceed the higher yielding genotypes when grown under different conditions. Under extreme G × E interactions, the genetic progress apparently achieved in one season might be wiped out in terms of performance during the next one.

G × E interactions affect traits throughout all stages of a plant breeding selection programme, and affect conventional breeding programmes as well as programmes driven by molecular technology. Typically, it is not usually possible to fully evaluate or investigate the G × E phenomenon until the advanced stages of breeding programmes.

G × E interaction based upon only two genotypes is shown diagrammatically in Figure 7.10. It should be noted that a significant interaction in an analysis of variance can be obtained even when there is no change in ranking of genotypes under study (no cross-overs in performance). Consider the two genotype cases in Figure 7.10. In the upper diagram both genotypes perform relatively similarly (i.e. the two response lines are parallel). Interactions can, however, appear significant if the lines converge, but do not cross (i.e. no change in ranking of the genotypes). If there is no change of ranking, the interaction can usually be designated as a **scalar effect** and is not considered by many as a true interaction. Some would argue that the lines would converge and eventually cross either above or below the range of the dataset collected, but there is no evidence that this is the case. Interactions caused by scalar effects can often be removed by transformation of the data. True interactions, where there are changes in ranking of lines in different environments, cannot be eliminated by data transformation.

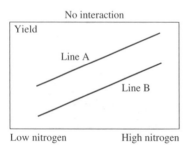

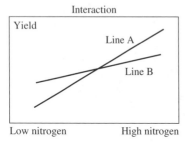

Figure 7.10 Genotype by environment relationships: top, where there are no genotype by environment interactions, and bottom, where the two genotypes (A and B) respond differently to different environments (nitrogen levels).

Environments can be classified into three different forms:

- Semi-controlled environments, where they are controlled by the grower or where there may be little change over years or seasons – for example, soil type, seeding density, fertilizer application.
- Uncontrolled environments, where there is often no chance of predicting conditions from one year to another – for example, rainfall, temperature, high winds.
- Managed stress environments, which are carefully managed locations, where abiotic stresses such as drought can be imposed under controlled conditions.

Obviously, even with the uncontrolled environmental conditions there can be some degree of prediction. For example, there will always be very low temperatures in North Dakota (US) in January and February, while it will tend to be continually warm in Death Valley, California (US), during the summer months.

The early and intermediate selections have been carried out at a single (or few) location, and so any surviving line should have at least been tested over more than a single year (albeit at a common location). It is difficult to carry out actual G × E studies on data where a large proportion of lines have been selected according to those data. For example, if there are three years of data available from an intermediate selection stage which begins with 1000 lines and reduces these to 20 lines based on phenotypic data collected over the years, then all the remaining genotypes will (by definition) have high phenotypic expression in all three years.

Multiple location trials are therefore necessary for three reasons:

- To identify particular genotypes that perform well over a wide range of environmental conditions. These lines are said to have **general adaptability**.
- To identify genotypes that perform to a high degree at specific locations or under particular conditions. These lines are said to have **specific adaptability**.
- In addition, they allow insight to be gained about the sensitivity of the germplasm to environmental variation, and thus the potential for its use as parental material for specific target areas.

7.4.5 Analysis of location trials

Various methods have been proposed for the statistical analysis of interactions in general, and G × E interactions in particular. The existence of interactions between genotypes and environments was recognized by R.A. Fisher and W.A. Mackenzie in the UK, even before the formation of the analysis of variance.

There are several conditions that are assumed when carrying out an analysis of variance (and indeed also some other analyses). These include: randomness; normality; additivity and homogeneity of error variance. The latter of these is the one that usually causes most concern when carrying out analysis of variance of multiple location trials. Regrettably, often the formal statistical assessment of error variance homogeneity is not carried out.

If there are only two experiments (i.e. two locations, two years, etc.), then the plant breeder can simply perform an analysis of variance on each experiment separately and from each obtain an estimate of the error mean square (σ^2). The larger of the two σ^2s can be used as the numerator and the smaller the denominator to carry out an F-test with the appropriate degrees of freedom. The resulting statistic can be compared with expected values from tables to determine whether the two σ^2 values are indeed different. But note that the probability given in the variance ratio (F) tables needs to be multiplied by two before interpretation, since we are simply taking the biggest variance over the smallest and hence involving both tails of the probability distribution – it is a two-tailed test.

In cases where more than one experiment is being considered (e.g. an experiment is carried out over three different years), a different approach needs to be considered. The F-test can still, however, offer a simple test where the largest σ^2 from the experiments is compared by dividing by the smallest σ^2 value. However, a more accurate method is available called a *Bartlett test*.

Bartlett test

There are two forms of the Bartlett test, depending upon whether the variances (σ^2) to be compared all have the same number of degrees of freedom, or whether they have different degrees of freedom. The first of these two situations will be considered first.

When all σ^2 values are based on the same degrees of freedom, the Bartlett test takes the form:

$$M = \text{df} \left\{ n \ln \left(\underline{S} \right) - \sum \ln \sigma^2 \right\}$$

where $\underline{S}$ is:

$$\underline{S} = \sum \sigma^2 / n$$
$$C = 1 + [(n+1)/(3n \ \text{df})]$$

where n is the number of variances to be tested, df is the degrees of freedom that **all** variances are based on, and ln refers to natural logs.

Now the Bartlett test in this case reduces to a chi-square (χ^2) test with $n - 1$ degrees of freedom, where:

$$\chi^2_{n-1} = M/C$$

Consider this simple example involving four variances (σ^2), all based upon 5 degrees of freedom:

df	σ^2	$\ln \sigma^2$
5	178	5.182
5	60	4.094
5	98	4.585
5	68	4.202
Total	404	18.081

$$\underline{S} = 100.9; \quad \ln(\underline{S}) = 4.614$$
$$M = (5)[(4)(4.614) - 18.081]$$
$$= 1.88 \text{ with 3 df}$$
$$C = 1 + [(5)/(3)(4)(5)] = 1.083$$
$$\chi^2_{3 \ \text{df}} = 1.88/1.083 = 1.74 \text{ ns}$$

As the χ^2 value in this case is smaller than the corresponding value from tables shown for the 5% probability level with 3 degrees of freedom, we would say that the four variances were *not significantly different* at the 5% level.

Consider now the case where each of the variances under test is based on different degrees of freedom. Now the Bartlett test is based on:

$$M = \left(\sum \text{df} \right) \ln(\underline{S}) - \sum \text{df} \ln \sigma^2$$

where $\underline{S}$ is equal to:

$$\underline{S} = \sum \text{df} \left[\sum \text{df} \cdot \sigma^2 \right] \Big/ \left(\sum \text{df} \right)$$

and

$$C = 1 + \{(1)/[3(n - 1)]\} \left[\sum (1/\text{df}_i) - 1 \Big/ \left(\sum \text{df} \right) \right]$$

and

$$\chi^2_{n-1} = M/C$$

To further examine this, consider the simple example involving five variances, each based on a different number of degrees of freedom:

df	σ^2	$\ln \sigma^2$	1/df
9	0.909	−0.095	0.1111
7	0.497	−0.699	0.1429
9	0.076	−2.577	0.1111
7	0.103	−2.273	0.1429
5	0.146	−1.942	0.2000
Total 37			0.7080

$$\underline{S} = \sum df\sigma^2 \Big/ \sum df = 13.79/37$$

$$= 0.3727$$

$$\left(\sum df\right)\ln(\underline{S}) = (37)(-0.9870) = -36.519$$

$$M = \left(\sum df\right)\ln(\underline{S}) - \sum df \ \ln \sigma^2$$

$$= -36.519 - (-54.472) = 17.96$$

$$C = 1 + [1/(3)(4)](0.7080 - 0.0270)$$

$$= 1.057$$

$$\chi^2_{4\,df} = M/C = 17.96/1.057$$

$$= 16.99^{***}, \quad \text{with 3 df}$$

As this value of chi-square exceeds the value given in the statistical tables for probability less than 1%, with three degrees of freedom, we can say that the five error variances show significant ($p < 0.01$) **heterogeneity**, and so conclude there are significant differences between the error variances.

It should be noted that the Bartlett test tends to be *over*-sensitive to deviations, and it is usual under practical situations only to consider heterogeneity of variance occurring where we have probability levels of 0.01% or lower.

Detecting significant treatment × environment interactions

The most common method, by far, of detecting genotype × environment interactions is to carry out an appropriate analysis of variance.

Various methods have been proposed for the statistical analysis of interactions in general, and genotype × environment interactions in particular. Consider first a simple example where a number of genotypes are each grown in a number of different environments. Variance components can be used to separate the effects of genotypes, environments and their interaction by equating the observed mean squares in the analysis of variance to their expectations in the random model. Therefore, in terms of a mathematical model, the yield y_{ijk} of the kth replicate of the ith genotype, grown in the jth environment, can be estimated by the formula:

$$y_{ijk} = \mu + g_i + e_j + ge_{ij} + E_{ijk} \qquad (7.1)$$

where μ is the average yield over all trials, g_i is the effect of the ith genotype, e_j is the effect of the jth environment, ge_{ij} is the interaction between the ith

Table 7.6 Degrees of freedom and expected mean squares from an analysis of variance with g genotype entries, grown at l locations, and r replicates at each location. In this analysis it is assumed that locations are random effects.

Source	df	EMS
Locations (l)	$l-1$	$\sigma_e^2 + g\sigma_{rwl}^2 + rg\sigma_l^2$
Replicates within locations (r)	$l(r-1)$	$\sigma_e^2 + g\sigma_{rwl}^2$
Genotypes (g)	$g-1$	$\sigma_e^2 + r\sigma_{gl}^2 + rl\sigma_g^2$
Genotypes × locations	$(g-1)(l-1)$	$\sigma_e^2 + r\sigma_{gl}^2$
Error	$l(r-1)(g-1)$	σ_e^2

genotype and the jth environment, and E_{ijk} is an error term. Also:

$$\Sigma_i g_i = \Sigma_j e_j = \Sigma_{ij} ge_{ij} = 0$$

The expected mean squares from an analysis of variance are dependent on whether main effects are fixed or random. In the case of genotype × environment trials it is usually assumed that both effects are random. From this, given a set of trials with g genotype entries, r replicates at each location and l locations, the expected mean squares obtained from an analysis of variance would be as shown in Table 7.6.

In this case the genotype × location interaction would be tested for significance against the pooled replicate within locations × genotypes interaction. The effect of genotypes would be tested in an F-test against the interaction G × E. Replicates within locations would be tested against the replicate error and the effect of locations would be tested against the replicates within locations mean square.

When both years and locations are included, often different locations (or different numbers of locations) are used in each year. In these cases the analysis reduces to genotypes × environments interactions where environments include a combination of years and sites (i.e. y years and s sites would result in ys environments). Environmental differences are often greater over years than over locations, and it can be informative to separate year and location effects in the analysis of variance.

More complex models and expected mean squares can be derived for location trials grown over more than one year which produce year ×

Table 7.7 Degrees of freedom and expected mean squares from an analysis of variance with g genotype entries, grown at l locations over y years, and with r replicates at each location. In this analysis it is assumed that locations and years are random effects.

Source	df	EMS
Years (y)	$y-1$	$\sigma_e^2 + gy\sigma_{rwly} + rgy\sigma_{lwy}^2 + rgl\sigma_y^2$
Locations within year (l)	$y(l-1)$	$\sigma_e^2 + g\sigma_{rwly}^2 + rg\sigma_{lwy}^2$
Replicates within loc. and years (r)	$yl(r-1)$	$\sigma_e^2 + g\sigma_{rwly}$
Genotypes (g)	$g-1$	$\sigma_e^2 + r\sigma_{glwy} + rl\sigma_{gy}^2 + rly\sigma_g^2$
Genotypes × year	$(y-1)(g-1)$	$\sigma_e^2 + r\sigma_{glwy}^2 + rl\sigma_{gy}^2$
Genotypes × L within Y	$y(g-1)(l-1)$	$\sigma_e^2 + r\sigma_{glwy}$
Error	$yl(r-1)(g-1)$	σ_e^2

genotype, location × genotype and year × location × year interactions. In this case the expected mean squares (EMS) would be those shown in Table 7.7.

Given that a significant interaction is detected in the analysis of variance, and this is not due to a scalar effect, then it can often be useful to have additional information regarding the partition of the interaction variance.

The analysis of multiple location data has benefited from progress with the implementation of linear mixed models to breeding data. There are several software packages available that allow the mixed model-based analysis of data at the pace that breeding programmes require.

Worked example

Consider the example where 20 spring canola (*B. napus*) cultivars or breeding lines were tested for yield potential in nine different environments throughout the Pacific Northwest (Idaho, Montana, Oregon) region of the US. The analysis of variance was found to be that shown in Table 7.8.

From this we have detected highly significant differences between sites (locations) and cultivars, and a highly significant interaction between sites and cultivars.

However, when yield data from each location were analysed separately we find that the error mean squares from each analysis of variance differ. This might have been expected given the large difference between the average yield at each location. The question is whether these differences are significant.

Table 7.8 Degrees of freedom, sums of squares, mean squares and F-ratio values from an analysis of variance where 20 spring canola (*B. napus*) breeding lines were tested for yield potential at nine different environments throughout the Pacific Northwest region of the US.

Source	df	SS	MS	F
Sites	8	220,698.7	27,581.3	133.6 ***
Reps within sites	27	5,575.5	206.5	58.8 ***
Cultivars	19	2,602.5	89.7	9.0 **
S × C	151	1,499.4	9.9	2.8 ***
Error	513	1,801.7	3.5	

A significant χ^2 value ($p < 0.001$) is found from a *Bartlett analysis,* indicating that these error variances are indeed significantly different. In this case, therefore, we are violating one of the least robust assumptions upon which the analysis of variance is based.

This is not uncommon, and the question would then arise '*well, what can I do about a situation where there is heterogeneity of variances?*' The only practical solution to overcome this is to transform the data. The most common transformations are square root, log, or to transform the raw data to a standardized normal distribution with mean of 0 and variance of 1. Alternatively, a variance/covariance structure could be used in the analysis that accommodates heterogeneity of variances. This can be accomplished through a linear mixed model analysis framework,

Given that a significant interaction is detected in the analysis of variance, and this is not due to a scalar effect, then it can often be useful to have additional information regarding the partition of the interaction variance.

Interpreting G × E interactions

The idea of breaking the G × E interaction into several components is entirely missing from the simple analysis of variance table shown above. In the G × E context a method of partitioning this interaction was suggested by Yates and Cochran in 1938, although this was largely neglected for 20 years until it was re-discovered in a paper by Williams in 1961. In their paper, Yates and Cochran stated that "the degree of association between varietal differences and general fertility (as indicated by the mean of all varieties) can be further investigated by calculating the regression of the yields of the separate varieties on the mean yields of all varieties".

This is when ge_{ij}, in the model equation (7.1) above, is regressed onto e_j. Therefore:

$$ge_{ij} = \beta_i e_j + \alpha_{ij} \qquad (7.2)$$

where β_i is a linear regression coefficient for the ith genotype and α_{ij} is a deviation from regression. Using a combination of equations (7.1) and (7.2) we can write:

$$y_{ijk} = \mu + g_i + (1 + \beta_i)e_j + \alpha_{ij} + E_{ijk}$$

An analysis of variance can be used to partition the genotype × environment interaction into **heterogeneity of regression** (i.e. that the regression slopes of the different genotypes have different slopes) and **deviation from regression** (i.e. that the relationship between genotypes and environments is not explicable by linear regression). In an analysis of variance with g genotypes and l environments the partition of interaction sum of squares would give $(g - 1)$ degrees of freedom for heterogeneity of regression and $(g - 1)(l - 2)$ degrees of freedom to deviations from regression.

In the analysis, each of these terms can be compared in an F-test on division by the error mean square. The heterogeneity of regression can be further compared with the deviations from regression to see whether it accounts for a significant part of the observed interaction.

This particular approach is called *joint regression analysis*. Despite some major theoretical difficulties with the analytical technique, it has been widely

used by plant breeders to determine the stability of genotypes in the advanced stages of selection.

From the analysis, each genotype has an average performance (mean over all environments) and a regression coefficient $(1 + \beta_i)$. The regression coefficient can be used to determine the stability of different genotypes over environments. Genotypes which have high $1 + \beta$ values are said to be more responsive to environment than those with low $1 + \beta$ values. From the analysis there can be four types of genotype:

- high average performance and high $1 + \beta$ value;
- low average performance and high $1 + \beta$ value;
- high average performance and low $1 + \beta$ value;
- low average performance and low $1 + \beta$ value.

The four types of genotype are shown diagrammatically in Figure 7.11. Obviously all plant breeders would like to develop cultivars which have high average performance and low $1 + \beta$ values as these genotypes would produce high yield in all environments. Unfortunately, performance and regression coefficient are usually related, such that it is usually high expression associated with high sensitivity.

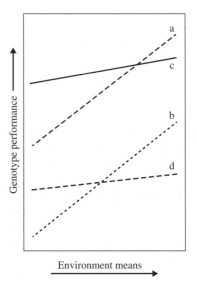

Figure 7.11 Different phenotypic responses (yield) to environment as measured by average performance and regression coefficient of performance against average environment performance, where (a) has high average yield and high **1 + β** response, (b) low average yield and high $1 + \beta$, (c) high average yield and low $1 + \beta$, and (d) low average yield and low $1 + \beta$.

A major criticism of joint regression analysis is that the regression is carried out between the performance of each genotype at each site onto the average performance of all genotypes at that site. Obviously these two are not independent variables, as the genotype performance is *a factor in determining the site mean*. To avoid this it has been suggested that the genotype being regressed should be omitted from calculating the site mean, and this has been carried out but has been shown to make little difference to the slopes obtained. Another way to avoid this is to carry out regression of the breeding lines onto the performance of one (or preferably more) control cultivar.

Other methods are available to examine G × E interactions in breeders' trials. Genotypic stability over environments can be examined by comparing the variance of each genotype over the different environments. This method can produce very similar results to joint regression analysis with increasing variation over environments related to less stability.

Genotype means and variances over environments can also be used simultaneously to predict the frequency with which test entries would exceed specified target values (often the value of controls in the trials). One advantage of the probability prediction method is that it provides a single datum for each test entry to be evaluated (i.e. the probability that a genotype will produce a yield greater than 2000 kg/ha). A second, perhaps more important feature is that this method can be expanded to cover several traits simultaneously. This is achieved by estimating the average performance for each trait, the variation over environments for each trait, and the covariance between traits and environments. Probabilities are calculated by evaluating the area under a univariate or multivariate normal frequency distribution (see cross prediction later).

Finally, another form of analysis called **Additive Main Effects and Multiplicative Interaction** (AMMI) has been suggested as suitable by several researchers. An AMMI analysis partitions the residual interaction effects between genotypes and environments by principal component analysis. The technique therefore involves a multivariate transformation of the residuals (after the main effects of genotypes and environments have been removed) within a two-dimensional table of genotypes and environments. Provided the first and second eigenvectors (transformations) account for a large proportion of the residual data, two-dimensional graphic inspection of eigenvalues from the first and second eigenvectors for different locations and genotypes can be used to examine the interaction effect. One aspect of AMMI readers have to be aware of is that it considers all statistical factors in the model as fixed, whereas in reality breeding data consist of a combination of both fixed and random factors. The theory of AMMI is outside the scope of this book, although readers should be aware that this option of interpreting data is available.

Selection

Understanding data and interpreting results from multilocation trials are a great aid in selection from such trials. Factors of importance in selection will include genotypic stability (or lack of) and correlation between characters over environments and between traits within a single location.

Finally, efficient selection with the very large datasets that are common in the advanced selection stages can be helped by inspection of performance ranking. Indeed, it has been shown that summing the rank of individual traits (or single traits over environments) will produce similar results to multivariate probabilities. Summing ranks can be considerably easier to compute than finding the area under an *n*-dimensional normal frequency distribution.

7.5 Cross prediction

The number of breeding lines discarded in a plant breeding programme is inversely proportional to the number of selection rounds. In the early stages there are many thousands of lines evaluated, with a low proportion selected for testing at the intermediate stage. At the advanced stage a few surviving lines are tested with great intensity, with only a few lines being discarded after each selection stage.

A number of researchers have shown that selection in the initial stages (that period where the greatest proportion of genotypes are discarded, and hence the greatest genotypic variation lost) is the most ineffective stage at identifying the most desirable lines. At the early generation stage, selection has been shown to result in, at best, only a random reduction in the number of genotypes within the breeding scheme. Some advances (particularly for qualitatively inherited traits) have shown a response to initial selection, although some have shown a negative response where the best phenotypes under conditions of early generation selection have been shown to be those least likely to become commercial cultivars.

Therefore, the early generation selection stage of a plant breeding programme is often very ineffective in terms of selection. This inefficiency is in part simply due to low heritability in performance characters in the early selection stages, compared with those in more advanced levels. Low efficiency is, in part, the result of:

- The inaccuracy of selecting in small plots (often single plants) because of the error variance and sampling variation along with environmental variation and competition effects of surrounding plots. This is not helped by the inability to adequately replicate and/or randomize the vast number of genotypes involved at the earliest selection stages.
- Selection under atypical conditions that do not mimic, for example, plant spacing that would be common in commercial production.
- Selection amongst highly heterozygous lines where genotypic worth can be severely masked by dominance effects (inbreeding species only).

Selection in the early stages is most ineffective for quantitatively inherited traits such as yield, quality and durable disease resistance. If early generation selection is purely a random (or near-random) reduction in the number of lines that are to be tested at the intermediate stage, then it is questionable whether this operation will merit the time and resources to complete the task. It has therefore been suggested that a more effective protocol would result from growing fewer breeding lines and doing no selection at the earliest stages. The reduced effort and resources at the early generation stage could therefore be used more efficiently to screen more genotypes at the intermediate stage (where efficiency due to replication and larger plots is more effective).

An alternative method of reducing the numbers involved in early to intermediate selection is available. This procedure involves the identification of the most attractive cross combinations from the many that would be possible, assuming that there is greater probability of obtaining a successful cultivar from the most desirable cross combinations. Having identified the '*best*' crosses, then maximum effort and resource can be directed to screening individual recombinants from within these specific crosses, while the '*poorer*' crosses are completely discarded. This process is called **cross prediction**.

7.5.1 Univariate cross prediction

As we noted briefly earlier, methods of predicting the properties and distribution of recombinant inbred lines (derived by inter-mating homozygous parents) using early generations of crosses have been proposed by Jinks and Pooni.

They showed that, for any continuously varying character, the expected mean and variance of all possible inbred lines, derived by inbreeding following an initial cross between two homozygous parents, can be specified in terms of the components of means and variances as specified by biometrical genetics. For example, if an additive–dominant genetic model of inheritance proves adequate, the expected mean is m, the mid-parent value, and the expected variance of the inbred sample is V_A.

From the predicted mean and variance, we can determine many of the properties of the recombinant inbred lines that can be derived in a pure-line breeding programme, based upon the performance of generations in the early generation stages. In addition, the relative probabilities with which different pairwise crosses will produce inbred lines with particular properties can also be predicted and hence used as a selection criteria for reducing the number of breeding lines in a plant breeding programme.

The crosses that show highest probability of producing desirable recombinants can therefore be identified from those with a lesser chance of producing desirable lines. Rather than selecting individual genotypes at the early generation stage, the number of surviving lines can be reduced by selection of the superior cross combinations. Similarly, if the probability of a desirable recombinant is known from a particular cross, then this value can be used to determine the number of recombinants that need to be evaluated to ensure that 'at least one' is found. When a single trait is examined,

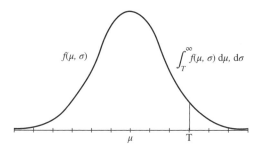

$f(\mu, \sigma)$

$\int_{T}^{\infty} f(\mu, \sigma) \, d\mu, \, d\sigma$

μ T

Figure 7.12 Illustration of univariate cross prediction technique.

this procedure of estimating and using genetic parameters is called **univariate cross prediction**.

Univariate cross prediction has been applied to a number of inbreeding species based on the initial work of Jinks and Pooni with *Nicotiana rustica*. Predictions of the proportion of recombinant inbred lines that will transgress a predefined target value (T) are based on the evaluation of the integral:

$$\int_{T}^{\infty} f(x_i) \, dx_i$$

where the trait of interest is normally distributed and the function $f(x_i)$ is based on m, the mean of all possible inbreds for a character and A, the additive genetic variance for the character (Figure 7.12).

Estimation of m and V_A

The additive genetic components of the expected variance (V_A) can be estimated from a number of different sources initiated by a cross between two pure-breeding lines. Methods that have proven reliable include:

- producing a sample of inbred lines from a pairwise cross between two parents by rapid-cycle single seed descent or doubled haploid techniques;
- the standard P_1, P_2, F_1, F_2, B_1 and B_2 families;
- a triple test cross, where a random sample of the F_2 from each parent cross-combination is backcrossed to P_1, P_2 and the heterozygous F_1 (the theory of the triple test cross is beyond the scope of this book);
- evaluation of a random sample of F_3 families.

If a sample of inbred lines from a number of different crosses are grown in properly designed assessment trials, it is possible to estimate m, the average

performance of all possible inbred lines from each cross, and A, the additive genetic variation for each cross. The average performances of the inbreds are a direct estimate of m, and the variance between inbreds (σ_g) in the sample is a direct estimate of A after error variance has been removed.

It is possible to calculate the proportion of lines expected to transgress a predefined target value (T) by using:

$$\frac{T - m}{\sigma_g} \quad \text{or} \quad \frac{m - T}{\sigma_g}$$

depending on whether the predictions are for values greater than (or equal to), or less than (or equal to) the target value set – where T is the target value, m is the mean of all inbred lines and σ_g is the genetic standard deviation.

Following the calculation of the probability integral from the predicted equations, the expected proportions of transgressive segregants can be obtained from tables of the normal probability integral.

In cases where it is easy to obtain a sample of inbred (or near-homozygous) lines from a large number of crosses (i.e. by using a doubled haploid techniques or rapid-cycle single seed descent), then this method will produce excellent predictions as there are no dominance effects to complicate estimation.

In many instances, however, it is not possible to produce inbred lines quickly and cheaply on a practical level in a breeding programme, and so cross prediction will involve estimating genetic means and variances from early generations of crossing designs.

Although the triple test cross will provide breeders with the best estimate of additive genetic means and variances, it requires a great deal of time and effort to complete. A similar effort will be required to obtain these estimates using the standard P_1, P_2, F_1, F_2, B_1 and B_2 method. Both these mating designs, therefore, have merit for genetic investigation but may have limited use in a practical plant breeding situation where many hundreds of cross combinations need to be screened.

Evaluation of a random sample of F_3 families from each cross under investigation offers a more practical approach. Approximate genetic parameters can be estimated from the mean of a random sample of F_3 families and from variation between families derived from a common cross.

For example, consider a single cross $(P_1 \times P_2)$. Then F_1 seed would be grown to produce F_2 single plants. A random sample of these would be harvested and a single plot grown from each of the (say 20 to 25) single plant plots. These plots would be evaluated to obtain the average performance of the families, and variation between families would also be estimated.

The mean (average performance of families) of the F_3 plots would be:

$$m + 1/4d$$

and the true variance of the F_3 family means $(\sigma_{F3}{}^2)$ would be:

$$1/2V_A + 1/16V_D$$

assuming that the additive–dominance model of inheritance is adequate to describe the character of interest. Therefore the following approximations can be made:

$$\text{mean of } F_3 \text{ families} = \text{mean of all possible inbred lines}$$

$$2\sigma_{F_3}^2 = V_A$$

Both estimates will of course be accurate only if dominance effects are relatively small in comparison to additive effects. In cases where $[d]$ is large, then the average of all possible inbred lines can be estimated by growing the parental lines in the prediction trial and estimating m as $(P_1 + P_2)/2$. Alternatively, when $[d]$ and D are large then they can be estimated by including a bulk sample of the F_1 or F_2 in the prediction trial. This latter option will of course also offer a better estimate of m.

If the parental lines are included in the cross prediction trial in which the F_3 families are evaluated, then it is also possible to carry out a crude scaling test to determine the dominance components and effect from:

$$[(P_1 - P_2)/2] - F_3$$

Using these predictions of the additive genetic mean and the additive genetic variance, we can predict the probability that a single inbred line selected at random will equal or exceed a predefined target value (T) by:

$$\frac{T - \text{mean } F_3}{\sqrt{(2\sigma_{F_3}^2)}}$$

Similarly, the probability that a single inbred line, taken at random, would be equal to or less than a predefined target value (T) would be:

$$\frac{\text{mean } F_3 - T}{\sqrt{(2\sigma_{F_3}^2)}}$$

Following the calculation of the probability integral from the predicted equations, the expected proportions of transgressive segregants can, as above, be obtained from tables of the normal probability integral.

Setting target values

A number of different options are available in setting target values upon which the predictions are based. These include:

- To include a set of control cultivars in the same experiments where the genetic parameters are estimated and using the means (and variances) to set target values which would be needed by a new cultivar.
- To include all parents used in crosses and setting target values of either the mean of P_1 (the parent with highest expression) or $(P_1 + P_2)/2$, the average parent value or m.
- If the F_1 progeny are included in the evaluation trial, then the means of the heterotic F_1 values are often used as target values in cross prediction.

Predicted from number in sample

When a sample of inbred lines is produced, then a second option for determining the frequency of particular recombinants has also shown promise. This method simply involves counting the number of inbreds within the sample which exceed the given target value.

Obviously, the accuracy of this method will be directly proportional to the sample size of inbreds used upon which the counts are made. Several researchers have shown, however, that even relatively small samples (around 25 lines) can still provide useful prediction results.

Use of cross prediction in clonal crops

Initially cross prediction was not considered as a tool in the breeding of clonal crops. This may perhaps have been due to the fact that heterozygosity is not a problem in the selection procedure. In other words, although the initial seedlings in, say, a potato breeding programme are all genetically unique and highly heterozygous, they are subsequently multiplied clonally and so are fixed in the sense that they are the genotypes that can be commercially exploited. However, it has been shown in a number of clonal crops that early generation selection suffers from all the inefficiencies found in inbred cultivar development.

In clonal crops there are just as many difficulties incurred in trying to identify desirable lines in the early generation stages (i.e. seedling stage and first clonal year stage) where only single plants are evaluated.

At the Scottish Crop Research Institute, research has shown severe inefficiencies in the traditional method of selection used in the early generations. Work resulting from this prompted an examination of cross prediction methods which might prove an effective alternative to the recurrent phenotypic selection used. A sample of 25 seedlings from the 200 grown from each of 8 cross families (C1 … C8) was evaluated for breeders' preference (a visual assessment of commercial worth on a 1 to 9 scale with increasing value attributed to increasing commercial worth) by four breeders independently. Table 7.9 shows the progeny means and within-progeny variances (σ_p) used to determine the frequency of clones that would exceed a preference score of greater than 5 on the 1 to 9 scale.

Table 7.9 Progeny means of breeders' preference ratings and within-progeny variances (σ_p), estimated on 25 progeny from each of eight potato crosses (C1 to C8), and the univariate probability that a genotype chosen at random from each cross will exceed a breeder's preference rating greater than 5, on the 1 to 9 scale.

Cross	Mean	σ_p	Predicted > 5
C1	4.36	1.52	0.337
C2	4.01	1.65	0.274
C3	3.61	1.50	0.176
C4	4.17	1.23	0.251
C5	3.04	0.91	0.015
C6	3.68	1.52	0.192
C7	4.21	1.36	0.281

Table 7.10 Relative ranking of eight potato crosses (C1 to C8) based upon cross prediction of breeder's preference that genotypes chosen at random from each cross will exceed a breeder's preference rating greater than 5, on the 1 to 9 scale, and the number of selected breeding lines from each cross that was selected in the 4th, 5th and 6th selection stages in the breeding programme at the Scottish Crop Research Institute.

Cross	Rank	Selected to stage:		
		Four	Five	Six
C1	1	15	3	2
C2	3	9	3	2
C3	6	1	0	0
C4	4	2	0	0
C5	8	1	0	0
C6	5	11	6	1
C7	2	12	7	3
C8	7	0	0	0

Seed tubers from a sample of 200 clones from each cross were increased without selection to a stage where a large amount of field-grown seed tubers were available and the 1,600 genotypes were evaluated for three years in a breeding programme. Selection of the populations was based on all the characters (yield, quality and appearance) that are normally assessed in the breeding scheme. Table 7.10 shows the number of selected clones that survived four, five and six rounds of selection from each cross. Also shown is the rank of each cross based upon the 25 seedlings and cross prediction of breeders' preference.

Obviously, there were more highly desirable clones from crosses C1, C2 and C7 which were ranked first, second and third on the univariate cross prediction of glasshouse grown seedlings.

In the potato, cross prediction was investigated with higher numbers of crosses (204) in a similar way as a result of this first study. Results from the larger study were in agreement with those shown above. This has prompted several other breeding organizations to change the means by which they reduce clonal numbers in the early generations of potato and sugarcane breeding schemes.

Use of normal distribution function tables

The area under a unit normal distribution (a normal distribution with mean of zero and variance of one) is frequently tabulated in statistical tables. It is common for the area to be given from $-\infty$ to the required target value (T). The whole area from $-\infty$ to $+\infty$ is of course equal to 1, so the area from T to $+\infty$ can be obtained by subtracting the table value from 1.

For example, from a given cross (A × B) estimates were obtained of the mean ($m = 12.0$) and genetic variance ($s_g = 16.0$, therefore $\sigma_g = \sqrt{\sigma_g^2} = \sqrt{16} = 4$). What would be the probability that a recombinant will exceed a set target value of 14? To solve this we have:

$$\frac{T - m}{\sigma_g} = \frac{14 - 12}{4} = 0.5$$

Using the unit normal distribution tables and a value of 0.50 we have the probability of $-\infty$ to T to be equal to 0.6915. The actual probability we want is $1 - 0.6915 = 0.3085$.

Therefore given the above genetic parameters we would expect that 30.85% of all possible recombinants from the cross will have a greater (or equal) value than the target value.

Consider the same set of parameters ($m = 12$, $\sigma_g = 4$) but now with a target value of 11 (i.e. a target value less than the progeny mean). We now have:

$$\frac{T - m}{\sigma_g} = \frac{11 - 12}{4} = -0.25$$

Looking this value up from the tables we have a probability value of 0.5987. In the example above we then subtracted this value from one to obtain the correct probability. In this case however, this $(T - m)/\sigma_g$ has a negative value, and so our required probability is $1 - (1 - \text{table value})$ which is in fact simply the value obtained from tables.

In summary, four possibilities exist:

- If the probability that a recombinant is to **exceed** a target value and $(T - m)/\sigma_g$ is **positive**, then the probability is $1 -$ the tabulated value
- If the probability that a recombinant is to **exceed** a target value and $(T - m)/\sigma_g$ is **negative**, then the probability is simply the tabulated value
- If the probability that a recombinant is **less than** the target value and $(m - T)/\sigma_g$ is **positive**, then the probability is simply the tabulated value
- If the probability that a recombinant is **less than** the target value and $(m - T)/\sigma_g$ is **negative**, then the probability is $1 -$ the tabulated value.

Univariate cross prediction example

To consider further possible problems involved in selection of the 'best' cross combinations, consider the following example.

Below are shown the means and genetic variances of crop yield (t/ha) of four barley crosses (A, B, C and D). Also grown in this prediction trial were five control lines. These controls are all commercially grown cultivars predominating in the region where new varieties are to be grown. The average yield of the controls was 21 t/ha and the variance of the controls was 7.5 t/ha. Which of the crosses should have greatest emphasis in a breeding programme where high yield is the major selection criteria?

Cross	Mean	Genetic variance
A	20.0	24.135
B	22.0	8.111
C	21.5	19.245
D	18.0	26.051

First it should be decided whether selection is based only on the mean performance of the crosses. If this is the case then the answer is quite simple. Greatest emphasis should be placed on cross B, followed by cross C. The remaining two crosses perhaps should be discarded as their average performance is less than the average of the control cultivars.

However, it should be noted that the four crosses have different mean yield values, but there are also large differences in the genetic variance (σ_g^2). Would our decision now change if we consider the 'best' cross based on the mean and variance?

First it is necessary to set a target value upon which the prediction is to be based. As there were a number of commercial cultivars included in the cross prediction trial, it may be useful to use as the target value the average performance of the controls (21 t/ha).

When this target value is used, the four crosses were estimated to have A = 42.4%, B = 63.7%, C = 54.4% and D = 27.8% of their progeny to be greater (or equal) to the mean of the control entries. Again, if these were the criteria used then the greatest emphasis should be put on cross B, followed by

Table 7.11 Progeny mean, genetic variance and genetic standard deviation of progeny from four different parent cross combinations, and the probability that genotypes chosen at random from each cross will exceed a specific target yield.

Cross	Mean	σ_g^2	σ_g	$T = 21$	$T = 24$	$T = 26$
A	20.3 (3)	24.13	4.91	0.424 (3)	0.209 (3)	0.111 (2)
B	22.0 (1)	8.11	2.85	0.637 (1)	0.242 (2)	0.081 (3)
C	21.5 (2)	19.24	4.38	0.544 (2)	0.284 (1)	0.153 (1)
D	18.0 (4)	20.05	5.10	0.278(4)	0.119 (4)	0.058 (4)

cross C, A, and lastly D (the same order as when only the means were used).

If this were an actual breeding scheme, however, it may be several years before a selected genotype from any of these crosses would become a commercial cultivar in agriculture. It would therefore be wise to set our target higher than the controls, as it might be expected that in several years' time newer and higher-yielding lines would be available. As the variance of the controls is available, we can use this to set a target value which is the mean of the controls plus the standard error of the controls (i.e. $21.0 + 2.74$), which would be approximately 24 t/ha.

With this target value we have A = 20.90%, B = 24.20%, C = 28.42% and D = 11.90%. Now there has been a change in the cross rankings (in parentheses in Table 7.11), with cross C now giving the highest probability of lines exceeding 24 t/ha. Cross B is now ranked second, but there is little difference between the probabilities of cross B and cross A.

If the target value is further increased (say to the control mean plus twice the control standard error), we would have a target value approximately equal to 26. With this target value the ranking of the four crosses is C, A, B and D. Now, while cross C has the highest probability, cross A has a higher probability of producing a genotype exceeding the target value (11.12%) than cross B (8.08%).

In conclusion, therefore, it is clearly important that univariate cross prediction is based upon the mean and genetic variance of a cross. When target values are relatively close to the progeny mean values, then not surprisingly the mean of each cross will be a large factor in the cross prediction. As target values are increased then the genetic variance becomes a more important factor in determining the probability of desirable recombinants. Finally,

in the above example it should be noted that it is the genetic standard error (σ_g) that is used in the estimate and not the genetic variance (σ_g^2). Even when there are large differences in genetic variance (compare cross B and cross D), the cross with highest mean value was always the better choice for further breeding work, despite the high variance of cross D. However, if the target was taken to a greater extreme, then the relationship would cease to hold true, and, of course, it is extremely '*good*' genotypes that breeders are usually trying to identify.

7.5.2 Multivariate cross prediction

Despite the usefulness of univariate cross prediction in determining the frequency of desirable recombinants that would transgress a given target value, its use is limited because, by definition, only a single character is being evaluated. As we noted many times already, a new cultivar will not be successful because of high expression in a single character, but rather it needs to express an overall improvement in a number of morphological, pathological and quality characters combined with high productivity.

The problem of selecting the most desirable cross combinations can partially be overcome by considering a trait such as breeders' preference, which is based on a visual assessment of several characters simultaneously by a breeder. Indeed breeders' preference scores have been shown to give very similar results to multivariate index selection schemes.

Visual inspection of several characters simultaneously, to result in a single overall rating for each individual, has several limitations. In potatoes this form of assessment has been shown to have advantageous features when used in a plant breeding selection scheme. Breeders' preference scores in potato breeding are highly related to actual yield, number of tubers per plant, tuber size, tuber conformity, tuber disease and absence from defects. It has been shown that this type of evaluation does not have such a good agreement with other important characters such as seed size, disease resistance, yield, and so on. Similarly it is not possible to combine characters that are expressed at different times in the growth cycle. For example, it is difficult to consider pre-harvest characters such as flowering time, plant height or maturity if preference scores are recorded at harvest. In addition, it is difficult to combine morphological characters such as yield along with quality characters that may be assessed in a laboratory at a later stage. Thus it is usually necessary to consider selection for more than a single trait.

If more than one trait is to be considered in cross prediction studies, it is possible to treat each independently, carry out univariate cross prediction on each character, and examine the probabilities obtained to make decisions on the 'best' crosses. This would of course ignore the fact that the different traits are interrelated (correlated) and that the relationship between the traits is constant over all crosses involved. This may cause problems, and so it may be necessary to expand the univariate procedure to cover several different traits simultaneously.

Univariate cross prediction is based upon evaluation of the normal distribution function determined by the mean and genetic variance of each cross and a chosen target value (T):

$$\int_T^\infty f(x_i)\mathrm{d}x_i$$

Suppose that two characters are to be considered. The bivariate normal distribution of the data from these two traits can be described by the mean of each character (m_1 and m_2), the genetic variance of each character (σ_1^2 and σ_2^2) along with the correlation between the characters (τ). Given these five parameters it is possible to estimate the proportion of recombinants from the cross that will transgress a given target value for character 1 (T_1), and simultaneously transgress a second target value (T_2) for the other trait. This probability is given by:

$$\int_{T_1}^\infty \int_{T_2}^\infty f(x_1,x_2)\mathrm{d}x_1,\mathrm{d}x_2$$

where the function $f(x_1,x_2)$ is a bivariate normal distribution function based on the mean of both traits, the variance of both traits and the correlation between traits.

It is easy to extend this to cover n different traits by evaluation of the integral:

$$\int_{T_1}^\infty \int_{T_2}^\infty \cdots \int_{T_n}^\infty f(x_1,x_2,\ldots,x_n)\mathrm{d}x_1,\mathrm{d}x_2,\ldots,\mathrm{d}x_n$$

In this case the function $f(x_1,x_2,\ldots,x_n)$ is a multinormal distribution function based on the means

($m_{1,n}$) of all n traits, the genetic variances ($\sigma_{1,n}^2$) of all n traits and the genetic correlations ($\tau_{i,n}$) between all n traits.

Given the various means, variances and correlations, it is possible to obtain bivariate, trivariate and multivariate probability estimates from statistical tables. These tables are, however, not commonly presented in standard statistical tables (as for example the ones that usually show unit normal distribution function, t-tables, χ^2 tables or F-tables). In addition, use of the tables that do exist can be complex and would require detailed description.

Parameters used in multivariate prediction are estimated using the same design types (i.e. triple test cross, F_3 prediction) that were explained previously for univariate predictions.

When it is necessary to estimate multivariate probabilities, computers offer an easier alternative. Computer software is available (although not commonly commercially) which projects a probable value, when the means, variances, correlations and target values are entered. To our knowledge there is software that can handle up to seven traits simultaneously – how the software manages this need not detain us here!

Similarly, it is beyond the scope of this book to try to explain in more detail the theory of estimating these probabilities. It is sufficient to understand the basic concept and to be aware of the usefulness of the procedure as applied in cross prediction techniques. You should, however, be aware that the procedure exists and that multivariate cross prediction can offer a powerful tool for selection in plant breeding.

Example of multivariate cross prediction

The eight crosses (C1, ... , C8) that were evaluated for breeders' preference (see earlier in univariate prediction) also had tuber yield, tuber size and the number of tubers recorded for the 25 progeny from each cross. Tuber shape was also visually assessed. The means and variances of each trait were estimated along with the correlations between traits for each cross. Based on these statistics the probability that genotypes would exceed target values for each character simultaneously was estimated using a computer software package called POTSTAT. The relative rankings of the multivariate predicted values (MV.rank) are shown in Table 7.12 along with

Table 7.12 Relative rankings of eight potato crosses (C1 to C8) based upon multivariate cross prediction (MV.rank) and univariate cross prediction (UV.rank) of breeder's preference that a genotype chosen at random from each cross will exceed a breeder's preference rating greater than 5, on the 1 to 9 scale, and the number of selected breeding lines from each cross that was selected in the 4th, 5th and 6th selection stage in the breeding programme at the Scottish Crop Research Institute.

Cross	MV.rank	UV.rank	Selected to stage		
			Four	Five	Six
C1	2	1	15	3	2
C2	2	3	9	3	2
C3	6	6	1	0	0
C4	4	4	2	0	0
C5	8	8	1	0	0
C6	5	5	11	6	1
C7	1	2	12	7	3
C8	7	7	0	0	0

the rankings of the univariate cross prediction of breeders preference (UV.rank) and the frequency of desirable clones selected from a large sample in the fourth, fifth and sixth round of selection.

There is good agreement between the multivariate predictions, based on four traits and the univariate prediction based on breeders' preference. Therefore the preference scores were highly related to yield, number of tubers, tuber size and tuber shape. There was also very good agreement with the predicted worth of each cross and the number of clones that indeed show commercial value in the advanced selection stages.

Observed number in a sample from each cross

It is possible to obtain good multivariate probability estimates by observing the frequency of individuals in a small sample that exceed given target values. The difficulty in using observed frequencies is related to sample size. The accuracy of the predictions will be directly related to the sample size examined. When the frequency of desirable recombinants is low (i.e. when high target values are used), then larger samples will need to be examined.

Similarly, if there are low correlations between traits of interest, sample sizes will need to be relatively large to carry out prediction effectively.

Use of rankings

It has been noted above in the univariate case that the relative importance of the different selection pressures imposed can affect the results of prediction. In multivariate prediction, three types of parameter are used: means, variances and correlations. In cases where the progeny means predominate in the prediction equations, which is often the case, then very good estimation of progeny worth can be obtained by summing the relative rankings (based on the phenotypic mean of the cross) for each of several traits.

Consider the potato example shown above, where four traits, yield, tuber size, tuber number and shape, were used to assess progeny worth of eight potato crosses. The ranking of the eight crosses based on the multivariate normal probability (MVP) and those obtained by summing the relative rankings of each character were:

Cross	Sum rank	MVP
C1	1	2 =
C2	2 =	2 =
C3	5	6
C4	4	4
C5	8	8
C6	6	5
C7	2 =	1
C8	7	7

As can be clearly seen, ranking each individual trait and then summing the rankings of each cross can be a good estimate of the commercial worth of different cross combinations.

7.6 Parental selection

Selection in a plant breeding programme takes two forms:

- Selection of superior parents.
- Selection of desirable recombinants which have resulted from inter-mating chosen parents.

Selection of the desirable recombinants has been covered in the foregoing sections of this book. We will now consider parental selection.

Parents used in plant breeding programmes are chosen from a wide range of possible genetic material. In general, however, parents are of three different types:

- Unadapted, or relatively unadapted, genotypes that possess one (or more) character which is not available within more cultivated types (i.e. parents from plant introduction accessions and germplasm gene-banks).
- Adapted genotypes which may be new (or old) cultivars from other breeding programmes.
- Genotypes selected from within the breeding programme. Often these lines will become new cultivars, but it is not uncommon that advanced selections (which have only a few slight defects that would render cultivar introduction infeasible) are used as parental lines.

It may appear strange that recombinant selection was discussed prior to parental selection (i.e. putting the cart before the horse). In actual practice there is no definite order of either selecting parents or selecting offspring. A large majority of parents used in plant breeding scheme are derived from selections within the breeding programme.

Parental selection is therefore a cyclic operation where parents are selected, inter-mated, recombinants screened from segregating populations and these, in turn, are used as parents in the next round of the scheme.

In deciding which parents are to be used in a breeding scheme, there are basically two types of evaluations possible:

- phenotypic evaluation;
- genotypic evaluation.

This information could have been derived from experiments or assessment trials carried out within the breeding scheme or by other organizations (e.g. available in germplasm databases).

7.6.1 Phenotypic evaluation

Phenotypic evaluation is often the first stage of parental selection. New genetic material is continually being added to the available parental lines within a plant breeding programme.

It can be of great benefit to a breeder, and will add increasing knowledge of possible new parental lines, to grow parental evaluation trials. When a potential new parent is made known, often the information of commercial worth is lacking. Information may

be available from a database management scheme, although often these data are related to performance in different geographical regions to the target region of the breeding programme.

Phenotypic parent evaluation trials can be carried out at relatively low cost. When many new parents are to be assessed then specific trials can be arranged. These trials should be organized with the same criteria of good experimental design that any evaluation requires. In cases where only one or two new parents are to be considered, it is often useful to include these genotypes as controls in one of the breeding trials.

7.6.2 Genotypic evaluation

Although it is often only after a new parent has proven to have some merit on its phenotypic performance that the more time-consuming and detailed examination of genotypic worth is carried out, it must be pointed out that the possibility that a valuable genotype (in terms of becoming a parent) might hide within a poor phenotype still exists. Nevertheless, because of limited resources and a lesser probability of a poor phenotype proving to be a good genotype, most effort is devoted to further evaluating proven material to determine the true value of the parent in cross combination.

The most common means to determine the genetic potential of new parental lines is to examine a series of progeny in which the new parent features as one of the parents. From these studies it is possible to determine the *general combining ability* of a genotype and to use this information to select the most desirable parental lines.

General combining ability is an indication from the progeny of how a particular genotype, when crossed with a range of other genotypes, responds. The most effective means of determining general combining ability is by diallel crossing designs, where the variation observed in the diallel table is divided into general combining ability of the parents used, and specific combining ability (all variation that cannot be explained by an additive model of parental values). But as noted before, this does limit the number of lines that can be examined.

General combining ability can be estimated from other crossing designs. The simplest of these involves evaluating the progeny that are produced by crossing the potential parent with one or more tester lines. Tester lines are chosen because of past experience in producing worthwhile results. For example, a new parent may be crossed to a genetically productive genotype and also to one with little genetic worth. The contribution of the parent can be observed by examination of the offspring from the crosses.

General combining ability can also be estimated using North Carolina crossing designs. These, as noted earlier, are of two forms:

- *North Carolina I designs*, where a number of parent genotypes are crossed to one or more tester lines. In these designs it is not necessary to hybridize each parent to a common set of tester lines. From the design an analysis of variance can estimate general combining ability of parents, which is tested for significance against the testers within parents' mean square.
- *North Carolina II designs*, where a number of parent genotypes are all crossed to one or more tester and each parent is crossed to the same tester lines. In this case an analysis of variance can partition the total variation into differences between general combining ability of the parents, general combining ability of the testers and an interaction term (parents × testers) that indicates specific combining ability.

In addition to the statistical analysis of diallels and other crossing designs, more information can be obtained from genetic analysis. The most common means to achieve this is from a Hayman and Jinks' analysis where within-array variances (V_r) and between-array covariances (W_r) are used to estimate the proportion of dominant to recessive alleles for a given character. Hayman and Jinks' analysis can be used therefore to choose parents with high phenotypic performance and with a known degree of dominant alleles.

When a suitable cross prediction scheme is employed in the early generations of a plant breeding scheme, it is possible to use the cross prediction data to indicate which specific parents have the highest probability of producing desirable recombinants. A potential new parent is hybridized to a number of different genotypes and the progeny are examined to estimate the mean of all crosses in which the parent is used and the genetic variance of all crosses in which the parent appears. These data can be used in the same way as illustrated earlier in cross prediction.

Similar probabilities based on several traits simultaneously can provide useful indicators of the exact worth of a new parent without waiting several years to determine this potential from survivors in a selection scheme.

At the Scottish Crop Research Institute, cross prediction at the seedling stage of the potato breeding programme became standard practice. Each year between 200 and 300 crosses were evaluated in cross prediction trials. The hybrid combinations with the highest probability of producing a new cultivar were increased, while the less desirable cross combinations were discarded.

This scheme, in addition to providing information on the commercial potential of each cross combination, was also used to determine the suitability of individual parents. The progeny mean and genetic variance of each parent was used in the prediction estimation. Shown in Table 7.13 are the rankings of nine parents based on this system along with the number of desirable recombinant lines that resulted from crosses involving the parents. Despite one or two changes in rank order, there was good agreement between the predicted and observed indicators. The differences that were observed could be explained by morphological characters (i.e. Cara has a pink eye and there was positive emphasis to select these types) or pest preferences (i.e. Maris Piper has nematode resistance and only clones that possessed the resistance were continued, irrespective of other characters).

7.6.3 Parental combinations

Having decided on a set of parental lines, the next decision to be made is *how many crosses should be made* and *which combinations will yield the best results?*

If there is a means by which large numbers of crosses can be evaluated, then many crosses will yield better results than if only a few are tried. However, it should be noted that there is little to be gained by making more crosses than can be screened in an effective manner.

In a straightforward commercial context, and for a short-term objective, only a limited number of crosses are to be considered; one simple and effective strategy is to cross the best with the best. Therefore identify the phenotypically and genetically best parents, intercross these and select amongst their progeny.

Table 7.13 Univariate probability that a genotype taken at random from a segregating progeny with a common parent will have a breeders' preference greater than 4, on a 1 to 9 scale, relative ranking of that probability, along with the proportion and ranking of genotypes that survive the 4th selection stage at the Scottish Crop Research Institute.

Clone	Cross prediction of preference > 4	Rank	Percentage of year 4 clones selected in year 7	Rank
Maris Peer	69.17	1	17.69	1
3683.A.2	62.57	2	11.76	2
Pentland Ivory	60.40	3	7.11	4
G.6755.1	59.74	4	6.29	5
Cara	57.34	5	10.95	3
8204.A.4	54.42	6	5.13	6
Pentland Squire	49.25	7	3.18	7
Dr Macintosh	47.37	8	0.00	8 =
Self crosses	37.99	9	0.00	8 =

Many breeders use the strategy of combining complementary parents, for example, to inter-mate a high-yielding, poor-quality line with a low-yielding but high-quality line. In theory this type of combination could allow the selection of a high-yield, high-quality recombinant. However, what is often achieved is an average yield with average quality. It is usually necessary to use some form of pre-breeding where the high-yielding line is first crossed (or backcrossed) to a high-quality line, parents are selected several times, and these are in turn used in final cross combinations.

Similarly when a character is introduced from a wild or unadapted genotype, it may take many rounds of backcrossing to get the desired character into a commercial background before the trait is introduced into a new cultivar. This is, of course, where some of the newer techniques of genetic transformation and marker-assisted selection offer other alternatives to the breeder.

7.6.4 Germplasm collections

Germplasm is the basic raw material of any plant breeding programme. It is important that genetic diversity is maintained if crop development is to continue, so that new characters can be introduced into already existing cultivated genotypic background.

Why is genetic variability so important? Well, it has been continually stated in this book that without genetic variability, there can be no gain from selection. A further need is related to the appearance of new forms of pest or disease or new husbandry techniques, or new environmental challenges. If a new disease became important in an agricultural area to which all known cultivars were susceptible, then it may be possible to identify new sources of disease resistance from closely related wild or weedy species.

There is a growing awareness of reduced germplasm resources throughout modern agriculture. The greater use of monoculture crops and homozygous cultivars has greatly reduced the genetic variability within our agricultural crop species. For example, at the turn of the century, farmers growing cereal crops were propagating land races that were a collection of genetically different types grown in mixture. Land races have been replaced, in most countries, by homozygous lines or hybrids, and much of the variability that existed has already been lost. Disease epidemics can also greatly reduce genetic variability within a crop species. The potato blight that affected Western Europe (not just Ireland) had the effect of greatly reducing the genetic variability within European potato lines (in addition it triggered a famine killing around one million people in Ireland alone in the nineteenth century). Worldwide organizations have been formed with the specific aim of conserving germplasm which is accessible to breeders to search for new traits that are not available within the cultivated crops. Bioversity International is one such organization which coordinates germplasm collection activities on an international level. Bioversity International is part of the Consultative Group on International Agricultural Research (CGIAR) Consortium.

In addition to the national germplasm collections and Bioversity International other organizations in the CGIAR Consortium centres, such as the International Potato Research Center (CIP, Peru), the International Center for Maize and Wheat Improvement (CIMMYT, Mexico), the International Rice Research Institute (IRRI) and the International Crops Research Institute for the Semi-Arid Tropics (ICRISTAT, India), have remits to maintain germplasm collections on specific crop species.

Germplasm is available within the US from the Plant Introduction System. Genotypes are made available from the location that maintains plant introduction material, or from one of the regional stations. Some of the major crop responsibilities of each station are as follows:

- **North Eastern Regional Plant Introduction Station, Geneva, New York**: perennial clover, onion, pea, broccoli and timothy;
- **Southern Regional Plant Introduction Station, Georgia:** cantaloupe, cowpea, millet, peanut, sorghum and pepper;
- **North Central Regional Plant Introduction Station, Ames, Iowa**: corn, sweet clover, beets, tomato and cucumber;
- **Western Regional Plant Introduction Station, Pullman, Washington:** alfalfa, bean, cabbage, fescue, wheat, grasses, lentils, lettuce, safflower and chickpea;
- **State and Federal Inter-regional Potato Introduction Station (IR-1), Sturgeon Bay, Wisconsin**: potato.

Germplasm in itself is of little use to a plant breeder unless there is information regarding the attributes or defects of different genotypes. Most germplasm collections have associated data banks detailing and classifying material within the collection. For example, the Germplasm Resources Information Network (GRIN) is a computerized database containing information on the location, characteristics and availability of accessions within the plant introduction scheme. This information is available to any breeder through the Database Management Unit of the Agricultural Research Service, Plant Genetics and Germplasm Institute, Beltsville, Maryland.

Think questions

(1) Selection in a plant breeding programme can be divided into three different stages: early generation selection, intermediate generation selection and advanced generation selection. Briefly, state the major differences between these three stages.

(2) A 10×10 half diallel crossing design was used to examine the potential of each of ten parents in a wheat breeding programme. F_1 families along with each of the parents were grown in a properly designed field trial. Yield (kg per plot) was recorded and from the data, array means, within-array variances (V_rs) and between-array covariances (W_rs) were estimated. These statistics are shown below. From this information, determine which three of the ten potential parents would be best suited for use in a cultivar development scheme designed to increase wheat yield. Explain your choices.

Parent	Mean	V_r	W_r
1	32.3	234.0	215.2
2	15.2	45.2	19.2
3	21.3	150.4	298.1
4	24.5	17.3	19.2
5	29.3	100.1	90.1
6	17.4	210.9	250.3
7	16.3	199.0	99.1
8	19.1	26.9	15.6
9	17.1	292.8	211.2
10	22.3	379.5	403.1

(3) In a bean breeding programme, it is desired to produce new cultivars that are short (dwarf) in stature and have oval shaped beans (rather than round). Both bean shape and dwarfism are controlled by single loci with two alleles at each. The allele for oval beans is dominant to the alternative allele for round beans. The dwarfing allele is recessive to the non-dwarf (tall) allele.

A cross is made between two homozygous parents where parent 1 is dwarf with round beans (*ttrr*) while parent 2 is tall with oval beans (*TTRR*). A number of F_1 plants are selfed to produce F_2 seeds from which 1,600 F_2 plants are grown. Assuming independent assortment of genes, outline a selection scheme that will result in harvesting F_4 seeds that are homozygous for oval beans and dwarf stature. Indicate the number of plants selected at each selection stage.

(4) You are a potato breeder working in a publicly funded organization. Due to the breakup of the former Soviet Union, you have inherited 500 potato lines from the Siberian Potato Research Centre. Briefly outline (using diagrams if necessary) how you would screen these genotypes for their potential as new parents in your breeding programme.

(5) In a winter rapeseed breeding programme, 3,000 near-homozygous breeding lines were evaluated for yield and oil content from a properly designed field trial. The correlation coefficient between yield and oil content was found to be $r = 0.41$. Using independent culling you want to select the highest-yielding 10%, and select for oil content retaining the best 15%. How many genotypes would you expect to be: (1) selected for high yield and high oil content; and (2) selected for high yield but discarded for high oil content?

(6) Independent culling can be a very effective means to select for two or more characters simultaneously and can be easily applied to breeding data. Under what circumstances would independent culling not be very effective? If index selection is used, list two methods that could be used to weight the traits.

(7) 50 progeny from ten potato crosses were evaluated for yield in a properly designed field trial. From the results the following cross means and genetic variances (σ_g^2) were obtained:

	Mean	σ_g^2
Cross 1	25.60	27.34
Cross 2	19.33	19.40
Cross 3	27.71	13.31
Cross 4	12.06	10.39
Cross 5	13.11	15.63
Cross 6	26.56	14.21
Cross 7	27.45	25.69
Cross 8	19.21	15.21
Cross 9	23.21	39.13
Cross 10	19.32	17.31

Also grown in the same trial were ten commercial cultivars. The average performance of the commercial cultivars was 20.14 and the standard deviation was 3.26. Using univariate cross prediction procedures, determine which three crosses should be used in breeding for cultivars that would have high yields. Rank your choices as first, second and third and include the probabilities used for your decision, under the following criteria:
- Greater than average controls plus one control standard deviation.
- Greater than the average control plus twice the control standard deviation.

Using the data presented above, how many clonal lines would need to be raised from Cross 6 to be 90% certain of having one line that would have a yield potential exceeding 3 standard deviations from the control mean?

(8) A half diallel crossing design is carried out involving 30 homozygous parents. F_3 progeny from each of the 435 possible cross combinations were raised and grown in a two replicate yield trial (planting F_3 and harvesting F_4 seed). Also grown within the trial were all 30 parental lines. Based on the yield results from the trial it was found that the regression of progeny performance (y) onto mid-parent value (x) was:

$$y = 0.832x + 0.002$$

Also from this trial, the phenotypic variance of the F_3 families was found to be $\sigma_p^2 = 65.216$, and the average yield of all 435 F_3 lines was 301.5 kg. What would be the expected gain from selection if the F_3 families were selected at the 10% level (i.e. discard 90% of families), and what would be the expected yield of these 43 selected lines at F_4, according to yield performance? Would you expect the same response to selection if the 43 selected F_4 families were further selected for yield the following year? (Explain your answer).

(9) In chickpea breeding it is assumed that the correlation between yield performance of F_4 families in one year and F_5 families in the following year is $r = 0.57$. 4,000 F_4 breeding lines of chickpea were evaluated for seed yield in a properly designed field trial. All

4,000 lines were re-evaluated at the F_5 stage the following year. If the highest yielding 10% were selected based on F_4 performance and the highest yielding 15% were selected based on their F_5 performance, how many of the original 4,000 lines would you expect to be selected based on both F_4 and F_5 performance? Explain what Type I and Type II errors mean in the context of selection. Estimate the Type I and Type II errors expected by selecting F_4 families for yield at the 10% level and selecting F_5 families at the 15% level.

(10) Parental selection can generally be divided into two different types. What are these types and, briefly, indicate the differences between them.

(11) Describe the main features of North Carolina I, North Carolina II and diallel crossing designs. Explain the terms in the model for the analysis of a diallel according to the method described by Griffing:

$$Y_{ijk} = \mu + g_i + g_j + s_{ij} + e_{ijk}$$

and indicate the importance of these terms and the use of Griffing analysis in selecting superior parental lines.

(12) You have been appointed as Assistant Professor/Plant Breeder in the Crops Division of McDonalds University in Frysville, MD, United States. It appears that the breeding programme has been trying to select improved genotypes with decreased sugar content in the tubers. However, in the 10 years previous to your appointment, there appears to have been no genetic improvement resulting from breeding. Outline three reasons that could individually, or in combination, have caused this lack of response.

(13) In a series of properly designed experiments, a team of plant breeders produced estimates of narrow-sense heritabilities (h_n^2) for plant height and plant yield from two different segregating families of spring barley (95.BAR.31 and 95.BAR.69). Both families were grown on two farms (Moscow and Boise, in Idaho, United States) in three successive years (1993, 1994 and 1995). The h_n^2 values from each year and site are summarized below.

Character	Year	95. BAR. 31		95. BAR. 69	
		Moscow	Boise	Moscow	Boise
Plant	1993	0.20	0.22	0.83	0.86
height	1994	0.01	0.21	0.31	0.74
	1995	0.21	0.30	0.52	0.79
Plant	1993	0.11	0.25	0.52	0.57
yield	1994	0.02	0.15	0.12	0.46
	1995	0.21	0.24	0.21	0.48

Which family is likely to give better responses in a breeding programme, given equivalent selection intensities, and why? In such a breeding programme, which of the two characters (plant height or plant yield) is likely to give the better response to equivalent selection intensities, and why? Consistently higher average values of h_n^2 were obtained at the farm near Boise. Which site, if either, is likely to provide the more accurate estimate of h_n^2, and why? The heritability estimates (particularly those for plant height for 95.BAR.69 in Moscow) varied greatly over the 3-year period. What could be the cause?

(14) 4,000 F_3 lentil breeding lines were evaluated for yield and the highest yielding 500 genotypes were selected. The average yield of the 500 selections was 1,429 kg/ha, while the average yield of the discard genotypes was 1,204 kg/ha. A random sample from the discards and the 500 selected lines were grown in a properly designed F_4 trial where the average yield of the random genotypes was 1,199 kg/ha and the yield of the selected 500 lines was 1,362 kg/ha. Determine the narrow-sense heritability for yield at the F_3 of these lentil progenies.

(15) In a barley breeding programme for "high diastase power", it is known that the correlation between yield and diastatic power is $r = 0.60$. In this breeding programme 3,000 doubled haploids are evaluated for yield and diastatic power in a properly designed field trial. If you wish to retain approximately the 'best' 150 lines, what proportion would you discard based on yield and what proportion would you discard based on diastase power to achieve this selected number?

(16) In 1998, 10,000 F_3 head-rows were grown in an unreplicated, but randomized, design from *Bobby Z's Wheat Breeding Program*. At harvest, seed from all rows were thrashed and weighed. The average yield of all 10,000 lines was found to be 164.24 kg. After weighing, 200 rows were taken at random and retained irrespective of their yield performance. From the remaining lines, the 'best' 1,000 rows for yield were selected for further evaluation in the breeding scheme. The average yield of these selected lines was 193.74 kg. In 1999, the bulk F_4 seed from the 200 random lines and the 1,000 selected lines were grown in a randomized complete block design. At harvest, each plot was harvested separately and the yield recorded. From this 1999 trial it was found that the average yield of the selected lines was 168.11 kg, while the average yield of the 200 randomly chosen lines was 133.41 kg. Using the above information, determine the narrow-sense heritability for yield in this material. Explain how this estimated value of heritability might influence your selection strategy for future F_3 head-row selection stages.

(17) In a Douglas fir breeding programme you have only sufficient resources to screen segregants from two crosses each year. However, in 1995, you have been provided data of family mean (MEAN), phenotypic variance (VAR) and narrow-sense heritability (h^2) from each of six crosses. From these statistics, determine which two crosses you will concentrate your efforts on in 1995 – assuming that you will subsequently select at only the 10% level. Explain, briefly, your choices.

Cross code	MEAN	VAR	h^2
DF.33.111	22.3	16.7	0.45
DF.66.123	24.6	14.3	0.51
DF.97.37	28.1	6.3	0.47
DF.97.332	26.1	15.3	0.11
DF.99.1	22.5	10.2	0.84
DF.99.131	18.9	26.1	0.75

(18) 25 progeny from ten winter wheat crosses were evaluated for yield in a properly designed F_3 cross-prediction study under field conditions.

From the results, the following means and additive genetic standard deviations (σ_A) for seed yields were obtained for each cross:

	Mean	σ_A
92.WW.46	236.60	127.34
92.WW.53	199.33	191.40
92.WW.54	241.82	91.43
92.WW.61	142.00	119.10
92.WW.71	133.11	125.46
92.WW.74	236.73	102.14
92.WW.93	233.55	281.77
92.WW.108	201.22	106.63
92.WW.111	229.37	299.39
92.WW.116	169.11	119.32

Also grown in the same trial were five commercial cultivars. The average performance of the commercial cultivars was 211.10 kg and the standard deviation was 41.83 kg. Using univariate cross prediction procedures, determine which three crosses should be used in breeding for cultivars that would need to produce yields:

- Greater than the average of the controls plus **one** control standard deviation.
- Greater than the average control plus **twice** the control standard deviation .

Rank your choices as first, second and third and include the probabilities used for your decision.

(19) 500 dry pea F_6 breeding lines were evaluated for yield potential at two locations (Hillside and Nethertown). After harvest and weighing, the 15% highest yielding lines were selected at each location. When results from this selection were examined it was found that 39 lines were selected in the top 15% at each site. From this information, estimate the narrow-sense heritability.

In this same study, the following phenotypic variances and site means for yield (over all 500 F_6 lines) were:

	Hillside	Nethertown
Phenotypic variance	96 kg^2	124 kg^2
Site mean	27 kg	29 kg

Given these data and the heritability estimated above, determine the expected response to selection at 10% level.

(20) Describe the difference between general combining ability and specific combining ability.

The following are average plant heights (over four replicates) of a 5×5 half diallel (including selfs) between sweet cherry cultivars.

Golden Glory	112				
Early Crimson	72	53			
Sweet Delight	102	64	99		
Dwarf Evens	56	41	65	49	
Giant Red	130	100	109	107	115
	Golden Glory	Early Crimson	Sweet Delight	Dwarf Evens	Giant Red

Estimate the general combining ability of each parent and calculate the specific combining ability of each cross. Which parents would be 'best' in a breeding programme to develop cultivars with short heights?

(21) What is the difference between a cultivar with general environmental adaptability and one with specific environmental adaptability?

Eight yellow mustard F_8 breeding lines were grown at 20 locations throughout the Pacific Northwest region. Significant genotype $\times$ environment interactions were detected for yield by analysis of variance, and when a joint regression analysis was carried out. The following are line means and environmental sensitivity $(1 + \beta)$ values from the analysis.

Line	Overall mean	Environmental sensitivities
90.EW.34.5	1,234	0.55
90.JB.456	1,890	1.05
90.JB.562	2,345	1.34
91.HG.12	1,897	0.76
91.HH.145	1,976	0.52
92.22.12	2,567	1.42
92.AE.1	2,156	0.83
92.HK.134	2,152	1.72

Select the two '*best*' breeding lines that show general environmental adaptability. Select the two '*best*' breeding lines that show specific environmental adaptability.

8

Broadening the Genetic Basis

By far the most common procedure for creating genetic variability in plant breeding is to make controlled cross-pollinations between two (or more) chosen parents (Parental Selection in Chapter 7). However, there are other techniques available to plant breeders, which have been further developed over the last 40 years, that have made, or are starting to make, significant contributions to the production of new cultivars. These include: induced mutation; interspecific species hybridization; protoplast fusion; and plant genetic transformation – all of which have been used to increase the genetic variability available to breeders, therefore increasing their likelihood of success.

8.1 Induced mutations

The variation that is displayed between all living plants and animals, including all crop species, is the result of natural mutations at the DNA level, with subsequent recombination and selection occurring, much of it over millions of years. But this has also been accompanied by changes at a structural level of the genetic material, such as rearrangement within and between chromosomes.

Mutations result in the generation of additional genetic variation within plant species. It has been estimated that mutations occur naturally with a frequency of 1 in 1,000,000 (one in a million, 10^{-6}) per generation per locus. Most of these mutations are recessive and deleterious and so these new alleles

usually do not survive at anything other than a very low frequency in nature. However, if the mutation results in an advantageous effect, the genotype possessing the mutation may thus be more adapted to the environment compared with the non-mutant types, and hence will tend to leave more offspring. Over generations this therefore leads to an increase in the frequency of the new allele within the population. Unlike other more recent developments described in this chapter, mutations have been used explicitly as an additional source of genetic variation during the last 90 years or so.

Obviously, the extremely low rate of natural mutation and even lower frequency of *desirable mutation* events are such that natural mutation has had little impact on modern plant breeding strategies. However, natural occurring mutations are still utilized and commercialized in clonal crops like apple. One such example are "spur" apple cultivars developed from previous existing varieties, which appear as spontaneous mutations in the field and which can be propagated by apple breeders. Another, more famous example would be the potato cultivar 'Russet Burbank'.

In the mid-1920s it was discovered that X-rays could be used to induce high mutation rates, first shown in the fruit fly and later in barley. Plant breeders were quick to realize the potential of induced mutation and, with the inevitable over-enthusiasm for a new idea, mutation breeding became common practice in almost all crop species and in many ornamental flower breeding programmes.

Plant Breeding, Second Edition. Jack Brown, Peter D.S. Caligari and Hugo A. Campos.
© 2014 John Wiley & Sons, Ltd. Published 2014 by John Wiley & Sons, Ltd.
Companion Website: www.wiley.com/go/brown/plantbreeding

The aim of mutation breeding is to stimulate an increase in the frequency of mutation events within a crop species and then to select desirable new alleles from amongst the mutants produced. More basically, mutation breeding has been utilized to make minor, advantageous genetic changes in already established and adapted cultivars through induced mutation treatments, for example, by inducing mutations in a highly adapted crop cultivar and then screening the resulting mutated lines for a specific character of interest. By doing this it is hoped to retain the existing cultivar's adaptability while adding the mutated *advantageous* trait.

Mutagenesis derived lines in a plant breeding scheme are labelled according to the number of generations after mutagenesis has taken place. For example, the mutated generation is called M_0, while the generation immediately after the mutation treatment is termed the M_1. In appropriate species, these plants can be self-pollinated to produce an M_2 generation, and so on (in line with F_1, F_2, F_3, etc., for the more usual sexual generations).

8.1.1   Methods of increasing the frequency of mutation

In general terms there are two methods that have been used to produce an increased frequency of mutations in plant species: radiation and chemical induction, with the highest frequency of mutation derived cultivars being from radiation-induced mutants.

Mutations following exposure to radiation are produced by a variety of effects from physical damage through to disturbing chemical bonds. Two main types of radiation have been utilized to induce mutation in crop breeding schemes.

- **Gamma rays** are the most favoured radiation source in plant breeding and have been used to develop 64% of the radiation-induced mutant derived cultivars. Gamma rays are electromagnetic radiation with a high energy level and are produced by the disintegration of radioisotopes. The two main sources of gamma radiation for induced mutation are from cobalt-60 and caesium-137. Plant breeding programmes have used gamma ray radiation treatments applied in a single dose, or have treated whole plants with long exposure to gamma radiation.

- **X-rays** were the original radiation mutagen, yet have been responsible for only 22% of the cultivars released worldwide from mutation-induced breeding programmes. X-rays are produced when high-speed electrons strike a metallic target. X-rays are high-energy ionizing radiation which have wavelengths ranging from ultraviolet to gamma radiation. Mutations are induced by exposing seeds, whole plants, plant organs, or plant parts to X-ray radiation of a required frequency for a specific time. X-rays have to be handled by trained radiologists, and so they are not always easily accessible to plant breeders, who often have to rely upon medical facilities (e.g. hospitals) for mutagenic treatment of plants.

Other forms of radiation that have been used to induce mutation include neutrons (an electrically neutral elementary nuclear particle produced by nuclear fission by uranium-235 in an atomic reactor), beta radiation (negatively charged particles that are emitted from radioisotopes such as phosphorus-32 and carbon-14) and ultraviolet radiation (used primarily for inducing mutations in pollen grains).

Many of the mutagenic chemicals are alkylating agents that bind to cellular DNA and interfere with chromosome division, such as: sulphur mustards; nitrogen mustards; epoxides; ethylene-imines; sulphates and sulphones; diazoalkanes; and nitroso-compounds. The most commonly used chemical mutagens have been ethyl-methane-sulphonate (EMS) and ethylene-imine (EI). It should be noted that all mutagenic chemicals are highly toxic, as well as being highly carcinogenic, so at all times they must be exclusively handed by qualified personnel in appropriate facilities.

The frequency of mutation can also be influenced by the oxygen level in plant material (higher oxygen related to increased plant injury and chromosomal abnormalities), water content (also related to oxygen content) and temperature.

8.1.2   Types of mutation

Mutations can be conveniently classified into four types:

- **Genome mutations** where there are changes in chromosome number due to either addition or loss of whole chromosomes or sets of chromosomes.

- **Structural changes in the chromosomes** involving translocation (a chromosomal aberration involving an interchange between different non-homologous chromosomes), inversions (changes in the arrangement of the loci, but not in their number), deficiencies, deletions, duplications and fusions (reduction or increase in the number of loci borne by the chromosome).
- **Gene mutations**, often termed point mutations, where the change is in a single gene, and often the result of a single base-pair change at the DNA level. Because of the redundant nature of the genetic code, some point mutations do not affect the amino acid encoded by the triplet of nucleotides where the mutation took place. These are silent mutations.
- **Extranuclear mutations** where the mutational event occurs in one of the cytoplasmic organelles. The DNA involved includes plastids and mitochondria and means that the mutation will usually be transmitted from one generation to the next through just one of the sexes, usually via the egg cells (i.e. maternally inherited). One example of this form of mutation is cytoplasmic male sterility, common in many crop species.

8.1.3 Plant parts to be treated

Mutagenic agents can be applied to different parts of plants and still produce effects. Seeds are the most obvious vehicle for treatment, and can be treated with either chemical or radiation mutagens. Seed have been preferred by many breeders because seeds are more tolerant to a wide range of physical conditions, such as being desiccated, soaked in liquid, heated, frozen or maintained under varying oxygen levels, which allows their exposure to the various mutagens to be rather easily carried out. Seed treated with an induced mutagen will result in plants that are:

- Non-mutants, the same as the parent plant.
- True mutants, where a mutation event has occurred throughout the whole plant.
- Chimera, where only a portion of the resulting plant has been mutated.

Another attractive possibility in mutation breeding is to treat pollen grains with radiation or chemical mutagens. The major advantage of treating pollen grains is that they are easily collected in large numbers and can easily be presented to a radiation source. Pollen grains are effectively single cells, so induced mutation of pollen avoids the occurrence of chimeras. Pollen grains are also generally haploid, in terms of their genetic composition, and so this opens the possibility, in an increasing number of species, of using tissue culture treatment to lead to their direct development into plantlets – which, with suitable treatment, can be induced to double their chromosome number and give true breeding, homozygous lines that will express both recessive and dominant mutated alleles.

Treatment of whole plants is less common but can be achieved using X-rays or gamma rays. This is often carried out using small plants or plantlets but, for example in Japan, they have built a large facility (resembling a sports arena) with a large gamma source at its centre, and a large number of fully grown plants are exposed over varying periods.

The treatment of cuttings and apical buds with radiation or chemical mutagens can be effective in developing mutant types in new shoots and plantlets. An important factor is whether the meristematic region forms mutations, since this is the region from which the new propagules develop. Treating cuttings and apical buds has been particularly important in developing mutants in clonally propagated cultivars.

It is now becoming more popular to combine mutation with *in vitro* cell and plant growth. The idea centres upon mimicking the possibilities developed with micro-organisms and so often involves treating single cells with a chemical or radiation mutagen and screening regenerating plantlets. One very desirable approach is the use of selective media, which only allows the growth of specific mutant types. This has been useful in developing herbicide-resistant cultivars where a low concentration of the selected herbicide is added to the media, and so only herbicide-resistant cell lines multiply.

8.1.4 Dose rates

It is apparent that all mutagenic treatments are basically damaging to plants. When too high a dose rate is applied, all the plant cells may be killed. Conversely, if the applied mutagen dose rate is too low, then very few mutant types will be induced.

It is therefore first necessary to determine an appropriate dose rate to use. The optimal dose rate will change according to crop species, plant part exposed to the mutagen and its physiological state. Indeed, the first stage of most mutagenesis-based breeding is to determine the most appropriate dose rate to minimize adverse effects, yet still produce sufficiently high levels of mutation.

In simple terms, dose rate is equal to mutagenic intensity × time applied. In chemical mutagenesis this involves the concentration of mutagenic chemical and time that plant cells were exposed to the chemical solution. Intensity of radiation can be altered by varying the distance from the radiation source or by varying the radiation form. The dose received can also be adjusted by changing mutagenic intensity or exposure time (or both). It is common to investigate different dose treatments until one is found that allows about 50% of plants to survive the treatment. These tests are called lethal dose 50 or LD_{50} tests.

8.1.5 Dangers of using mutagens

Mutagenic chemicals and radiation are effective because they alter the genetic makeup of plants and create variation. They will, of course, similarly affect the DNA of plant breeders who are exposed to them! It is not therefore possible to overemphasise the importance of using appropriate safety procedures when handling any mutagen. As already mentioned, the facilities for applying mutagenic treatments (in this case mainly radiation) are not always directly available to the average plant breeder; specialized operators or personnel (e.g. hospital radiologists) usually carry out the actual exposure to the radiation.

To use chemical mutagenic agents safely requires a number of safety features, spelt out in many countries (and by most suppliers in safety/hazard assessments) by specific safety protocols. Staff using these chemicals should be aware of the advised risks and safety procedures. Minimum safety will probably require suitable gloves, protective clothing and safety glasses combined with compulsory '*Good Laboratory Management Practice*'. It is also important that procedures and equipment are in place to deal with appropriate disposal of chemicals, and to contain and clean up any accidental spills of mutagenic chemicals.

8.1.6 Impact of mutation breeding

Mutation-derived cultivars have been released as a direct result of mutagenesis, or have mutant genotypes as parents in traditional breeding programmes. Since the inception of mutation breeding, over 2,250 cultivars have been released worldwide (FAO/IAEA [Food and Agricultural Organization/International Atomic Energy Agency] Mutant Varieties Database). It should be noted, however, that a high proportion of mutation-derived cultivars that have been released have been ornamental plants and flowers rather than agricultural crops. Over 70% of these cultivar releases were developed directly from mutant breeding lines and many of them were developed and released in Asia, with 27% being developed in China and 11% developed in India. Mutation-induced cultivars are not quite as common in other countries, although over 125 mutant-induced cultivars have been released in the US and 32 in the UK (31 of which were barley cultivars) in the past 70 years.

The highest numbers of mutant cultivar releases were in rice (433), followed by barley (269), wheat (220), soybean (89), groundnut (47), maize (32), pea (32), cotton (24) and millet (24). Only 46 mutant fruit cultivars were released over the same period. Mutant genes were developed mainly for dwarf stature, improved disease resistance, stress resistance, herbicide resistance, and improved grain or oil quality.

Although many have argued that these released cultivars have made little impact on our agricultural crops, some major positive impacts cannot be denied. Semi-dwarf rice derived from mutation breeding has been cultivated on millions of hectares. The barley cultivar 'Diamant', developed as a gamma-ray mutant of 'Valticky', was selected to have the *ert* dwarfing gene. It has been estimated that over 150 cultivars have been released in Europe that have Diamant in their pedigree. In addition, the Scottish barley cultivar 'Golden Promise', which also has this mutant dwarfing gene, is arguably the best cultivar ever released in the country and has been a major contributor to the Scottish brewing industry. Similarly, mutant durum wheat occupied over 25% of the Italian wheat acreage in the mid-1980s. Finally, health concerns about '*trans*' fats' in our diets has prompted many food processors and others in the food industry to use non-hydrogenated vegetable fats, which are low in polyunsaturated fats. The first canola cultivar with low linolenic acid, 'Stellar', inherited the fatty acid desaturase gene from a German EMS-induced mutant line coded as M47. Other low linoleic acid mutants have subsequently been developed using microspore mutagenesis, and now ultra-low polyunsaturated canola oil cultivars with highly elevated oleic acid contents are commercially available.

8.1.7 Practical applications

Having decided which mutagenic agent to use and a suitable mutagenic treatment strategy (i.e. rate, time, and which plant part to treat), in reality the physical treatment of plant cells by a mutagenic agent to induce mutation is the easy part of mutation breeding. By far the most difficult aspect of mutation breeding relates to selecting desirable mutants while avoiding the subsequent detrimental effects of mutagenesis.

Mutagens are non-discriminatory and inevitably produce a complex mixture of mutations. Mutants selected as having the trait of interest may also have undergone some chromosomal rearrangements and structure changes as well as exhibiting non-genetic (or at least non-nuclear) aberrations. It is also rare that selected mutants have been genetically altered for the single gene of interest, and there may be multiple mutation events, most of which have a negative impact on the normal growth of the plant. In seed-propagated crops, sterile segregants need to be sorted out and discarded in the first round of selection. But even having done this, selected

mutants will have been altered in a number of different ways – most of them bad!

Consider the following example of using mutation breeding techniques in yellow mustard. The aim of the breeding programme was to screen a large number of mutants (derived by chemical mutagenesis using EMS) to produce lines that were low in seed meal glucosinolate content. After screening several thousand lines over a four-year period (using glucose-sensitive test tape), eight lines were identified that showed lower glucosinolates compared with the non-mutated parent genotype ('Tilney').

However, when these eight lines were grown in replicated field trials it was found that all mutant lines were considerably lower in yield, produced smaller and less vigorous plants and matured later than the parental genotype (Table 8.1). So although exposure to the mutagen had produced mutations that had affected glucosinolate content in the manner hoped for, it had also affected other aspects of the genotype such that the selected lines all appeared to have mutated for other important traits. Further crossing and selection was clearly necessary before an adapted cultivar could be developed.

It has to be pointed out that experience suggests that, apart from loss of function alleles, it is usually easier to *find* a new allele than to *create* one. It may be significant to note that several years of intense mutagenesis breeding effort to develop yellow mustard lines with very low glucosinolate content in the seed meal failed to achieve this objective. Interestingly, the year after the mutagenesis programme was stopped, a gene that almost eliminated seed

Table 8.1 Seed yield, oil content, ground cover, days from planting to flowering and days from planting to flower ending of the yellow mustard (*Sinapis alba*) cultivar 'Tilney', and eight EMS mutants selected from Tilney with modified seed quality.

Identifier	Seed yield (kg ha^{-1})	Oil content (%)	Percentage ground cover, 1–9 scale	Days to flowering (days)	Days to flower end (days)
Tilney	2,166	30.1	8.7	54	80
Til.M3.A	555	30.5	3.0	59	86
Til.M3.B	1,797	29.2	4.7	58	85
Til.M3.C	1,004	30.4	2.0	60	86
Til.M3.I	1,347	29.7	4.7	59	86
Til.M3.II	1,242	29.7	3.0	60	87
Til.M3.III	766	30.1	2.3	59	86
Til.M3.IV	1,480	31.0	3.3	58	86
Til.M3.V	1,215	31.0	5.0	58	86

glucosinolates in yellow mustard was identified within a wild *Sinapis alba* population from Poland.

Initially many plant breeders believed that mutation breeding would have a revolutionary effect on cultivar development. Although there are multiple examples, like those above, of success in mutation breeding, for many this revolution has not happened and it is unlikely it will do so. Indeed, many believe that mutation breeding is too unpredictable and should only be considered as a *last resort technique* when all other avenues have tried, because of its limitations.

The question must therefore be asked as to what are the circumstances in which mutation techniques can be useful? To address this question the following points should be considered.

- Mutagenesis is an indiscriminate process and generates large numbers of undesirable variants along with those that are wanted.
- Most successful mutant cultivars and mutation-derived cultivars have resulted in selection for characters controlled by single (or at best a few) genes. Quantitatively inherited traits are more challenging in mutation breeding and will require large efforts to achieve success, if indeed success is ever achieved.
- The frequency of desirable mutants is likely to be low, and it is essential that there are suitable selection techniques to screen the thousands of mutants generated to identify the few that have the required mutated trait. The most common example of mass selection in mutation breeding relates to herbicide tolerance. Most herbicide tolerance in crop species is qualitatively controlled, and many thousands of mutated breeding lines can easily be screened for tolerance by simply spraying the mutants with the herbicide of choice, on the premise that all those that survive carry a mutated form of the gene conferring tolerance. Other examples would include selecting mutants that exhibit morphological changes in plant structure (e.g. dwarfs), maturity, flower colour, qualitative disease resistance, or enhanced end-use quality (e.g. fatty acid profile or starch types).
- When desirable mutants are selected they are usually adversely affected by the mutagenic treatment, and it will be necessary to 'clean up' the mutant genotype either by recurrent

selfing and selection, or more likely by recurrent back-crossing and selection. Most mutations are recessive, and identification of recessive mutations is difficult due to them being masked by their dominant counterpart alleles. In diploid seed-propagated crops, recessive mutations can be identified by selfing or inter-mating mutated lines, albeit the frequency of lines that are homozygous recessive for the desirable gene mutation will be very low. For obvious reasons, selection of desirable recessive mutations in clonally propagated crops is considerably more difficult, and breeders require excessive effort in selfing mutant lines, or hope for the extremely rare event where both (all) alleles at a given loci have the same recessive mutated gene.

The unpredictable nature of mutagenesis raises the question as to whether it is possible to 'direct' the mutational effects towards changing only characters of interest and to affect them in rather particular ways. The first possibility for 'direction' of the effects that can be exploited is that different mutagens have different forms of action, as noted earlier. Some induce point mutations and so are more likely to produce a particular array of effects, while others are likely to induce grosser structural chromosomal changes. An even more specific array of possibilities arises from the potential to induce site-specific mutagenesis. Consider our increasing ability to identify particular genes, to clone or synthesize these and introduce them back into the plant species (or, of course, to another) – this clearly raises the potential to 'mutate' these DNA sequences and reintroduce them.

One lesson that was learned from the early efforts of mutation breeders is that it is necessary to have clear objectives, which are biologically reasonable, if success is to be achieved. However, even in cases that have been well organized with realistic objectives, the effort that was required to sort out the desired products in a useable form was often greater than would have been required to achieve the same results from a more traditional hybridization breeding scheme.

In plant breeding the need for new sources of variation and mutations (natural or induced) will feature as part of future breeding efforts. Therefore mutation breeding has a very real place in cultivar development, but it would be unwise to base a complete variety development programme on mutagenesis.

In summary, a mutagenesis breeding programme must deal with large numbers of mutated lines so that the low frequency of desirable mutations, in an acceptable genetic background, can be selected. Similarly, a mutation scheme must offer a quick, cheap and effective selection screen to identify the few desirable mutants.

8.2 Interspecific and intergeneric hybridization

Another method of increasing the genetic diversity of a crop species is by interspecific or intergeneric hybridization. When sources of variation for a character of interest (e.g. disease or pest resistance) cannot be found within existing genotypes in a species, it is logical to look at related species or genera and examine the possibility to introgress traits from them into the one of interest.

Interspecific hybridization refers to crosses between species within the same genus (i.e. *B.rapa* × *B.oleracea*), while intergeneric hybrids are crosses between different genera (*Triticum* × *Secale*).

The probability of developing a successful new cultivar is related to the frequency of desirable (or undesirable) characteristics in the parents used in hybridization. The most commonly used parental lines will be adapted cultivars or highly desirable genotypes from within breeding programmes. When a character of interest is not available within this gene pool, then the obvious next step is to screen other lines that are not as adapted, or indeed in wild genotypes, in an attempt to identify expression of the desired trait in them. If the character cannot be identified within this wider germplasm source, then breeders spread their search wider and will screen related species in an attempt to find a natural genetic source.

Successful interspecific or intergeneric hybridization should therefore be considered when:

- the desired expression of a character of interest is not available within the gene pool of adapted genotypes, or their unadapted counterparts from the same species;
- acceptable expression for this trait has been shown to exist within a related species or genera;

- it is possible to introgress alleles from the related species into the cultivated species.

Successful interspecific crossing depends on two factors: obtaining viable seeds from plants in the first generation of the cross between the species (and later generations), and eliminating undesirable characters also introduced from the donor species. One, or both, of these factors may be the major determining factor in the actual success of gene transfer between species by this approach (see later for the possibilities using genetic transformation).

8.2.1 Characters introduced to crops from wild related species

A high proportion (over 80%) of genes introduced to our crop species through interspecific or intergeneric hybridization relate to pest and disease resistance. This trend continues today, whereby wild species related to our crop species are continually being screened and evaluated to identify new genes for resistance to crop pests and diseases. Resistance to grassy stunt virus was introgressed from *Oryza nivara* to cultivated rice. A number of late blight resistance genes have been transferred from *Solanum demissum* into potato cultivars. In addition, most new potato cultivars released in the European Union contain the H_1 gene conferring resistance to potato cyst nematode (*Globodera rostochiensis*) transferred from *S. verni*. Cabbage seedpod weevil resistance has been transferred to rapeseed through intergeneric hybridization between *Brassica napus* and *Sinapis alba*. More recently, genes conferring resistance to Hessian fly, a major insect pest of wheat in the US, have been transferred from *Aegilops tauschii*. Several rust-tolerant genes have been introgressed into wheat bread from related *Triticum* species.

In addition, significant efforts to develop genetic tolerance to wheat stem rust race Ug99 through the introgression of gene Sr39 existing in goatgrass (*Aegilops speltoids*) have been carried out. Sr39 provides tolerance to multiple stem rust races. The use of molecular markers helped to eliminate over 95% of the goatgrass DNA from wheat × goatgrass genotypes, as well as to monitor the introgression of this gene into elite wheat breeding lines.

Other traits that have been transferred through interspecific or intergeneric hybridization include abiotic stresses, drought tolerance, heat tolerance and salinity tolerance. Examples of enhanced yield or quality characters from wild relatives into crop plants, not surprisingly, are very rare because many different loci would need to be introgressed, which increases the probability of also passing undesirable loci. Nevertheless, there are successful stories about the use of wild relatives to enhance the pool of genetic variation existing in crops; using wild *Brassica oleracea* resources, scientists at the John Innes Centre (Norwich, UK) have been able to significantly enhance in broccoli (*Brassica oleracea*) the accumulation of glucosinolates – secondary metabolites that play a cancer-preventive role in humans. The first commercial cultivars with these enhanced levels of glucosinolates have already been released to the market in several countries.

8.2.2 Factors involved in interspecific or intergeneric hybridization

In order to make a successful hybrid involving two different species, there needs to be some degree of compatibility between the parents used. A number of factors need to be addressed to ensure successful gene introgression.

The first stage, which must be achieved, is that the male and female gametes from the different genotypes must unite to form a zygote. Failure at this stage can result from:

- inability of pollen grains to germinate on the receptive stigma of the female parent;
- failure of pollen tubes to develop successfully and grow down the style, or non-attraction of the pollen tube towards the ovary;
- inability of male gametes that do reach the embryo sac to actually fuse with the egg cell;
- inability of the nuclei from pollen and egg to fuse.

All of these aspects are related to fertilization barriers, and a number of techniques can be used to overcome the problems at each stage. *In vitro* fertilization (i.e. using excised organs) can sometimes be used to overcome problems in the first two barriers listed above.

Success in interspecific hybridization may be unidirectional (i.e. style length differences that mean that pollen tubes fail to reach the ovary) and these can be overcome by attempting the reciprocal cross. Therefore successful hybrids might be possible from the mating A × B, but difficult or unsuccessful when tried as B × A. A good example of this is seen in the cross *Brassica napus* × *B.oleracea*, which will produce viable hybrid seed if the cross is carried out in this direction (i.e. *B. napus* as female). If, however, *B. oleracea* is the female, then very few or no seed is produced (without using tissue culture techniques).

Cross incompatibility resulting from failure of pollen grains to germinate and develop pollen tubes is associated with proteins on the pistil that interact unfavourably with proteins in the pollen. In some cases this reaction has been overcome by mixing pollen from the donor species with compatible pollen from the female species (so called "mentor effect").

Pollen tubes often fail to reach the ovary (or 'miss' the ovary) due to the physical differences in style lengths between the different species. This can sometimes be overcome by mechanically reducing the style length of the longer style parent, although this will only be successful if the shortened pistils remain receptive (i.e. as in maize). In the extreme case the complete pistil can be removed and pollen applied directly into the ovary, but this usually requires *in vitro* techniques.

Once fertilization is achieved the problems are not necessarily over. When two species differ in ploidy level the resultant offspring are usually sterile, and it may be necessary to first reduce or increase the ploidy of the parents prior to crossing. In potato, potato cyst nematode resistance was identified in *Solanum verni* (a close relative to cultivated potato, *S. tuberosum*). *S. verni* is diploid while *S. tuberosum* is tetraploid. Two methods of successful hybridization have been achieved:

- Doubling the ploidy level of *S. verni*, using colchicine, and then carrying out interspecific hybridization at the tetraploid level.
- Producing dihaploids from *S. tuberosum* and crossing the two species at the diploid level. Progeny from the hybrid cross are then doubled to the tetraploid level using colchicine or by spontaneous doubling arising during growth *in vitro*.

A similar manipulation of ploidy in interspecific crosses in potato was used to introgress late blight resistance (*Phytophthora infestans*) into cultivated potato cultivars. The source of blight resistance was found in a wild relative of the cultivated potato (*S. demissum*), which is a hexaploid. A small proportion of tetraploid progeny can be obtained by crossing dihaploid *S. tuberosum* (see above and in haploid section) with the hexaploid *S. demissum*.

If attempts to obtain hybrid seed by means of sexual crossing fail, then somatic fusion (protoplast fusion) can provide a realistic possibility. Isolated wall-less cells (protoplasts) can be induced to fuse and affect the production of a somatic hybrid, and hence facilitate genetic transfer between two species if it can be followed by regeneration of whole plants. The resulting somatic hybrids will have the combined chromosome number of both parents (e.g. as in allotetraploids), so it may be necessary to first reduce the ploidy of parental lines or reduce the ploidy of hybrid combinations produced. However, the most difficult aspects of this technology are:

- being able to regenerate plants from protoplasts, even without fusion;
- selecting fused heterokaryons from unfused or self-fused parental protoplast.

8.2.3 Post-fertilization

After fertilization, failure of seeds to develop and/or to reach maturity can result from embryo and/or endosperm abortion, or failure in the stages of embryo or fruit development to complete their necessary development to give mature seeds.

Successful fruit and flower retention after fertilization can be a simple function of a dependency on having a sufficient number of developing embryos. In some interspecific or intergeneric hybrids the number of fertilized ovules is too low to stimulate mature fruit development, and growth regulators (e.g. gibberellic acid) have been used as a means to encourage fruit retention. It has also been suggested that increasing the frequency of developing seeds in fruits (by applying a mixture of compatible and incompatible (mentor) pollen) can be used to avoid flower or fruit abscission.

Many interspecific or intergeneric hybridizations fail as a result of post-fertilization factors, which cause embryo or endosperm abortion. It may, however, still be possible to obtain hybrid plants despite abortion by using *in vitro* techniques such as:

- ovule culture, where the complete ovary is removed from the plant and aseptically transferred to a growth media chosen where the

nutrients are provided and thus seed development can proceed;
- embryo rescue, where immature embryos are excised from the ovary and transferred to growth media; the chosen media therefore replaces the natural endosperm (which may not have developed or has aborted).

Sometimes a combination of both techniques is necessary to successfully achieve interspecific hybrid seed production. Early in the embryo development the ovary is removed and cultured *in vitro* to achieve embryos, which are then transferred when of a suitable size in order to allow successful rescue and culture.

Finally, it is not uncommon in hybrid crosses that rather than resulting in hybrid combinations the resulting seeds develop as matromorphic plants, which are thus derived exclusively from the maternal genotype. This characteristic has been developed to advantage in producing homozygous lines (doubled haploid lines) (e.g. *Hordeum vulgaris* × *H. bulbosum*). The seeds from interspecific crosses should thus be checked to ensure that the matromorphs are discarded if the desire is to produce hybrids – but retained if this feature is being used to produce haploids of the maternal genotype!

8.2.4 Hybrid sterility

In many cases the F_1 plants resulting from interspecific crosses are completely (or partially) sterile. A common technique used to overcome sterility, caused by lack of chromosome pairing, is to induce chromosome doubling in the hybrid, and hence develop alloploids. When doubled, it allows each chromosome to have a homologue with which to pair at meiosis, and thus reduce the infertility problem.

8.2.5 Backcrossing

After interspecific hybridization the resulting progeny will generally contain a large proportion of undesirable characters from the donor species, along with the character it was wished to introduce. In such circumstances it is necessary to carry out several rounds of backcrossing to the host species, with selection for the new character, to obtain genotypes that will have commercial value. Any programme involving interspecific or intergeneric hybridization is therefore likely to be long term, but, of course, once you have a well-adapted genotype you can use it as a parent for further variety production.

8.2.6   Increasing genetic diversity

Many crop species have a relatively narrow genetic base, and it is often advantageous to broaden genetic diversity by introgressing traits from related weedy species. Several crop species (i.e. rapeseed and wheat) have evolved as allopolyploids, whereby they contain complete chromosome sets from two or more diploid ancestors. Greater genetic diversity and variation can be achieved in breeding by resynthesizing the crop species from its ancient ancestors. One such example has been the use of alleles from *Brassica oleracea* wild accessions to increase the levels of glucosinolates, in broccoli.

8.2.7   Creating new species

It is even possible to create new crop species by intergeneric hybridization, but despite the possible attraction of this there are very few instances where successful new crops have resulted. However, three notable examples are:

- **Triticale**, which resulted from intergeneric hybridization between wheat (*Triticum*) and rye (*Secale*).
- **Tritordeum,** which resulted from intergeneric hybridization between wheat (*Triticum*) and barley (*Hordeum*).
- **Raphanobrassica**, which resulted from the intergeneric cross between kale (*Brassica oleracea*) and radish (*Raphanus sativus*).

When each of these new species was created there was great hope that they would almost immediately have high potential commercial value. However, in none of the cases has this full commercialization at a global scale occurred – at least not yet. Only in the case of triticale have a limited number of cultivars been released; nevertheless, on a worldwide scale its impact, in terms of food/feed production, has been negligible thus far.

8.3   Plant genetic transformation

The stable, heritable introduction of foreign genes into plants represents one of the most significant developments affecting the production of crop species in a continuum of advances in agricultural technology relating to plant breeding. The progress in this area has depended largely on two main components: the tissue culture systems having been

Figure 8.1   Winter biennial forms of yellow mustard (*Sinapis alba*) (left) and Indian mustard (*Brassica juncea*) (right) produced through intergeneric hybridization.

developed that provide an amenable vehicle for the transformation induction; and the use of molecular biology to isolate, modify and clone agronomically relevant genes and the transfer of these into plant species.

The term 'genetic transformation' comes from that used for a much longer period, bacterial transformation in which DNA has been successfully transferred from one isolate to another or between species of bacteria, and integrated into the genome. It was shown that the stably transformed bacteria then expressed the new genes and had appropriately altered phenotypes. In eukaryotes, transformation has a further complicating dimension, at least in many plant breeding contexts. The transforming DNA must not only be integrated into a chromosome, but it must be a chromosome of a cell or cells that will develop into germ-line cells. Otherwise the 'transformation' will not be passed on to any sexual progeny. Transient transformation refers to the event where the introduced gene is not fully integrated into the host genome.

Using plant transformation techniques has made feasible the transfer of (mostly) single genes (i.e. simply inherited traits) into plants, to have such transgenes properly expressed and to function successfully. Theoretically, at least, specific genes can be transformed from any source into developed cultivars or advanced breeding lines in a single step. Plant transformation, therefore, allows plant breeders to bypass barriers that limit sexual gene transfer, and exchange genes (and traits) from unrelated species where incompatibility does not allow sexual hybridization. These recombinant DNA techniques therefore apparently allow breeders to transfer genes between completely unrelated organisms. For example, bacterial genes can be transferred and expressed in plants. Nevertheless, up to this point the genes or gene components used to develop transgenic plants only include plant, bacterial and/or virus origins or genes modified in the laboratory. None of the transgenic plants currently commercially available to farmers in many parts of the world contain genes, or components thereof, of human and/or animal origin.

However, as we learn more and more about the DNA and hence the genes involved, the perspective of the picture somewhat changes. Increasing direct evidence of the presence in different species of the same basic gene, or clear variants of it,

demonstrate the greater conservation of genetic material during evolution than we expected. Also, we are being reminded of the existence of parallel natural processes for much of what we regard as novel. For example, bacteria, viruses and phages have already successfully evolved mechanisms to transfer genes, in just the same way that is regarded by some as being so alien! But clearly, the new techniques enable modern plant breeders to create new variability beyond that existing in the currently available germplasm on a different scale and in a different time-frame from that which was possible previously. Each plant carrying, at a specific chromosomal location, a given DNA construct introduced through genetic transformation is referred to as arising from a specific transformation event. Events with unique, stable chromosomal insertions are commonly selected for commercial development. Many such events have been developed, such as DAS-1507-1, MON-89034-3 and SYN-IR162-4 created by Dow AgroSciences/Pioneer, Monsanto and Syngenta, respectively; they all encode tolerance to Lepidopteran insects in maize. There are also an increasing number of commercial events developed in the academic/public sector, such as EMBRAPA in Brazil and USDA/Cornell University in the US. For instance, EMBRAPA (Brazil's National Institute of Agronomic and Livestock Research) has just developed event EMB-PV051-1 in common bean in Brazil, which renders this key crop tolerant to the bean golden mosaic virus.

In addition, a myriad of universities and research institutions in most countries now develop transgenic plants on a more or less routine basis for diverse research purposes. Although plant transformation has added (and some say dramatically) to the tools available to the breeder for genetic manipulation, it does, as with all techniques, have its limitations. Some of the limitations will reduce with increased development of methodologies, while others are those that are inherent to the basic approach.

At present, recombinant DNA techniques tend to be restricted to the transfer of a single or a few gene(s). This means that they are very effective where the trait is determined by one, or a few, gene(s). However, some agronomic traits showing continuous variation are actually controlled by a few loci showing rather large effects, but others by a myriad of genes with much lesser effects. So, for example, it is not clear how much yield itself, which could be argued is one of the most important characters of interest to farmers, can be manipulated by discrete steps of individual transformation events. Interestingly, however, recent reports do indicate the potential to transform with a number of genes (constructs) in one go with a reasonably high level of co-transformation. In addition, it is feasible to pile up, or stack, several transformation events through breeding or molecular approaches. In the case of maize, at the time of writing there are hybrids developed for several countries that carry up to eight genes, and soybean varieties are being developed that carry up to three stacked events.

It may seem obvious, but another restriction currently imposed is that the techniques are only readily applied to genes that have been identified and cloned. Despite the development of genomics, bioinformatic and high throughput sequencing approaches, the number of such desirable genes is still relatively modest, although increasing rapidly. What is becoming clear is that there is a deficiency in the knowledge of the underlying biochemistry or physiology of most agronomically relevant traits. Another feature is the rather limited availability of suitable promoters for the genes that are to be introduced. Optimizing the expression of transgenes for a particular developmental stage or in specific tissues has been fully recognized, and so the search for promoters now equals that for the genes themselves. In addition, it has been recognized that because of the random nature of the incorporation of transgenes into the host's genome, a large number of transformed plants need to be produced in order to allow the selection of the few that have the desired expression of the transgene without any detrimental alteration of all the characters of the host. Breeders have played a critical role in the development of novel breeding schemes and selection approaches, so enabling the selection of the best events which have then been used in breeding programmes.

8.3.1 A glimpse at the genetic transformation of plants

Before plant transformation can be used successfully in a plant breeding programme and cultivars are developed using recombinant DNA techniques, the following have to be in place.

- A desirable, agronomically relevant gene must be available for insertion into the target host plant. Therefore a DNA construct containing the gene that will confer a particular trait of interest must be developed and introduced to a plant, which then could be used in breeding programmes.

- There must be a suitable mechanism to transfer the gene to the target plant. In dicots (and also now increasingly cereals and grasses) the most commonly used vector has been *A. tumefaciens*. *Agrobacterium tumefaciens* is the causal agent of crown gall disease and produces tumorous crown galls on infected plants. The utility of this bacterium as a gene transfer system was first recognized when it was demonstrated that crown galls were actually produced as a result of the transfer and introgression of genes from the bacterium into the genome of the host plant cells. This natural mechanism for the introduction of genetic material that *Agrobacterium* has developed is an exquisite example of a naturally evolved mechanism to transfer genes from one species into another, also known as horizontal gene transfer. The crown galls produced by *Agrobacterium* induce in its natural hosts the synthesis of plant growth regulators and other compounds like opines which are used by *Agrobacterium* as a source of nitrogen and carbon. Vectors based on *Agrobacterium* used to develop transgenic plants are known as disarmed, since they lack the set of genes inducing crown galls. Therefore they retain the ability to transfer DNA, yet lack any genetic component that might represent a pathogenic, disease-causing factor.

Several physical or mechanical DNA delivery systems have also been developed, and have been particularly popular for monocots. The most common, at least initially, of these systems involved electroporation of protoplasts, although now particle bombardment is regarded as the method with the widest applicability. Particle bombardment involves the use of gold or tungsten microparticles (0.5 to 5.0 μm), carrying the DNA to be introduced on their surface, which are accelerated to speeds of up to several hundred metres per second. Particle bombardment has been used for gene transfer into a variety of target tissues including pollen cells, apices and reproductive organs.

Because transformation events created through particle bombardment tend to produce less precise or multiple insertions, *Agrobacterium tumefaciens*-based approaches are generally the methods of choice in current research efforts aimed at creating transgenic variation.

- A suitable construct has to be created that includes:
  - a promoter region that is recognized by the host, which drives the expression of an introduced gene and therefore its pattern of expression. These may be:
    - **Constitutive promoters** such as the 35S of cauliflower mosaic virus or *nos* from the nopaline synthase gene, which switches the gene on in all plant tissues.
    - **Tissue specific promoters** such as those from α-amylase (specific to the aleurone in some cereal species), patatin (specific to tubers) and phaseolin (specific to cotyledons).
    - **Inducible promoters** such as those from alcohol dehydrogenase I (induced by anaerobiosis); and chlorophyll *a/b* binding protein (induced by light).
  - a structural gene that encodes a protein involved in a trait of importance to farmers, such as insect resistance, herbicide tolerance, drought tolerance, virus resistance or quality.
  - a transcription terminating sequence at the 3′ end of the gene which signals the end of the transcription of the structural gene.

It should also be noted that molecular biologists might have to redesign the gene of interest, for optimal codon usage (i.e. codons most frequently used by the protein synthetic machinery in the host plants). This is because the codon usage varies among species, and it might be the case that a codon frequently used in the donor species to encode a given amino acid is less frequently used in the host species. In addition, it is sometimes necessary to include a 'transit' peptide in the genetic construction, when the expressed protein functions in a particular cell compartment (for instance the chloroplast). In addition, optimization of gene expression might be achieved by including short DNA sequences ('enhancers') in the promoter region.

Regardless of the delivery (vector) system used to transform plants, foreign DNA will be inserted into relatively few cells. A means, therefore, must be available to select, or at least significantly enrich, the cells that have been transformed. This is achieved through the use of a selectable gene marker along with the gene of agricultural, nutritional and/or industrial relevance that needs to be introduced into the host plant. In the early years of genetic transformation, the most usual

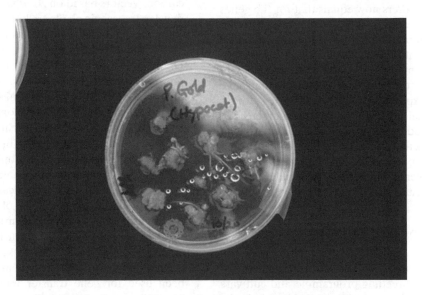

Figure 8.2 Regeneration of Indian mustard (*Brassica juncea*) plantlets from callus tissues.

selection agents used were **antibiotics** (mainly kanamycin or hygromycin) or **herbicides** (i.e. glufosinate or glyphosate). As the few cells effectively transformed also carried a selectable gene marker rendering them tolerant to such selection agents, they were able to thrive in selective culture media and give rise to adult, fertile plants. A growing number of transgenic plants lack selection markers, since depending on the genetic transformation strategy employed, they could be segregated out from the structural gene using routine breeding procedures.

Thus, genetic transformation enables the insertion of foreign DNA into the genome of a few selectable cells. A method must, of course, be available to develop intact mature, fertile plants from these single transformed cells. One of the major early barriers in transformation of a number of crop species was the inability to regenerate whole plants from single cells *in vitro*. However, over the years efficient tissue culture systems have been developed, enabling the generation of full, fertile plants from genetically transformed tissues in many species. In many dicots, leaf disks are transformed by infection with *A. tumefaciens*. Plantlets are then regenerated by tissue culture methods from the leaf disks. In many monocots, cultured cells or embryos are transformed by a suitable DNA delivery system (e.g. the particle gun) and intact plants are then regenerated from transformed cells, again in tissue culture. Other methods that have had success are the transformation of embryogenic cell cultures or protoplasts, followed by regeneration of whole plants.

Once whole, fertile transformed plants have been produced they need to be thoroughly characterized. This may be achieved by some or all of the following, often in series:

- Polymerase chain reaction (PCR) techniques can be used to detect the presence of transgenes, although these techniques alone cannot fully characterize insertions into the host genome.
- Southern blots are used to assess how the gene is integrated into the plant genome, and to estimate the number of gene copies that have been inserted. It is likely that current Southern blots will be replaced by electronic Southern blots in the years to come.
- Northern blots to assess gene expression through RNA transcribed from transgenes. Northern blots have been increasingly replaced by high throughput approaches such as quantitative reverse transcriptase-PCR.
- Western blots and enzyme immunoassays to characterize protein expression of the transgene in plant tissues

However, this is just the start, since after such tests have shown that transformed plants have the gene of interest, that it is integrated and that it is functional, this simply means it is worth proceeding further in the development of the GM crop. It then needs to be demonstrated that the gene (and expression of the trait it encodes) is stable; this first means demonstrating Mendelian transmission of the transgene through clonal or sexual generations in order to show that any progeny will inherit and express the gene as expected. Subsequently, extensive field and molecular testing begins to ascertain whether the expression of the gene has the desired effect on the phenotype, whether any other characters have been affected directly or by the transformation process and, of course, what the actual field performance is. Once field performance has been demonstrated, plant transformation events must undergo the thorough, extensive and detailed characterization analyses requested by regulatory agencies before their commercialization is allowed.

Genetic transformation does not constitute a breeding approach by itself. Instead, it is a very powerful source of novel genetic variation which, like any other source, must be channelled through an effective breeding programme, if successful cultivars carrying an agronomically relevant transformation event are to be developed and commercialized.

8.3.2 Some applications of genetic engineering to plant breeding

Weeds represent a significant constraint to crop production in most countries. Weed control is based on the use of herbicides, although there are still countries where it is based on human labour. If a plant was able to withstand herbicides while its associated weeds do not, it would address the needs of many farmers. Engineering herbicide tolerance into crops represents a novel alternative for conferring selectivity of specific herbicides. Two general approaches have been taken in engineering herbicide tolerance:

- altering the level and sensitivity of the target enzyme to the herbicide;
- incorporating a gene that will detoxify the herbicide.

As an example of the first approach, glyphosate, the active ingredient of herbicides such as 'Roundup', acts by specifically inhibiting the enzyme 5-enolpyruvylshimate-3-phosphate synthase (EPSPS). Tolerance to glyphosate has been engineered into various crops such as maize, soybean, canola, cotton and wheat by introducing genetic constructs for the production of a tolerant variant of the EPSPS from *Agrobacterium* strain CP4.

Genetic transformation has also been used to develop plants tolerant to other herbicides such as gluphosinate. In addition, novel systems of herbicide tolerance are being developed which would enable the use of herbicides such as 2,4 D and dicamba in weed control programmes based on the use of herbicide tolerant plants developed through biotechnology.

The production of plants that are tolerant to insect damage has been another application of genetic engineering with important implications for crop production. Transgenic plants with insect tolerance are widely used in maize and cotton in several countries already, and there are already transgenic soybeans able to withstand the attack of several Lepidopteran species.

Insect damage not only can create significant yield reduction by itself, but additionally the wounds it creates can open the way for opportunistic fungi to attack plants already damaged by insects. One route by which progress in engineering insect resistance in transgenic plants has been achieved is by using the genes of *Bacillus thuringiensis* (*B.t.*) which produces insect-specific endotoxins (so-called *B.t. toxins*). *B.t.* is a bacterium that produces a crystalline protein (also called Cry proteins) during sporulation which, when cleaved to the mature toxin peptide, produces paralysis of the intestine of specific target insects and so leads to their death. Thus it provides a useful and selective means of insect control. It is interesting to note that diverse microbial formulations of *B.t.* have been used for many years to control insects in organic food production systems. There are other sets of *B.t.* proteins that have also enabled the development of commercial insect

resistance transformation events, such as VIPs (vegetative insecticidal proteins).

However, with these resistance genes, an overall strategy is needed to prevent (or at least significantly reduce) evolution of resistance to the toxin in the insects that are being controlled. Insect resistance management programmes involve the use of refuges – field sections cultivated with an insect-susceptible version of the same cultivar – in order to reduce the likelihood of selection of insects resistant to *B.t.* toxins.

Another trait of interest is virus resistance, where significant resistance to tobacco mosaic virus (TMV) infection, termed 'coat protein-mediated protection', has been achieved by expressing only the coat protein gene of virus in transgenic plants. This approach has produced similar results in transgenic tomato and potato, although in some other cases a similar approach seems not to be as effective. It has been suggested that other genes, such as viral replicase genes among others, might provide an effective mechanism to control virus infection in plants. Recent work has shed light on the mechanism involved in such examples, and it has been demonstrated that these transgenic phenotypes are the result of RNA interference, an extremely specific DNA-homology based, naturally occurring mechanism, that enables the switching-off of the expression of genes.

There is a wide list of dicot crop species that have proved successful hosts for transformation, including alfalfa, apple, carrot, cauliflower, celery, cotton, cucumber, flax, horseradish, lettuce, potato, rapeseed, rice, rye, sugarbeet, soybean, sunflower, tomato, tobacco and walnut. In monocots, maize is leading the way, but is being followed by wheat, barley and rice.

Initial cultivar development using recombinant DNA techniques has focused upon modifying or enhancing traits that relate directly to the traditional role of farming (i.e. agronomic traits). These, as noted above, have included the control of insects and weeds, and mostly have relied on the expression of single bacterial genes in plants. What can be termed the 'first generation' of genetically engineered crops have been released into large-scale agriculture (including maize, tomato, canola, squash, potato, soybean and cotton) during the last 15 years, and other crops are already 'in the pipeline' as again mentioned above. More recent work has focused on modifying end-use quality (especially fatty acid, starch and vitamin precursors). In addition, transgenic maize plants aimed at industrial applications are already commercially available in the US. For instance, Syngenta developed transgenic corn hybrids carrying an enhanced form of the enzyme alpha amylase, which increases the efficiency of the corn-to-ethanol production process. Farmers growing such corn hybrids receive a premium price when delivering the corn to ethanol-producing facilities. The first transgenic maize plants able to successfully withstand certain levels of drought at the farmer's field level have been made available to farmers in the US by Monsanto.

It is expected that the availability of novel genes and genetic transformation methods will allow the development of transgenic plants carrying traits of increasing complexity and agronomic relevance. Nevertheless, much research is still needed before the availability of transgenic plants able to carry complex traits like grain yield or disease tolerance becomes a reality.

The genetic transformation of plants has had a significant and positive impact on global agriculture. The first transgenic cultivars were made available to farmers in 1994, and since then, over 1,500 million hectares have been grown with transgenic cultivars, an area almost 50% larger than the total land mass of Europe. Not only farmers running large farming operations have benefited from these, but also millions of small farmers around the world.

8.3.3   Cautions and related issues

There have been a number of concerns that have arisen over the past few years as the application of plant transformation technology has expanded, and particularly as new transgenic crops have been released into commercial cultivation. Plant breeders need to be aware of the concerns, as well as the local and international regulations, that apply to plants derived using recombinant DNA techniques, as well as to other forms of gene manipulation (e.g. induced mutations). In addition to the general social and environmental concerns, the breeder must consider the following:

- Is the level of expression of the genetically engineered crop plant sufficiently useful to agriculture to merit the time and resources that have gone into its development, and what will need to be done in terms of any further development?
- Is there a gene dose effect that will further optimize the results? In most cases what breeders require are single copy insertions which are inherited like any other Mendelian loci in their breeding programmes.
- What would be the best way to deploy the technology? Plant breeders are aware of the constant evolution occurring in organisms causing diseases and pests, which can often overcome single gene resistance when introduced into commercial cultivars. In this regard, transgenic crops must be managed within schemes encouraging good agricultural practices, such as the use of refuges already described in the case of crops resistant to insects, and the rational use of herbicides and agronomic practices such as crop rotation in the case of plants resistant to herbicides.
- It was at first, naïvely believed that plant breeders would be able to take an adapted cultivar and simply transform it with a specific gene to give an 'instant' new cultivar – one that had all the previous desirable characteristics but also with the transformed trait. It is now known that this is not in fact the case. New cultivars derived through plant transformation require the same rigorous field-testing prior to release that traditionally developed cultivars do. Multiple transformation events are necessary to ensure that one such event would give a transformed plant that has the desired level of expression for the altered trait, plus no deleterious epistasis interaction with the transformed gene or background changes. Therefore, genetic transformation represents a previously unforeseen source of jobs, professional development opportunities and research objectives to plant breeders.

Think questions

(1) Under what circumstances would a plant breeder choose to use mutation breeding?
(2) List two methods that have been used by plant breeders to induce increased frequency of mutagenesis in plants.
(3) List three problems that a plant breeder may encounter when using mutagenesis in a breeding programme.

(4) What is possible with plant transformation (recombinant DNA) techniques that is not possible with more traditional breeding techniques.

(5) You have been asked to use recombinant DNA techniques in your alfalfa breeding programme to increase profitability. Describe all the processes you will have to complete to achieve this goal.

(6) List three vectors that have been used to transform plants with genes from other organisms.

(7) Briefly describe three problems that plant breeders may encounter in developing transgenic cultivars.

(8) Name three types of mutation that can occur in plants, and briefly describe the features of each type. Outline two difficulties that might be problematic when using mutagenesis in a plant breeding programme.

(9) Interspecific and intergeneric hybridization can sometimes be useful techniques in plant breeding by introgression of characters and genes from different species. In an interspecific cross between *Brassica napus* and *B. rapa*, there was no evidence that any *B. napus* egg cells had been fertilized. List three factors that could have caused this non-fertilization.

9
Contemporary Approaches in Plant Breeding

9.1 Introduction

In Chapter 7, selection was introduced and the different forms of selection that are imposed in plant breeding described. However, it is increasingly common that plant breeders use biotechnology protocols to help select and multiply plants, including at the molecular level, and there are now a range of techniques available to plant breeders. These have been developing over the last 40 years and have made, or are starting to make, significant contributions to the production of new cultivars. These include *in vitro* propagation and genetic markers (mainly molecular markers) which are being increasingly used in breeding programmmes, to supplement the more traditional selection methods already described.

9.2 Tissue culture

A variety of techniques have been developed under the broad umbrella of tissue culture. It is not the intention to cover the details of these techniques but to briefly consider a couple of them, enough to give an idea of their application.

9.2.1 Doubled haploids

Establishing true-breeding, homozygous lines (as noted earlier) is an essential part of developing new cultivars in many crop species. These homozygous lines are used either as cultivars in their own right or as parents in hybrid variety development. Traditionally, plant breeders have used inbreeding, the process of selfing or mating between close relatives, to achieve homozygosity, a process that is time-consuming. Therefore the opportunity to produce plants from gametic, haploid cells has been the goal of many plant breeders, since this technique that can produce 'instant' inbred lines once the chromosomes of the haploids are doubled. The time needed to produce truly homozygous lines can be significantly reduced, which can result in reduced time to develop new cultivars. The additional costs and complexity involved in producing double haploid lines can often be justified by the speeding up of the breeding cycle they enable.

The genetic phenomenon critical to obtaining homozygous lines is the formation of haploid gametes by meiosis. During this type of cell division, the chromosome number is halved and each chromosome is represented only once in each cell (assuming the species is basically a diploid one). If such gametic, haploid cells can be induced to develop into plantlets (i.e. we encourage the development of the sporophyte – *note*: lower plants often have this as a specific phase of the lifecycle), a haploid plant can develop which can then be treated (usually with a chemical called colchicine) to encourage its chromosomes to double, to produce a completely homozygous genotype (a doubled haploid).

Plant Breeding, Second Edition. Jack Brown, Peter D.S. Caligari and Hugo A. Campos.
© 2014 John Wiley & Sons, Ltd. Published 2014 by John Wiley & Sons, Ltd.
Companion Website: www.wiley.com/go/brown/plantbreeding

From a quantitative genetics perspective, the main advantage of doubled haploid plants is based upon the elimination of heterozygous individuals, as the duplication of haploid cells would develop homozygous individuals (i.e. either *AA* or *aa*, using a single locus as example). By doubling the frequency of recessive homozygous individuals from the 25% found in an F_2 population, to 50% in a double haploid population, accordingly the probability of identifying superior breeding materials can be increased.

Techniques used for producing haploids *in vitro*

Although the generation of double haploids is a very attractive technique to many plant breeders, the natural occurrence of haploid plants is rare. However, the use of plant tissue culture has allowed the production of plants from gametic cells cultured *in vitro*.

Even though haploid plants can be regenerated from both male and female sex cells, it is generally the male cells (microspores or pollen) that have proven most successful in the regeneration of large numbers of haploid and doubled haploid lines. This is partly because of the ease with which pollen, as opposed to eggs, can be collected, and partly because it is simply that, in general, many more pollen grains than eggs are produced.

There are, of course, exceptions, and some examples include:

- The relative ease by which haploid barley plants can be produced from female sex cells. Interspecific crosses between cultivated barley (*Hordeum vulgare*) and the wild species *H. bulbosum* followed by *in vitro* culture of rescued immature embryos results in haploid plants as a result of exclusion of the *H. bulbosum* chromosomes during embryo development.
- Dihaploids from tetraploid potatoes have been produced in large quantities, using interspecific hybridization between cultivated potato (*Solanum tuberosum*) and a diploid relative (*S. phureja*). The cross of the tetraploid female *S. tuberosum* with the diploid male *S. phureja* would be expected to produce only triploid offspring – but it does not. Instead, the numbers of seeds obtained are relatively few and are predominantly tetraploid (as a result of the production of unreduced ($2n$) pollen from

S. phureja). Among the rest are some of the expected sterile triploids, but also some maternal dihaploids arising from the egg. Lines of *S. phureja* have been selected that produce a high frequency of dihaploid seed, greater than 70%. In addition such pollen parents have been selected to include a homozygous dominant embryo spot marker, which makes visual identification of the non-dihaploid seed easy.

- In bread wheat, its interspecific cross with maize pollen induces the production of haploid plants from female sex cells which can be recovered into full, fertile plants through the combination of tissue culture and chromosome duplication through colchicine.
- In maize, the crossing of inbreds with haploid inducer genotypes has recently enabled the cost-effective development of doubled haploid lines in an increasing number of both private and public breeding programmes.

There are other haploid induction mechanisms, but the most widely applicable are via anther or microspore (immature pollen grains) *in vitro*. The anthers, of course, are flower organs in which microspores mature into pollen grains under normal conditions (i.e. *in vivo*). The production of haploid plants from anther culture has been reported for over 200 species of higher plants. However, although the technique offers great potential for use in plant breeding programmes, the current examples of its application on a large, practical scale are restricted, but some are provided by commercial programmes in rice, wheat, barley, rye, canola, tobacco, potato, pepper and maize.

Probably the most successful examples of the use of doubled haploid plants are commercial programmes of canola and maize, which use double haploid plants on a regular and increasing basis. The most common approaches deployed to obtain double haploid plants in canola and maize are microspore culture and interspecific crossing with haploidy inducer stocks, respectively.

9.2.2 Some potential issues

Genotype dependence

One factor that has limited the use of anther culture in practical plant breeding programmes is that even the different variants of the protocol often show strong genotypic dependence. Therefore, if a protocol is identified which is effective for one genotype, that protocol often needs to be modified (sometimes to a large extent) to obtain success with another genotype, or, more appropriately, a range of genotypes.

Somaclonal variation

The techniques noted above involve producing plants that have been regenerated following *in vitro* culture. Variation can often be detected among such plants that are regenerated, and this variation has been termed somaclonal variation. The frequency of such variation has been suggested as reflecting the occurrence and length of the callus phase. In a haploid production scheme it is therefore essential that callus stages are kept to a minimum, so that any somatic variation is kept under an acceptable threshold.

Non-random recovery of haploid lines

An important need underlying the application of haploids in a plant breeding context is that the population of homozygous lines, derived from the chromosome doubling of the haploids produced, are a random representation of the gametic array possible. In other words, the possibility of unconscious selection occurring (effectively gametic selection) must be avoided. The genetic combinations recovered from haploid systems may be disproportionately composed of combinations from one of the original parents that were used to make the hybrid crosses from which the anthers were taken. An obvious possibility is that one of the parents showed a much greater propensity for regeneration in culture, or more responsiveness to plant growth regulators, and this would result in combinations with the genes that determined its response being represented more frequently in the population of gametes. In experimental studies it has been shown that non-randomness of the possible gametic combinations can occur and can be influenced by the culture protocols used.

Constraints on recombination opportunities

As already discussed in previous chapters, recombination processes are the main engine through which new genetic combinations and genetic variation is created, and therefore are of paramount importance to plant breeders. While a traditional, pedigree-based breeding programme offers several opportunities for chromosomes to recombine, in many cases double haploid lines are derived from gametes from F_1 populations. This could potentially create a constraint on the genetic variation creation ability of breeding programmes, which can nevertheless be addressed in several effective ways, such as inducing double haploid plants from heterozygous F_2 plants instead of F_1 generations.

Practical applications of haploids

Progress in evaluating gametic-derived plants under field conditions has been increasing dramatically; reports using numerous crops have indicated the importance of continued research in this area. A previously unforeseen advantage of using doubled haploid plants has been an improvement of the quality of field datasets in hybrid crops using testcross testing (Section 4.6.3), because it enabled testing of only homozygous genetic classes in the field. In the case of maize, for example, the traditional testing of testcross generations generated from F_2 populations was inherently based on a combination of homo- and heterozygous genotypes, whereas double haploid testcross generations derived from an F_2 population only carry two homozygous genotypes, lacking the heterozygous class.

It has been suggested that developing haploids in a practical breeding scheme will not be as effective as might be expected. In particular, concerns have been raised regarding:

- The cost of producing haploids.
- The inability to easily produce large numbers of homozygous lines through haploidy.
- The deleterious variation that is sometimes exposed as a result of recessive alleles in the original material or mutational/somaclonal variation induced as a result of the *in vitro* techniques.
- The dependence on the genotype of the parental material used in influencing the frequency of haploids produced – which often means that the very material the breeder most wants to use is non-responsive and haploids are not easily obtained.

Nevertheless, as refinements are made in methods and protocols, it is likely that it will become easier and cheaper to produce haploids on a routine basis. This will then mean that their impact on plant breeding programmes will be greater in the future. To date, very few cultivars have been introduced as a direct result of haploidy (perhaps those produced in China being the exception). However, there is little doubt that these techniques have added valuable information for plant breeders with regard to a number of aspects of genetics and tissue culture.

As noted earlier, one limitation to the widespread use of doubled haploids among many crops is the inability to produce large enough numbers of plants from culture. Regeneration frequencies are improving continuously, however, which will not only improve the applicability of the technique in a range of species, but will also increase the potential for their application in other ways. For example, the possibility of deliberately applying positive selection pressure during the culture phase for certain characteristics, that is, *in vitro* selection, will become even more attractive. Also this might be combined with induced mutagenesis during microsporogenesis, for example, allowing production of novel resistance to fungal or bacterial pathogens or to herbicides.

9.2.3 *In vitro* multiplication

In vitro multiplication of breeding lines can have two main benefits (particularly in clonal species) in relation to plant breeding programmes:

- Plants propagated *in vitro* can generally be initiated to be disease-free, and can be used to: help maintain stocks of breeding lines; facilitate long-term germplasm storage; facilitate international exchange of material; and reduce the length of quarantine periods.
- Short 'generation' times and fast growth means that rapid increases in plant number can readily be achieved.

Both the above have particular importance to clonal crops, which tend to have a relatively low multiplication rate as a result of their vegetative mode of propagation and which are particularly susceptible to viral and bacterial diseases that tend to be multiplied and transmitted through each clonal generation.

Good examples of maintaining high disease status and offering rapid plant regeneration potential include potato and strawberry. Other, perhaps less well-developed examples include *in vitro* propagation of date and oil palms. In these crops it was found that rapid plant regeneration would indeed offer an alternative to the slow and lengthy process of propagating side shoots in date palm and a more uniform planting material in the case of oil palm. However, in date palm the process is still very genotype dependent, and with oil palm there proved to be an unacceptably high frequency of sterile palms produced with initial protocols; however, these are now being revised and would appear to offer practical possibilities.

9.3 Molecular markers in plant breeding

Although plant breeders have successfully practised their art for many centuries, genetics is a subject that really only 'came of age' in the twentieth century with the rediscovery of Mendel's work. Since then research in genetics has covered many aspects of the inheritance of qualitative and quantitative traits, but plant breeders usually still have little, or no, information about:

- the locations of many of these loci in the genome or on which chromosome they reside;
- the number of loci involved in any trait;
- the relative size of the contribution of individual alleles at each locus on the observed phenotype, except where there is an obvious major effect (e.g. height and dwarfing genes).

9.3.1 Theory of using markers

The concept of associating easily visualized markers in plants with loci affecting qualitative and quantitative variation in traits of interest is not new, and was first proposed by Karl Sax in 1923 while studying the association between the colour and size of common beans. Since then a variety of contributions have been made to the general concept and theory of using mapped genetic markers for identifying, locating and manipulating genes of specific interest. The basic idea is straightforward. If a trait or characteristic is difficult to score for whatever reason (e.g. it shows continuous variation;

assessment is detailed and time-consuming; or the trait is only expressed after several years of growth), an easily scored marker that was determined by a locus genetically linked with that affecting the character would be an attractive alternative, surrogate way to monitor the locus of interest.

The concept, therefore, is to use the marker locus as a point of reference for the chromosomal segment in the vicinity of the gene that is really of interest. The approach requires that alternative alleles at the marker locus match the different alleles at the locus of real interest, thus effectively marking the sections of the homologous chromosomes containing the locus that determines the particular expression of the trait we are trying to select.

The association of these marked chromosome segments with the expression of specific quantitative characters can be evaluated while allowing other chromosomal regions in the same individuals to vary at random. The aim, therefore, is to obtain molecular marker that are closely associated with the locus determining the desirable phenotypic expression of polygenic characters such as yield or quality, and so selection procedures could be based upon these markers rather than on phenotypic observations.

The segregating nature of F_2 populations (resulting from selfing an F_1 produced by crossing two homozygous inbred lines) often makes this generation ideal for studying quantitatively inherited characters. Investigations have also been carried out using BC_1 generations, although the information obtained from this type of investigation is likely to be reduced (approximately half) compared with that obtainable from studies on F_2s. In species where double haploid populations are available, such as canola and maize, they have been used extensively in the quest for molecular markers to be used in selection schemes, because of the reduction in genetic complexity they represent.

With an adequate number of uniformly spaced markers, it is possible to identify and characterize linkage groups, which represent the chromosomes involved. It is also possible to construct such a detailed genetic map (a graphical array of ordered molecular markers along each chromosome as well as the genetic distances – expressed as centiMorgans – existing among these) so that the location of all major genetic factors associated with the quantitative trait might be linked rather easily and thus, by following the presence/absence of the different alleles, their individual and interactive effects can be described.

Markers in plants could assist plant breeders in the development of a better understanding of the underlying genes for characters of interest, as well as providing breeders and geneticists with a powerful approach for mapping and manipulating individual loci associated with the expression of these traits. In addition, if the marker genes are tightly linked to other qualitative or quantitative characters, then much of the selection in a plant breeding scheme could be carried out based on the identification of specific set of alleles at the marker loci.

The ability to identify loci that have effects on specific quantitative traits (termed quantitative trait loci – QTL) should lead not only to the ability to handle these loci in a much more deterministic manner, but also provide a more powerful means of investigating epistasis, pleiotropy and the genetic base of heterosis. So the effective use of mapped genetic markers enables advances in cultivar development and selection procedures.

Genetic markers in plants associated with expression of morphological characters have been used for quite a long time, and marker maps assembled. They have been quite well developed in a number of species (e.g. wheat, maize, potato, barley, peas and tomatoes, but also forest species) but generally had rather limited usage because of the problems in finding or generating such markers and their genotype-specific nature. This is changing very rapidly, though, because of:

- The dramatic cost reduction of molecular data as a consequence of the technological progress. The cost of a molecular datapoint today is at least 100 times cheaper than it was 10 years ago.
- The progress observed with developing breeding approaches based on molecular markers in animal breeding, which are being readily deployed in plants.
- The outcomes of the partial/complete sequencing of genomes in many important crop species by providing a massive number of molecular markers. For instance, the sequence analysis of a diverse set of inbred lines in maize yielded over 3 million SNPs (single nucleotide polymorphisms).

The characteristics of a 'good' marker system are:

- that the markers are easy, quick and inexpensive to score the phenotypes expressed;
- the markers are neutral in terms of their phenotypes, and so have no deleterious effects on fitness and no effects on any other traits, including undesirable epistatic interactions with any other traits;
- they reveal a high level of polymorphism (i.e. allelic richness);
- they are robust enough so they can be readily transferred between laboratories and researchers in diverse countries;
- they are stable in expression over environments;
- they can be assessed early in the development of the plant (seedling level), and/or in tissue culture, and require little plant material as source of DNA. Thus evaluation is possible without the need to grow a plant for months or even years before it can be scored. Indeed, a molecular marker that could be assessed in just a fraction of an embryo without compromising the viability of the remaining embryo would enable the selection of only those seeds carrying appropriate alleles, and remove the need to plant all those seeds carrying alternative alleles. The use of embryo fractions in plant breeding is not new, and in the 1960s and 1970s it was very successfully used in Canada to identify canola breeding lines lacking erucic acid in their fatty acid profiles;
- the scoring should be non-destructive, so that desirable individuals can be selected and grown to maturity;
- codominance in expression of the alternative alleles, so that heterozygotes can be differentiated from homozygous dominant genotypes. However, the increasing use of double haploid breeding approaches in crops like corn and wheat alleviates this otherwise important requirement.

9.3.2   Types of marker systems

Any type of genetic marker that has the above properties (or many of them) may be suitable for marker-based applications in the investigation and manipulation of quantitative traits, but the question is really: how closely do they conform to the ideal requirements given above?

The types of markers that can and have been used in plant breeding include:

- Morphological markers, which are basically those that you see by simply looking at a plant's phenotype, including characters such as pigmentation, dwarfism, leaf shape, absence of petals, and so on. It is possible, of course, to choose ones that are easily scored, but the difficulties with morphological markers include that they: cannot always be scored early in development (e.g. flower colour); are often associated with deleterious effects (e.g. albinism); are often relatively rare; their expression is not always independent of the environment in which they grow; and often show dominance/recessiveness.
- Biochemical markers, such as isozyme markers. Isozymes (an abbreviation for isoenzyme) are variant forms of an enzyme, which are functionally identical but can be distinguished by electrophoresis (in other words when placed in an electric field). Under these circumstances the different forms of the enzyme will migrate to different points in the electric field depending on their charge, size and shape. Isozymes have been used very successfully in certain aspects of plant breeding and genetics since they: generally appear to be nearly neutral in their effects on fitness; are rarely associated with undesirable phenotypic effects on other traits; are usually free of environmental influence; and can often be extracted from tissue early in development. So they have a number of inherent properties that allow them to be used effectively for characterizing, and selecting for, qualitative and quantitative characters. Unfortunately, the number of genetic markers provided by isozyme assays is not over-abundant, and they can be either co-dominant or dominant in expression. As a result, the use of isozymes as genetic markers did not allow the full potential of genetic mapping to be realized.

In reality, the practical impact of morphological and biochemical markers has been negligible because of their constraints, and molecular markers are the most appropriate option available for breeding programmes.

- Molecular markers, which are able to detect genetic variation directly at the DNA level. There are basically two systems by which molecular markers are generated, and these need to be described briefly to allow an understanding of their application, but their thorough description goes beyond the scope of this book. The two systems can conveniently be classified as non-PCR-based methods and PCR-based methods. Before briefly describing each it is worth pointing out that molecular markers are simply differences in the DNA between individuals, groups, species, taxa, and so on. Clearly the type and level of variation in DNA that we would want to examine is different depending on what level of distinction we are interested in and what questions we are answering.

Given the above characteristics of molecular markers, particularly their relatively unlimited numbers, it is no surprise that the advent of the possibilities for molecular markers in the 1990s was greeted with some excitement and is seen as providing a major change in the potential to exploit the ideas for using markers advocated some 70 years earlier.

9.3.3 Molecular markers

Non-PCR methods – DNA/DNA hybridization

The first and most widely known of these is restriction fragment length polymorphism (RFLP), originally developed in human genetics. Other non-PCR methods do exist, for example the use of tandemly repeated regions of DNA, known as mini-satellites or micro-satellites, but these will not be described here.

RFLP analysis involves digesting the DNA (cutting it at sites with very specific sequences – there are a number of different enzymes, called restriction enzymes, that cut different patterns of sequences) into fragments, which can then be separated out by gel electrophoresis (as for isozymes, separating them by their differing mobilities in an electric field). To visualize their positions, they are 'blotted' onto a filter, where they are hybridized with a labelled (usually radioactive) 'probe'. The probe is a short fragment of DNA, which may be from a known gene, an expressed sequence or an unknown fragment of the genome. When the 'blotted DNA', having first been denatured to reduce it to single strands (rather the usual double-stranded state of DNA), and the probe (also denatured) are brought together, where there is an exact match in the complementary sequences they will hybridize

(by hydrogen bonding) or bond. The filter is then washed to remove all the excess probe and leave only that which is now bonded with our sample DNA. If we expose the filter to X-ray film, when it is developed it will show where the probe still remains, hence where the probe has hybridized and so where there was a piece of the DNA we were investigating which had a complementary sequence. The pattern of bands obtained in this way is called the restriction fragment pattern. Using a varied combination of enzymes and probes gives a wide range of possibilities for exposing variation in the DNA sequences.

RFLPs are highly reproducible, they show codominance in their expression and are reliably specific. However, they are relatively time-consuming, rather expensive, not easy to automate, require fairly large amounts of 'clean' DNA, and tend to use radioactive probes for best results. Although they are extremely useful for detailed genetic analyses, they are not well suited to the needs of breeding programmes.

PCR-based methods – arbitrarily primed techniques – multi-locus systems

The most commonly used approach in the past was **randomly amplified polymorphic DNA** (RAPD). The technique basically involves using a single 'arbitrary' primer – a 10-nucleotide-long sequence of DNA in a PCR reaction. The basic ingredient of the PCR reaction is DNA polymerase, an enzyme that enables the copying of a duplicate molecule of DNA from a DNA template, and is commonly *Taq* polymerase, a thermally stable DNA polymerase. The primer anneals to its complementary sequences in the DNA sample being studied, and 'primes' the DNA polymerase to start DNA amplification. These amplification products can be resolved on agarose gels.

The advantages of RAPDs are that it requires only small amounts of relatively crude DNA; it requires modest, widely available equipment (thermal cycler and electrophoresis devices); and no prior knowledge of the gene or DNA sequence is required. It is fast and relatively inexpensive. However, the results can be rather variable depending upon slight changes of the PCR conditions or ingredients, and RAPD markers show dominance, and so their transfer between laboratories is less than ideal. Despite the early excitement and their very low cost, RAPDs are very rarely used on a regular basis in breeding applications.

A more reliable method developed is **amplified fragment length polymorphism** (AFLP), and this is not only more repeatable but also gives much higher frequency of markers, and *inter-simple sequence repeats* (ISSRs or anchored microsatellites). However, the details of these are beyond our present remit. AFLPs include a DNA digestion step with restriction enzymes, which tends to preclude their routine use as a high-throughput marker system in breeding programmes.

PCR methods – site targeted techniques – single locus systems

Rather than using arbitrary primers, it is possible to specifically design primers to be used in PCR. There are a number of possibilities to design primers, but one such approach is **single sequence repeats** (SSR), also known as microsatellites. Microsatellites are simple sequence repeats which are ubiquitous around the genome and are generally quite variable in exact DNA base-pair composition. If one pictures these at different places in the genome, the DNA 'flanking' these regions will be different depending on where they are (i.e. the site at which they are found will be unique). So you can 'fish' for these in genome libraries cloned in *E. coli* with simple repeats as probes, then sequence positive clones and design PCR primers with the main part being simple repeats but the ends being other unique 'tags'. This allows much more robust markers to be generated but with all the advantages of the PCR technology. SSRs have been extensively used in breeding programmes and genetic research, even though they are rapidly being replaced by **single nucleotide polymorphisms** (SNPs, pronounced "snips"). SNPs represent an outstanding source of molecular markers as they exploit variation existing directly at the DNA level: when directly comparing aligned DNA sequences of two individuals, nucleotide variants such as indels (insertion/deletions) or single point mutations can be detected that enable very specific PCR assays to be developed. SNPs are orders of magnitude more variable than any other known molecular marker; in maize, the analysis through sequencing of a diverse set of

inbreds determined that 1 base pair in every 44 was polymorphic.

SNPs are robust, amenable to automatization, and also able to uncover more genetic variation than any previously developed molecular marker systems. In addition, their codominant nature increases the genetic insight they provide. They are resolved on DNA sequencers rather than on gels, which further contributes to their reliability and transportability among laboratories. Nowadays there are thousands of publicly-available SNPs in the crops most important for humankind.

9.3.4   Uses of molecular markers in breeding programmes

Molecular markers can therefore be used to identify cultivars (DNA fingerprinting), to differentiate one cultivar from another (perhaps one already released), or to be able to prove proprietary ownership of specific cultivars. If you have a modest set of markers it is possible to produce a DNA fingerprint which is unique (or nearly so) and so can potentially be used to identify that particular genotype. Similarly, using the same principle it is possible to identify DNA that is not supposed to be there and so it can be used to ensure that a particular cultivar is pure and free from contaminants. A further possibility is afforded by the potential to assess how diverse genotypes are at the DNA level and hence assess their level of difference (genetic distance) if used as parents (e.g. parents of hybrid cultivars).

Marker-assisted backcrossing

When a gene of interest can be shown to be linked to a molecular marker, then assessment of the marker can help to accelerate the backcrossing process. Mature plants would not need to be grown to identify which backcross individuals carry the allele of interest. This is particularly helpful where, although determined by a major gene, the phenotypes are difficult or time-consuming to detect or are expressed later in development (e.g. fruit colour). Molecular markers can identify which of the backcross progeny have better restoration of the rest, or background, of the genome of the recurrent parent. Marker-assisted backcrossing has enabled the development of trait integration approaches where a transgene is introgressed quickly and accurately into elite breeding lines in breeding programmes. When marker-assisted backcrossing is combined with winter nurseries, allowing several crop generations to be planted within a calendar year, variety development can be dramatically accelerated.

They can provide breeders with vital information about the legitimacy of any cross, but particularly if a supposed wide cross (or interspecific cross) is a rare genuine event or the result of an unfortunate illegitimate pollination. Indeed, when used in conjunction with cytogenetic information, they can give very precise information about what chromosome or parts of chromosomes are present in interspecific hybrids or generations derived from such hybrids. In tree breeding programmes, this specific application is critical, as many years might pass between the time a cross is made and when the progeny from such a cross is field-tested.

When a number of markers have been generated then they can be used to build a genetic map of a species – an ordered array of genetic markers along chromosomes which also displays the genetic distances existing between these markers- and hence provide much clearer ideas of the positions on chromosomes of different genes and so determine the associations that might be expected between simply inherited traits. Thus helping to determine the selection strategy that will be most applicable.

QTL mapping

If a genetic map based on a mapping population is available, as well as phenotypic trait data collected from the same population, the genetic position (both in absolute terms, i.e. in which chromosome, and in relative terms, i.e. where in such a chromosome) of QTL involved in the genetic architecture of such a trait on the genetic map can be established. Subsequently, one could attempt to use neighbouring molecular markers to 'tag' such QTL, and so to follow their segregation and use such molecular markers as surrogates of the QTL. In breeding programmes this could be used, for instance, to conduct selection based on the presence/absence of molecular markers genetically linked to a given QTL, instead of running selection based on phenotypic assessments of those traits, thus reducing time and expenses. Also, it might assist the early identification in plantlets of individuals carrying a QTL for which expression would

take place during later stages of crop development. Even though conceptually this appears as appealing and straightforward, results of selection in breeding programmes based on QTL or their neighbour molecular markers have enjoyed very limited success, at best. This is because the difficulty remains in assessing the quantitative trait expressions accurately and in ways that are relevant to the agronomic circumstances in which the cultivars will finally be grown. Genotype × environment interactions could pose as large a problem in QTL as it does in traditional evaluation and selection. QTL, however, might offer plant breeders an opportunity to obtain a better understanding of the genetic basis of genotype × environment interactions, epistasis and heterosis. Also, it is clear that amongst the quantitative variations exhibited for many traits, there are some regions of DNA that determine rather large parts of the variation that we observe – if these could be handled effectively, the effort that was saved could be focused on the non-defined regions. Another significant issue encountered is that the expression of many QTL has turned out to be dependent upon genetic context, and so QTL discovered in a given population might not exert the same allelic effect in a different population, or might not even exert any allelic effect whatsoever. It is therefore fundamental that before embarking upon the deployment of breeding schemes based on QTL to enhance the efficiency of selection, those QTL must be thoroughly validated across genetic contexts and environments.

To a large extent the limited impact of QTL mapping in applied plant breeding relates to the quantitative genetics of agronomically relevant traits, controlled by a large number of QTL with small individual effect and expression strongly influenced by the environment. Another significant limitation is the way in which research around QTL has been organized: first QTLs are found, and subsequently their effects are estimated. The biparental populations used to map QTLs often lack breeding relevance and represent additional costs and time. A likely consequence of this is a biased estimate of QTL and the failure to observe QTL with small effects. These significant pitfalls can be mitigated by the use of association mapping approaches, as these rely on the use of populations closer to the context of 'real-life' breeding programmes. Nevertheless, the issue of biased genetic effects remains.

Association mapping

In order to overcome some of the pitfalls described in QTL mapping, mainly their lack of breeding relevance, association approaches have been developed that exploit linkage disequilibrium and enable the genetic mapping to be carried out in sets of genotypes rather than in mapping populations. For instance, an association mapping project might assemble several hundred individuals encompassing elite breeding lines, breeding germplasm and other sources of genetic variation. The molecular markers most often used in association mapping approaches are SNPs. Using statistical analyses, significant marker–trait associations are established. The typical outcome of association mapping efforts are haplotypes (a combination of SNPs encompassing a small chromosomic block) statistically associated with the expression of the trait under study, which could then be used as a selection tool.

Despite over 25 years of research with molecular markers and the publication of thousands of papers on the topic of QTL mapping (the first paper linking genetic maps based on molecular markers and the detection of QTL was published in 1988) and on a lesser scale association mapping efforts, there are very few documented cases of their successful or routine use in applied breeding programmes. Among additional reasons for such a less than ideal contribution to breeding programmes, several in particular could be mentioned, such as the use of small, biparental mapping populations lacking relevance to the populations used by breeders, sparse genetic maps, inappropriate data analysis, poor phenotypic datasets and lack of thorough validation – in the genetic and environmental context – before their routine use. In addition, although the allelic effects of most reported QTLs might be valuable from a research perspective, in many cases they do not support genetic gains large enough to offset the expenses incurred in detecting the QTL in the context of breeding programmes.

In order to attempt to overcome the pitfalls of molecular markers and marker-assisted selection approaches (briefly described above), a new approach has recently been developed known as *genomic selection*, or *genome wide selection*. It represents a good example where plant breeding borrows heavily from animal breeding concepts, being originally proposed by Theo Meuwissen and colleagues in Norway in 2001. Genomic selection (GS) does not refer to the use of molecular markers identified through QTL or association mapping approaches as an aid to plant breeding. Rather, it represents an entirely new perspective on the exploitation of molecular data in breeding programmes.

Genomic selection

Genomic selection builds on the availability of large numbers of molecular markers (mostly SNPs) to estimate marker effects across the entire genome, rather than at single QTL. So, GS aims to capture the entire set of both large- and small-effect QTL, and this in turn enables GS to fully comprehend the genetic variance existing for a given trait in a population, whereas QTL approaches can only capture a limited proportion of such genetic variance. GS integrates all those genetic effects into so-called *genomic estimated breeding values* (GEBV) which represent the genetic merit of individuals. It uses a training population of individuals for which a large amount of genotyping information (hundreds to thousands of SNPs) and high-quality phenotyping data is integrated to calculate GEBVs. Conceptually, they represent the ultimate selection criteria as they encompass the entire genome rather than just some markers or QTL/haplotypes. Subsequently, selection based upon GEBVs is imposed upon breeding populations for which a sufficient volume of genotyping information is available. It is important to stress that only genotypic, not phenotypic, information is required from those breeding populations, since this represents a paradigm shift in the way breeding programmes have successfully been run for many years (i.e. relying entirely on phenotypic information to make selection decisions).

As would be expected with any novel approach, there are many research issues that need to be addressed and resolved before GS could have a meaningful, significant impact in breeding programmes. One such area of research is the way training populations need to be created in order to maximize the capture of genetic variation while keeping breeding relevance. An additional area of work refers to the complex statistical models required to estimate the genetic merit of individuals. An often overlooked area of research is the phenotypic basis of GS efforts; because selection is imposed upon breeding populations lacking phenotypic data, it is imperative that the phenotypic data used to establish GEBV be of the utmost quality and accuracy. In other words, these phenotypic datasets must be based upon the best statistical designs available, the best agronomics and trait data collection procedures and the most appropriate linear mixed model-based statistical analyses, leading to datasets with large heritability values.

Even though genomic selection is still in its infancy, several research projects are already underway in wheat, corn, barley and the forest species eucalyptus. Particular research issues remain, such as our still rather limited ability to establish commercially relevant genotype to phenotype relationships. Regardless, it is envisaged that in the years to come GS will revolutionize plant breeding, by shrinking breeding cycles and increasing the overall efficacy of breeding programmes.

9.3.5 Issues with markers

In many instances, using molecular marker techniques (say for selection) is basically more expensive and more technically demanding than other selection options. It is therefore not really cost-effective to set the necessary laboratory facilities and trained staff to handle a few crosses or perhaps a situation where the profit returns on the breeding are low.

Finding a molecular marker that is associated with a major gene of interest or a QTL is not always too difficult, but ensuring that it is close enough not to be lost by subsequent recombination is more difficult. Also, the applicability of the marker combination over a range of crosses rather than just a specific one is a concern that takes considereable time and effort.

Nevertheless, the exploitation of QTL offers great potential that has yet to be realized in practical terms. Developing good and reliable QTL will require a great deal of well designed and accurate field-testing. As already noted, genotype × environment interactions may pose as large a problem in QTL as they do for traditional selection. Finally, there needs to be even better repeatability between results obtained by different research teams. Different researchers sometimes identify different loci to be responsible, in QTL analyses, for the major differences in expression of the phenotypes for the character of interest. Some of these differences will reasonably be ascribable to the fact that different alleles are segregating in different crosses or being expressed at different levels in different circumstances – note the similar problems with heritability estimates. But there are also technical differences that need to be corrected before the true potential of QTLs can be realized.

The only reported successful examples of deploying QTL information in large-scale, commercial breeding efforts represent private endeavours based in the US: Monsanto's maize breeding programme and Pioneer's soybean breeding programme.

A previously unforeseen positive consequence of the deployment of genetic markers in breeding programmes has been an increasing awareness of the need to have high-quality phenotypic datasets. Unless the field phenotypic data are of high enough quality, the promise and potential of molecular markers will not be realized, nor the delivery of the expensive investment in labs, molecular biology, bioinformatics and staff needed to set up marker-assisted selection approaches.

9.3.6 The increasing availability of genome sequences

The success of the human genome sequencing project, the continuous advancement of sequencing approaches, the dramatic reduction in sequencing costs led by the quest to achieve the sequencing of human genomes for under US$1000, reflected for instance in current claims of cost per megabase (10^6 base-pairs of DNA) sequenced below US$ 10 cents, and the availability of computing power and software to process, analyse and make sense of the deluge of genomic information, have all enabled the sequencing of crop genomes to become routine. A typical crop genome sequencing effort encompasses the following generic steps, although important modifications may occur both in terms of chosen steps and extent of analyses:

1. Production of sequencing libraries where the plant DNA to be sequenced is properly cloned and subjected to shotgun sequencing through high throughput approaches;
2. Assembly of the sequence reads in order to reconstruct the genome just sequenced;
3. Annotation, by assigning putative gene functions to the genic sequences unveiled. This is carried

out using computer simulation/analysis (*in silico*) using powerful software and gene identification algorithms; however, the final biological confirmation of putative gene sequence function is often achieved through further experimentation;

4. Further analyses such as synteny analyses, search for orthologous genes, patterns of genome duplication/rearrangement/loss, evolutionary analyses, and in some cases SNP identification.

The first plant genome sequenced, *Arabidopsis thaliana*, was made available in 2000, and since then the genome of over 20 plant species has been sequenced, including some of the major crops for mankind such as barley, maize, potato, soybean, sorghum, tomato and wheat. In addition, horticulturally important crops such as *Brassica oleracea*, cucumber, melon and watermelon have had their genomes already sequenced, as well as fruit species such as apple, cacao and diploid strawberry. It is particularly encouraging that additionally the genomic sequence of crops such as chickpea (*Cicer arietinum*), a key staple crop to many people and the second most widely grown legume globally, are also being reported, as this development reflects that cultivated species known as 'orphan' crops are also benefiting from genomic sequencing efforts.

Only a few years ago, sequencing a crop genome represented a major, expensive undertaking. However, the molecular and bioinformatic technology and expertise currently available allows the accomplishment of the sequencing and analysis of a crop species genome within a calendar year. It is foreseen that in the years to come, sequencing genomes will become even more widespread and available, and therefore genomic sequences will become a commodity.

The availability of genomic information in an increasing number of crop species would challenge the current paradigm of plant breeding – phenotype-rich, genotype-poor – into one that is phenotype-rich *and* genotype-rich. This in turn will positively impact plant breeding in a number of ways.

Increasing availability of molecular markers

The use of molecular markers in crop breeding is relatively new and began in the 1980s. Regardless of the chosen approach (QTL mapping, association

mapping or genomic selection), until recently a main hurdle to establishing agronomically meaningful marker–trait associations was the availability of genetic markers. The availability of cheap sequencing approaches provides plant breeders and researchers with thousands of SNPs, and has changed this situation dramatically. In some crops there are ten of thousands of SNPs, and even larger numbers available to establish marker–trait associations. Until now, even the most advanced genetic maps in crop species relied upon only hundreds of molecular markers; however, from this point on the number of genetic markers available should not represent a constraint in either developed or developing countries. In addition, not only the genome of specific genotypes could be made available, but also those of a myriad of individuals, which could be part of genetic mapping efforts or from genetic improvement programmes. At the time of writing an exciting new approach, called genotyping-by-sequencing, is being deployed in crop species such as barley and wheat, which enables marker discovery and genotyping at the same time.

Molecular basis of genetic variation

There is increasing evidence provided by sequencing efforts that there is significant structural variation among the genomes of diverse individuals within the same species. In the best reported crop example to date, maize, the comparative analysis of inbred genomic sequences has shown significant differences in gene copy number but also in presence/absence of genes. It was previously thought that the same set of genes existed in different individuals of a given species, and that phenotypic differences arising were due to the several allelic forms that could arise from those genes, or to non-allelic, epistatic interactions. The presence of genes in some maize inbreds and their lack in others might contribute to important biological mechanisms such as heterosis, and also shed light on the basis of quantitative variation and genotype by environment interactions.

Identification of agronomically relevant genes

It is known in cattle and chicken that traits selected by humans have lower levels of variation in 'improved' than in 'non-improved' individuals. These genomic regions with lower variation and skewed allelic frequencies might harbour important genes associated with domestication processes, and thus agronomically relevant genes or genes controlling the expression of other genes. This has been demonstrated in rice through the sequencing and thorough analysis of the genomes of 50 diverse wild and cultivated rice lines, enabling not only the discovery and identification of over 6 million SNPs, but also thousands of candidate genes that might have been artificially selected during the domestication of this important crop.

Another elegant example from rice has recently enabled the identification of functional nucleotide polymorphisms: a DNA polymorphism associated with the expression of a phenotypic trait, at the gene DTH2, which encodes the locus 'Days to Heading' on chromosome 2. These polymorphisms correlate with early flowering and explain the geographical expansion of rice cultivation in Asia, and likely represent a target of human selection for adaptation to long-day photoperiod conditions.

Think questions

(1) List four uses for molecular markers in plant breeding.
(2) Describe any advantages of using molecular markers in plant breeding.
(3) Describe one difficulty that might be encountered in utilizing QTLs in plant breeding selection.

(4) A cross is made between two parents in a hybrid wheat breeding programme, where P_1 is the female parent and P_2 is the male parent. When you run RFLP analysis you see electrophoretic patterns like those below.

P_1 is the lane pattern for the female parent, P_2 the male parent, and other lanes (a to l) were observed from a sample of seed of the F_2 progeny.

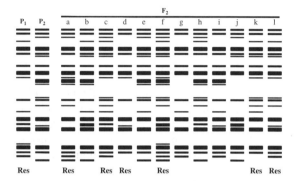

Explain what could have caused the pattern shown in lane f. If it is know that Parent 1 has a single gene conferring resistance to yellow stripe rust and that the stripe rust resistance of each genotype/lane is given below each lane (i.e. Res = resistant to strip rust) indicate on the diagram a possible molecular marker for the stripe rust resistance gene.

(5) Describe three applications involving tissue culture or *in vitro* plant propagation in a plant breeding scheme.

10
Practical Considerations

10.1 Introduction

As noted at the beginning of this book, plant breeding demands a range of skills, including good management and a multitude of scientific disciplines in combination, to achieve success. Plant breeding operations and evaluation of plant breeding lines will be conducted in laboratories, glasshouses and field situations. This chapter attempts to outline some of the practical realities in a plant breeding programme. Sections covered in this chapter examine: experimental designs, including the types of designs suitable for different parts of a plant breeding programme; glasshouse management and field management; and the applications that can be covered and managed using computers. Finally, this chapter considers some of the practical considerations of the actual cultivar release procedure.

10.2 Experimental design

It has been stressed previously that the basic operations of cultivar development can, for simplicity, be divided into three stages: producing genetic variation, selection among recombinants for desirable new cultivars with specific characteristics, followed by genetic stabilization, multiplication and seed certification. The following few sections are concerned particularly with the middle one of the three processes.

The aim in selection is to identify recombinants that are genetically superior to already existing cultivars. Superiority can be achieved by increased productivity (e.g. increased yield or better end-use quality), by making productivity less variable (e.g. reduced risk of crop failure by introduction of disease, insect or stress tolerance or resistance, earliness, dry down, and so on) or increased profit (e.g. reducing input costs by incorporation of disease resistance).

In each of the cases it will always be necessary to evaluate the performance of breeding lines for qualitative and quantitative characters. In some instances it is possible to select and screen for single gene characters without the complication of undue influences from environmental factors. However, it is widely accepted that virtually all quantitatively inherited characters (most often the ones with greatest commercial value, e.g. yield, quality and many durable disease resistances) are highly modifiable by the environment. Consider the observation of a single plant; the aim is to minimize the non-genetic effect from the equation:

$$P = G + E + (G \times E) + \sigma_e^2$$

where P is the phenotypic expression, G is the genotypic effect, E is the effect of environmental variables, $G \times E$ is the effects attributable to the interaction of the genotype with the environment, and σ_e^2 is a random error term associated with a single observation. In the evaluation of breeding material it is only possible to observe the phenotype

Plant Breeding, Second Edition. Jack Brown, Peter D.S. Caligari and Hugo A. Campos.
© 2014 John Wiley & Sons, Ltd. Published 2014 by John Wiley & Sons, Ltd.
Companion Website: www.wiley.com/go/brown/plantbreeding

(a combination of genotypic and environmental effects). The aim is to determine the genetic potential of each breeding line as accurately as possible, and hence it is necessary to either estimate or minimize the environmental and error effects on the phenotypes observed in the field. Achieving this demands careful use of a number of **experimental designs**.

Running a plant breeding programme is no different from organizing a whole series of scientific experiments, and therefore all aspects of the operation should be treated with the same care and detail that individual experiments require in terms of planning and handling. Good experimental design leads to knowledge of the accuracy of the data upon which evaluation and selection are based. The quality of information collected in a plant breeding programme is the key factor determining the success of the scheme, whether it is one based solely on traditional techniques or whether it incorporates molecular-based technology. The proper use of statistically valid designs can increase the heritability of the trait under selection, that is, it can enhance the 'genetic signal' relative to the 'experimental noise' ratio and therefore the rates of genetic gain, response to selection and likelihood of success. Unfortunately, often this is overlooked.

Sometimes, the set of trials that a breeding programme uses to gain insight into the performance of advanced breeding lines in a Target Population of Environments (the set of environments or geographical regions a breeder is breeding for) is referred to as **multiple environment testing**.

It is common to evaluate breeding lines (**test entries**) in comparison with existing cultivars (**controls** or **checks**) within the same trial. In some cases a single cultivar is used, but more often several cultivars are included in the evaluation trials. The choice and number of control entries is largely dependent upon the range and number of cultivars that are currently being grown in the target region for the new cultivars, the type of trial, and the number of evaluations that are to be made. For example, an evaluation trial may contain the highest yielding cultivar available to compare yielding performance, the best quality cultivar to provide a baseline for quality, a cultivar with disease resistance to evaluate response under disease pressure, and so on. When appropriate, it is often a good idea to include cultivars developed by competitors or those that

currently are in high demand by the farmers. It is always desirable to include as many control entries as is possible within the restrictions imposed by the extra land used and the effort involved. It should also be noted that evaluation trials can be costly, and that the cost of an evaluation trial is often directly related to the number of total entries that are included. If several thousand breeding lines are to be evaluated then it is unnecessary to include only a few control entries. If, however, only a few lines are under trial, then it would be unwise to include many hundreds of control plots. A simple rule of thumb, which is often useful, is that if between 1 and 200 breeding lines are to be tested, the number of control plots (not always entries) should be about a tenth of the total number of trial entries, while if more than 200 lines are under evaluation, up to 1/20th of the total number in the trial should be controls.

A wide spectrum of possible designs is available, but only a limited number will be detailed here, namely:

- unreplicated designs;
- randomized complete block designs;
- factorial designs;
- split-plot designs.

10.2.1 Unreplicated designs

Unreplicated designs, as their name suggests, are experimental designs where test entries are not replicated and so appear only once at each testing location. There are, however, several (or indeed many) different options even when single replicate designs are used. These include:

- Non-randomized designs without control entries, where genotypes that are to be evaluated are arranged in plots in a systematic order (e.g. numerical order, alphabetical order) (Figure 10.1a). Only genotypes under test are evaluated, and there are no control (check) cultivars grown at the same time. It is not possible to obtain any estimate of error from this type of design or make any direct comparisons with known cultivars.
- Non-randomized designs with control entries, where the test entries are arranged in plots in a systematic order (as above), but control cultivars are interspaced amongst the test entry

(a) No randomization, without controls

49	50	51	52	53	54	55	56
41	42	43	44	45	46	47	48
33	34	35	36	37	38	38	40
25	26	27	28	29	30	31	32
17	18	19	20	21	22	23	24
9	10	11	12	13	14	15	16
1	2	3	4	5	6	7	8

(b) No randomization, with controls

37	38	C.1	39	40	41	C.2	42
31	C.3	32	33	34	C.4	35	36
C.1	25	26	27	C.2	28	29	30
19	20	21	C.3	22	23	24	C.4
13	14	C.1	15	16	17	C.2	18
7	C.3	8	9	10	C.4	11	12
C.1	1	2	3	C.2	4	5	6

Figure 10.1 Non-randomized single replicate plot designs (a) without control entries, and (b) with control entries arranged systematically throughout.

(a) Randomization, no controls

5	22	50	47	34	36	52	24
31	40	18	38	28	3	49	25
43	32	7	41	10	37	19	1
14	56	29	55	12	54	35	30
26	9	45	48	27	44	2	42
53	20	51	17	6	23	21	13
11	39	46	15	16	4	33	8

(b) Randomization, systematic controls

38	13	C.1	41	29	5	C.2	32
33	C.3	1	24	4	C.4	18	30
C.1	21	39	23	C.2	36	26	9
40	8	2	C.3	15	16	27	C.4
42	12	C.1	35	3	19	C.2	11
20	C.3	34	28	37	C.4	25	31
C.1	7	22	14	C.2	6	17	10

Figure 10.2 Randomized single replicate plot designs (a) without control entries, and (b) with control entries arranged systematically throughout.

plots (Figure 10.1b). The control cultivars can be arranged in a systematic order (e.g. every 20 plots), or they can be allocated to plot positions at random. In most cases a number of control cultivars are included. Each control entry may also be replicated more than once in the whole design. Multiple entry plots of control cultivars can often be useful to determine an estimated error variance for the overall field trial.

- Randomized designs without control entries, where the test entries are arranged within the trial at random but no control entries are included (Figure 10.2a).
- Randomized designs with control entries, where the test entries are randomly allocated plot positions within the trial (Figure 10.2b). Control entries can also be randomized throughout the design (and often replicated in more than one plot) or they can be arranged in a systematic order (e.g. every 5th plot), again with the option of having replication only for the control entries.

The efficiency of evaluation trials will always be increased by randomization, and non-randomized trials should be avoided if at all possible and carefully considered before being used. Similarly, it is generally unwise to organize any breeding evaluation trials without including any control entries

against which the test lines will be compared. Without these considerations the trials are generally uninformative and can often be misleading.

In the early generations of a plant breeding scheme, there may be many hundreds or thousands of genotypes to be tested, each with only a very limited amount of planting material available. In many breeding programmes, the first 'actual' field trials are conducted using head-row plots, where each plot has arisen from a single plant selection in the previous year. Where thousands of lines are to be tested, it may be extremely difficult to completely randomize each individual head-row, but randomization at this early-generation stage can greatly increase efficiency. One option is to utilize nested designs. For example, say that a canola breeding programme has 200 cross combinations to be evaluated and that there are 100 individual single plant selections taken at the F_3 stage. Therefore there would be 2,000 F_4 head-row plots that would be planted in the field. A randomized complete block (with control entries) would be very large. In addition, from a practical aspect, it is often difficult to examine a single row plot on its own. As an alternative the 200 crosses could be randomized into five replicate blocks, and the 100 single plant selections are grown as rows within cross blocks. Each cross, therefore, would be represented by five sub-blocks (groups) of 20 head-row plots (grown adjacent), and replicated five times throughout the whole trial.

If control entries are arranged in a systematic order it will be possible to make direct comparisons of individual test entries with the nearest control plot, which can have advantages. For example, it makes possible the analysis of the data collected using **nearest neighbour** techniques, where plot values are adjusted according to the performance of appropriate surrounding test entries. In order to increase the value from unreplicated designs, they can be combined with powerful statistical analyses to allow the handling of any systematic heterogeneity existing among experimental units and/or the spatial autocorrelation that could arise between neighbouring plots. One example is the use of autocorrelation models such as AR1XAR1, a two-dimensional spatial model consisting of separable first-order autoregressive processes along both columns and rows, which deals with autocorrelation patterns running along both dimensions of a trial (columns and rows). The ability to account for spatial correlation patterns in two dimensions generally provides a more efficient analysis of breeding trials. The availability of more effective statistical analyses has recently encouraged a growing number of breeding programmes to include, at least in some stages of testing, unreplicated trials as they allow the testing of a larger number of breeding lines within a fixed budget.

10.2.2 Randomized designs

Although it is possible to obtain an estimate of error variance from unreplicated (single replicate) designs which have multiple entries (i.e. replicates) of chosen control cultivars, it is more common, where possible, to replicate both test lines and control cultivars in order to have a better estimate of the average performance of each entry, along with the variance in its performance, as well as obtaining a better and more representative overall estimate of error variance.

Completely randomized designs

If there is no knowledge of fertility gradients or other environmental variation existing within a test

(a) Four Replicates ~ No Blocking

16	16	14	9	13	16	15	3
1	13	6	11	5	12	10	4
10	12	2	7	6	15	11	2
2	5	13	7	1	7	6	12
4	14	3	13	9	6	9	1
11	4	3	8	7	14	15	10
8	12	14	4	2	5	16	8
3	9	15	8	10	11	5	1

(b) Four Replicates ~ Blocking

9	14	5	13	9	16	11	5
11	7	16	1	3	13	8	6
15	2	3	12	2	15	14	4
10	6	4	8	1	10	7	12
6	13	10	16	14	6	9	8
14	3	1	8	12	11	15	4
12	15	9	5	1	16	3	7
7	2	11	4	13	2	10	5

Figure 10.3 (a) Completely randomized block design, and (b) randomized complete block design.

area, many suggest that complete randomization be used to identify superior breeding lines. In such a design, each of the test and control entries are allocated at random to plot positions (Figure 10.3a). Each entry is repeated a number of times according to the required number of **replicates**. The error variance is estimated from the variance between replicate test entries.

Randomized complete block designs

Although there are often merits in choosing a completely randomized design, a more common design (probably the most common design) used by plant breeders is a randomized complete block design, or variations thereof. In these designs, the total area of the test field is divided into units according to the number of required replicates. Each unit is called a block. Each of the test and control entries are randomly assigned plot positions within each block (Figure 10.3b). In the cases where there are distinct fertility gradients or other differences between blocks, then these can be estimated and subtracted from the error variance. It is possible, therefore, to obtain a more accurate estimate of the error variance. Blocking does not necessarily need to be different areas within a field trial. Different blocks in a randomized complete block design could, for example, be different days of testing (where it is not possible to test all replicates in a single day).

Factorial designs

Single replicate designs are often referred to as single dimension designs, and randomized designs are called two-dimension designs. In many cases it is important to simultaneously evaluate a number of breeding lines with regard to their response to different treatments. These types of experimental designs are called multidimensional designs or **factorial designs**. To illustrate factorial designs, consider the example where there are only four breeding lines to be tested (L.1, … , L.4) and the performance of each is to be evaluated under three different treatments, or factors (T.1, … , T.3). Each genotype entry is grown with each of the different treatments. Overall, there are $4 \times 3 = 12$ entries. These are arranged at random as illustrated in Figure 10.4a. In the example, only two replicates are illustrated. In practice more than two plots of each test unit would be grown to ensure the necessary level of replication for a reasonable estimate of the error variation. Replicated factorial designs can be completely randomized, or each replicate can be blocked.

Analysis of factorial designs allows estimates of differences between test entries and between treatments compared with an estimated error. These designs also allow evaluation of any **interaction** that may exist between test entries and treatments (in other words, variation in the response of the lines to the treatment imposed). To illustrate this, consider the performance of two test lines (A and B) each evaluated under low- and high-nitrogen conditions. If the yield performance of the two lines follows the pattern where entry A is higher-yielding than entry B at both nitrogen levels, it is said that there is no interaction. If, however, entry A is highest yielding at high nitrogen levels but entry B is highest-yielding at low nitrogen levels, then there is said to be interaction between genotypes and nitrogen levels. The significance of the interaction is tested in the analysis of variance. It should be noted that in testing for genotype × treatment interactions, if there are no changes in genotype ranking then although formally an interaction may

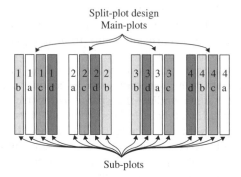

Figure 10.4   (a) Multiple factor factorial design, and (b) split-plot design.

be detected, the implications for plant breeding are minimal, unless other treatments (maybe outside the range tested) are envisaged as being likely.

Split-plot designs

In some cases a breeder is interested in estimating the difference between the effect of one factor and the interaction of that factor with a second factor, while having lesser interest in variation within the second factor in its own right. In plant breeding, for example, it is well established that higher nitrogen (within limits) applied to cereal crops will result in higher yield. Genotypes × nitrogen studies are routinely carried out, not to determine whether there is an average yield increase with increased nitrogen (as this has already been established). The primary goal is to determine the differences between genotypes and the interaction between genotypes and nitrogen levels. In these cases a special type of factorial design called a **split-plot design** is commonly used.

Split-plot designs divide the total area of a test into **main blocks**, **sub-blocks** and **sub-sub-blocks**. Firstly the main blocks are arranged at random, and then factors within sub-blocks are arranged at random within the main blocks, factors of sub-sub-blocks are arranged at random within sub-blocks, and so on. A simple split-plot design is illustrated in Figure 10.4b. The numbers 1, 2, 3 and 4 are main blocks and the letters a, b, c and d are sub-blocks. Only a single replicate is shown in the figure, but as before, other replicates would be similar in structure and would always be blocked.

The analysis of variance produces two errors for use in F-tests. The error #1 is estimated by the sub-blocks within main blocks × replicates effect, while the interaction effects would be tested against the interaction between main blocks × replicates.

10.2.3 The increasing role of linear mixed model approaches to analyse breeding data

Most breeding datasets represent not only fixed but also random sources of variation, and conventional analysis of variance approaches in software programs do not formally take such differences into account. Fortunately, the development of software enabling linear mixed model-based analyses such as ASReml, Genstat, SAS and S-Plus, among others, and the increasing computing power at the disposal of breeders, enable the use of linear mixed model platforms to analyze breeding datasets. This box provides only a coarse overview of linear mixed models and their use in breeding, but hopefully encourages interested readers to seek further support from more specialised books or from biometricians.

The basic linear mixed model includes both fixed and random factors as follows:

$$y = X\beta + Zu + e$$

where:
- y = vector of observations
- X = associated matrix design
- β = vector of fixed effects
- Z = associated matrix design
- u = vector of random effects
- e = random residual vector.

A factor is said to be random when it can be thought of as a random sample from a population (for instance, environments, genotypes, sampling units), or a randomization unit such as a plot. Otherwise factors are considered to be fixed, for instance

the treatments it has been decided specifically to apply. When the purpose of a breeding trial is selecting superior individuals the genotypes are often considered random, as this enables its estimation as BLUPs (best linear unbiased predictors). BLUPs maximize the correlation between true genotypic values and predicted genotypic values, which is typically the purpose of breeders.

The discussion of whether genotypes are to be declared fixed or random factors is beyond the scope of this book, but suffice to say here that any set of genotypes managed by a given breeding programme could be considered as a random sample of the entire germplasm collection available for a given species that could have arisen as a result of the selection process. This in turn allows the assessment of phenotypic performance as random effects through BLUPs.

Using BLUPs increases the amount of information used to assess phenotypic performance. It also enables a more effective allowance for spatial variation as it can be easily combined with statistical approaches resulting in a higher effectiveness in selecting 'truly' superior individuals. They also increase the predicting ability of future performance.

The seminal paper of Arthur Gilmour and colleagues published in 1997 demonstrated how linear mixed models could address frequent sources of variation existing in breeding datasets: large-scale variation across fields, natural variation or local trends existing within individual trials, and also extraneous variation sometimes induced by experimental procedures such as planting and harvesting of experiments or trimming of research plots. These sources of variation would go unnoticed in a regular analysis of variance, not only adding unwanted noise to datasets but also potentially impacting the ability to pinpoint truly superior individuals and therefore the ability to make meaningful genetic progress.

Linear mixed models also present a significant enhancement when it comes to analysing datasets from multi-environment trials where the performance of advanced breeding lines is evaluated across the target population of environments that a breeding programme focuses upon. Most breeders are unaware of the default variance/covariance structure available in many commercially available software packages, named compound symmetry, assuming an equal genetic variance over locations and the same genetic correlation/genetic covariance between any pair of locations. Seasoned plant breeders are aware this is rarely the case, and that often there are large differences among the set of locations they use to test their breeding materials. This can represent a significant loss of efficiency and accuracy to any breeding programme, particularly if some of its locations frequently induce biotic or abiotic stresses. Linear mixed models enable plant breeders (or biometricians) to analyse datasets with increasingly more realistic assumptions, such as a fully unstructured variance/covariance structure where genetic variances are specific for each location, whereas genetic correlations are typical for each pair of locations.

Currently, linear mixed model platforms can be organized and automated so that hundreds of experiments can be analysed at the speed needed by breeding programmes in order to make advancement selections on time. We do not necessarily advocate every plant breeder becomes a user of linear mixed model-based analyses; nevertheless, their deployment represents an effective approach to increase the efficiency of selection, to gain deeper insight in return for the large effort spent on multilocation trials, and to increase the likelihood of commercial success.

The use of linear mixed model approaches is strongly advised when molecular breeding or genomic selection approaches are pursued. Unless the best phenotype is used to establish marker–trait associations, the theoretical strength of molecular breeding or genomic selection might not translate into the improved cultivars that farmers, end-customers and humankind require.

Until recently, deploying linear mixed models was constrained by the lack of computing power, limited availability of user-friendly software, and the very small number of practitioners within the plant breeding community. Now it is feasible for any breeding programme located anywhere in the world to analyse its data through mixed models approaches as these three constraints have generally been overcome.

10.3 Greenhouse management

A large proportion of the tasks that are necessary in the early parts of a plant breeding programme can often be carried out in a greenhouse. An integrated greenhouse system is not essential for a successful varietal development programme. However, many of the operations can be carried out more effectively if a greenhouse system is conveniently available. This is particularly true for private breeding programmes where the speed of cultivar development is of critical importance.

Greenhouses come in many different shapes and sizes and can be constructed from many different materials, including wood, aluminium, glass, plastic and polythene. The actual design of these systems will not, however, be covered here. A greenhouse can simply be considered as a relatively large area where there is some control of environmental conditions such as soil type, irrigation management, nutrient management, lighting and temperature.

The operations that are often carried out in a greenhouse with regard to a plant breeding programme include:

- artificial hybridization;
- seed increases of breeding lines, including progressing to homozygosity;
- evaluation of traits that are difficult to control under field conditions.

10.3.1 Artificial hybridization

It is possible to carry out artificial hybridization under field conditions; however, many breeding schemes use greenhouse facilities for this task because it is easier to achieve the conditions necessary to ensure controlled pollination between chosen parents (Figure 10.5). Also, it can often be possible to achieve controlled pollination out of season in greenhouses, and usually it is easier to prevent unwanted illegitimate cross-pollination.

The method used for controlled pollination will be dependent upon the crop species involved, whether the crop is out-crossing or self-pollinating. The major goal in artificial hybridization is to ensure that the seed produced is in fact from the particular, desired parent combination. Therefore steps must be taken to ensure that seed has not resulted from an unwanted self-pollination or from an accidental cross-pollination that is not intended by the breeder.

Artificial pollination therefore demands that naturally inbreeding lines (or lines which are self-compatible) are **emasculated** to avoid self-pollina-

(a)

(b)

Figure 10.5 Artificial hybridization in (a) canola breeding and (b) wheat breeding. Note that pollination bags cover racemes and ears that have been pollinated so as to avoid unwanted crosses.

tion. Emasculation in most crop species can be achieved by manually removing the male plant parts (i.e. anthers) before they are mature and pollen is dehisced. In some cases it is possible to use chemical emasculation where specific chemicals applied at the critical growth stage will render the plants male-sterile. Chemical emasculation is, however, not widely used in routine breeding, and mechanical emasculation is used almost exclusively as a means of avoiding selfing in crossing programmes.

Once the chosen plants have been emasculated, within a few days pollen from the male parents can usually be applied to the receptive female stigma. Pollen can be transferred manually, often using a small paintbrush, or by removing dehisced male parent anthers and brushing pollen onto the female stigma. Cross-pollination can also be achieved simply by having emasculated females grown in close proximity to male flowers and allowing pollen to pass naturally from male to female. When this is to be done it is common to place emasculated female flowers and pollen-fertile male flowers together within a pollination bag to ensure that the desired hybridization occurs and to avoid the female being pollinated by stray pollen which may, for example, be blown in the air or carried by insects. If necessary, within these bags, suitable pollinating insects can also be placed to help pollination efficiency.

In a number of crop species there are self-incompatibility systems, which have evolved naturally to encourage cross- (rather than self-) pollination and

hence maximize heterozygosity of plants within the species (e.g. as exists in many *Brassica* species). Similarly, many crop species have male sterility systems (either nuclear or cytoplasmic in inheritance), which can be utilized in cross-pollination systems. In both these cases it is not necessary for the female parents to be emasculated to guarantee cross-pollination.

Regardless of the breeding system, it is common to place pollination bags over flowers either prior to pollination (to avoid unwanted crosses) and/or after pollination (to ensure that no further pollination takes place) (Figure 10.5). It should be remembered that bagging crosses is time-consuming, and that if not done carefully can have an adverse effect on the potential success of the artificial hybridization and the amount of seed produced. Physical damage can be caused during the bagging operation, or the bag may create an environment unsuitable for seed production. It is also important to label carefully at each stage, otherwise the origin of any seed produced may be in doubt.

With many crop species, particularly crops that are clonally propagated and where the end product does not involve the botanical seed (e.g. potato, banana, sugarcane), it is not always easy to have parental lines develop sexually reproductive parts. In a number of instances flower induction can be achieved by manipulation of environmental conditions, by adding or reducing nutrient levels, manipulation of day length, or by artificially controlling the natural source–sink relationship.

For example, in potato, many past breeders have specifically selected breeding lines that rarely produce flowers, with the idea that energy put into sexual reproduction would detract from tuber yield. Flowering can be induced in some genotypes by planting tubers under long day conditions and having plants develop to maturity in shorter days. Enhanced flowering in potato can also be achieved by 'growing on a brick', where parent tubers are planted on building bricks and covered with soil. At the stage when tuber initiation occurs the soil is washed from the mother tuber and newly initiated tubers are removed, hence offering greater resources for flower development. A similar effect can be achieved by grafting potato shoots onto tomato seedling rootstocks. Applying high levels of nitrogen at particular growth stages can also sometimes increase the duration of the flowering period.

In other crops (and sometimes also in potato) the opposite is true, and reduced levels of nutrients cause stress to parental plants thus inducing the plant to flower, which would otherwise not occur under optimum conditions.

Finally, irrespective of crop or breeding system, it is always desirable to have multiple and sequential plantings of parents that are to be used in crossing designs. Genetically different parents will flower and dehisce pollen at different times, and multiple plantings will increase the possibility of achieving all the hybrid combinations planned.

10.3.2 Seed and generation increases

If hybridization is carried out between two homozygous parents, then the F_1 plants will be heterozygous at all the loci by which those parent lines differ, and all plants will be genetically identical. It is therefore common practice to go from the F_1 populations to F_2 under glasshouse conditions. This tends to maximize the use of F_1 seed because of the high levels of germination and survival that can be achieved. If F_1 populations are grown under field conditions it generally requires greater quantities of hybrid seed. This is disadvantageous since the cost of producing F_1 seed is usually high, because it involves emasculation followed by hand-pollination, as opposed to simply bagging the F_1 to allow selfing in order to produce the F_2.

With many annual (and some biennial) crops it is possible to grow more than a single generation each year by utilizing greenhouses, so reducing generation times and hence increasing the speed to homozygosity. Single seed descent used in spring barley, where plants are grown at high density and with low nutrition, can be used to increase F_1 populations to F_3 populations within a single year (i.e. three generations in 12 months).

At the advanced stages of a plant breeding scheme, greenhouse growth can also be utilized to increase advanced selections under controlled conditions prior to producing breeders' seed. This can be particularly useful in crops that are grown as true-breeding, inbred lines, but in which a relatively high frequency of natural out-crossing occurs (e.g. *Brassica napus*).

Tissue culture techniques are becoming a routine part of many plant breeding schemes. Plants rarely can be transferred directly from *in vitro* growth to field conditions without involving an intermediate greenhouse stage. Here the greenhouse stage could involve an intermediate operation where plants are weaned from *in vitro* to *in vivo* sterile soil mix, allowed to develop and are later transplanted to the field. Alternatively the greenhouse can be used to produce seed (or tubers) from plants that have previously been grown *in vitro,* and hence provided with a more easily protected environment free of pests and diseases.

10.3.3 Evaluation of breeding lines

One advantage of growing plants under greenhouse conditions, rather than field conditions, is related to environmental control. Control of the environment can be critical to guarantee epidemics of pests or disease, or to evaluate stress factors, to allow resistance screening. There have been a number of studies that have resulted in protocols suitable for evaluating plants under glasshouse conditions.

Disease and pest testing involves subjecting segregating breeding populations to a disease or insect and selecting those plants that show resistance. Examples include spraying barley seedlings with a suspension of mildew spores and screening for resistant lines, or spraying potato seedlings with a spore suspension of late blight or early blight and recovering the seedlings that are not killed. These tests are often more effective if there is good environmental control, such as is provided in a greenhouse. This helps to guarantee that the results are repeatable and the particular pathogens are allowed to increase and indeed infect the plants. It also allows control of the disease when it is time to stop further infection

Screening breeding lines for abiotic stresses can also be achieved under greenhouse conditions if the environment can be controlled in a repeatable and relevant manner. Stress screening has been shown to be reliable for such factors as tolerance to nutrient deficiency, drought, salinity and heat, where it is not always possible or easy to control the relevant environmental factors involved under natural conditions in the field.

It should, however, be noted that evaluations designed to be carried out under greenhouse conditions must first be compared with results that would have been achieved under natural field condition. There have been numerous cases where selection has been carried out under controlled conditions and later found to bear little, if any, relationship to what subsequently is experienced under field conditions.

10.3.4 Environmental control

Artificial lighting (fluorescent and/or incandescent) is nearly always necessary to achieve maximum use of greenhouse space. Lighting is, however, expensive both to install and to maintain, particularly if different lighting regimes are required. When, however, lighting is available, it usually allows the greenhouse to be utilized throughout the whole year.

If plants are to be propagated in the greenhouse throughout the year it will also be necessary to have a suitable heating and/or cooling system. A range of different types of systems is available and these cannot be adequately covered here. However, it should be noted that all the types involve a relatively high cost to install and operate. Thus it is usual to expect to have to justify the costs in terms of likely returns of, for instance, increased numbers of generations, effectives of tests, and so on. Good control of temperature is of course important if healthy plants are to be propagated. A particular example in which temperature control is often needed is in biennial crops where plants require vernalization (chill treatment) before they will flower. Plants grown under greenhouse conditions can be vernalized outside the greenhouse (e.g. in a growth chamber or cold room) but this

involves moving plants between facilities, which can be time-consuming and expensive if the number of plants involved is large.

Growth within greenhouses requires artificial irrigation. Irrigation can be by hand, which allows for some flexibility but does not usually allow for complex management systems. Automatic irrigation is usually preferable and can be of three forms:

- Above-plant irrigation (or misting) where plants are sprinkle or mist irrigated from above. This can be relatively inexpensive but can cause problems if plants are tall. Above-plant irrigation can also increase the risk of plants becoming infected by fungal diseases where leaf moisture is necessary for infection to take place. It can also be a problem by generating leaf scorch in strong sunlight.

- Below-plant irrigation where plants are irrigated by capillary action from having moist or wet material below the plant pots. Below-plant irrigation avoids the above-ground plant parts becoming wet, although it can be difficult to establish young plants and maintain very large plants with such a system alone, and it is sometimes necessary to hand-water as a supplement to the system.

- Drip irrigation where each plant pot is individually irrigated directly into the soil by a drip line. There are several different forms of drip irrigation and this system offers greatest flexibility over all others. This system is, however, the most expensive to install and is not always available in all plant breeding greenhouses. Since the system requires that individual drip lines are located in each plant pot, there can be some restriction on the number of plant units that can be grown, so this needs to be carefully considered when setting the system up.

All methods of irrigation offer the possibility of applying nutrients along with the water and so they can be provided 'continuously', thus enabling more optimized growth of plants over other methods of nutrient/fertilizer application.

10.3.5 Disease control

Unless disease is to be deliberately imposed as a selection pressure, as in the case of a resistance screening scheme, it is desirable to avoid as many diseases and pests as possible in a breeding greenhouse. The best results are invariably achieved when the plants are as healthy and disease-free as possible. Crop failure in a greenhouse as a result of plant pests (mainly insects) or disease can carry a high cost and should, of course, be avoided if at all possible.

Disease and pest control can be achieved by adopting good management practices, including sensible breaks in production along with appropriate sterilization strategies. However, the application of chemical insecticides and fungicides (and sometimes nematicides) is also a frequently needed practice. Application can be by spraying plants or pests or by fumigating a whole area within the greenhouse. The main advantage of chemical controls of disease and pests are that they can be applied in anticipation of a problem appearing. Therefore they offer **preventative** disease and pest control. The disadvantage is that many of these chemicals are indeed harmful both to humans and other plants as well as insect life, and it is therefore always desirable to minimize their use.

There are now many types of biological controls that can be used to control insect pests within a greenhouse. A well-known example is the release of ladybugs (ladybirds), which are natural predators of aphids, into greenhouses. There are many other predator insects available that can offer effective control of other insect pests. A sample of specific predator types available and the pests they attack include: *Amblysieus cucumeris* against thrips; *Aphidoletes aphidimyza* against aphids; and *Encarsia formosa* against whitefly; while ladybugs and green lacewings are used as general insect predators. The major drawback to biological predatory control relates to the fact that the pest must in fact be present, even if at a low level, before the predators are released (otherwise how will they be able to survive!). It is therefore difficult to avoid some insect damage and almost impossible to achieve complete preventative control.

The risk of soil-borne diseases can be avoided (or at least substantially reduced) by using only sterile soil, or soil mixes, in the greenhouse. However, unless an inert, synthetic soil substitute is used (e.g. 'Perlite' or sand/gravel/Perlite) the possibility that disease will occur as a result of infected soil cannot be entirely avoided. Often the sterilization procedure fails to remove all disease or fails to kill microbial spores or weed seeds. In addition if peat moss is used in soil mixes it is almost impossible to ensure the mix is free from insect pests that have a reproductive cycle in the peat moss.

Achieving good disease and pest control in greenhouses also relies upon other factors. For example, good insect-proofing throughout the house will reduce the risk that insects will enter the greenhouse. However, it should be borne in mind that people are very effective spreaders of plant diseases in greenhouses. Personnel from the breeding programme are likely to be in contact with plants outside the greenhouse (i.e. will visit field plots) and so there is a great risk that these staff will transmit disease or carry in insect pests prevalent to the crop with them while visiting the glasshouse. Simple rules, such as carrying out any greenhouse operations first thing each day with other field tasks being done later, can help in reducing disease incidence and spread. Having a mat soaked with a disinfectant agent at each door of the greenhouse is a cheap yet effective barrier to keeping many pathogens out of a greenhouse.

Plant viruses can cause particularly serious problems in plant breeding schemes since many virus diseases are transmitted through the planting material (e.g. seed viruses in cereals and tuber-borne viruses in clonal crops). Many viruses can be eliminated by avoiding the virus vectors, which are often insects (particularly aphids). Workers in the breeding programme can also be responsible for carrying insect vectors into greenhouses on their hands or clothes. Again, the risks of infection can be reduced by applying simple rules (e.g. protective clothing, sterile gloves, etc.).

10.3.6 Economics

Despite the attraction of greenhouses as an integral part of any plant breeding programme, there is no doubt that this facility can be responsible for a high proportion of the overall cost of operating a breeding system. In addition, due to the high cost of building and maintaining greenhouse facilities, the actual space available will be limited. In the practical world (the one in which we unfortunately all live), economic use of greenhouse space is an ever-present major factor.

Plants in greenhouses are grown either in pots (or some other individual unit) or in beds (where many plants are propagated together). The size of pot used (or the plant density in seedling beds) will have a large influence on the number of plants that can be grown in a unit area. It is therefore necessary to choose a pot size/density that will allow good plant health and growth. If small pots are used, then more plants can be propagated at lower cost. If they are, however, too small, then plant health, root development or reproductive efficiency can be affected.

It is necessary to allow access to plants grown in greenhouses. Increased efficiency of greenhouse space can be achieved by using **rolling benches**, where plants are grown on benches that can be easily moved to allow access, but minimizes the greenhouse space that is allocated to walkways. Rolling benches can, however, cause problems in cases where plants are tall and require staking or tying to prevent them falling over and being damaged. In addition, rolling benches can increase the need for uniform lighting over the whole greenhouse area rather than only over designated growth areas or static benches.

10.3.7 Experimental design in the glasshouse

One final note on the use of greenhouses and plant breeding relates to experimental design. Many believe that the conditions in greenhouses are such that there is uniformity in soil type, lighting, irrigation, and so on. In comparison with conditions that may prevail in the field, there may indeed be less environmental variation in a greenhouse. Despite this, it should be noted that there will be differences nonetheless between, say, plants next to the glass and those in the centre of the house. Therefore **all** experiments grown in greenhouses should be treated with a clear understanding of the fact that variability in environmental conditions still exists, and therefore good experimental design, replication and randomization will be as important in greenhouse experiments as in other situations.

10.4 Field plot techniques

A large proportion of the work in a plant breeding programme is carried out using field trials. The aim of plant breeding is to develop superior cultivars that are genetically more adapted than the cultivars that are already available. New and old cultivars are grown within agricultural systems on a large scale. For example, wheat grown in the Pacific Northwest of the US is grown in fields that cover many hundreds of acres. Obviously it is not possible to evaluate the many thousands of potential new cultivars in a plant breeding scheme on the large field areas in which they will eventually be grown if successful. The aim, therefore, of field trialling is to **predict** how each genotype would perform **if they were grown on a large acreage basis**.

In order to grow accurate and representative field plot trials, it is first necessary to determine the way that the crop is grown in agriculture and to try to use this as a basis for the practices used in the plot trials. Factors that need to be determined include:

- land preparations and agronomic management;
- seeding rate, final plant density and depth of planting;
- nutrient levels and when nutrients are available (pre-plant and/or post-emergence);
- irrigation management;
- timing of operations such as planting and harvest windows;
- chemicals available, for example, which insecticides, fungicides or herbicides are registered for use on the crop, at what rates they are applied, and what the seed treatments are;
- regions where the new cultivars will be targeted.

Do not forget that the major aim of field trials is to **mimic what would happen in commercial agriculture**. Therefore field trials should usually be planted at the same time that the crop is normally planted. Planting depth, plant density, nutrient management, weed control, disease control, harvest time and method, and post-harvest treatment should all match commercial production as far as this can be achieved within the restraints of small plot management.

10.4.1 Choice of land

In order to choose a good area of land for field plots, it is necessary to identify the factors that magnify soil differences and to reduce, if possible, soil heterogeneity.

Fertility gradients are generally more common in sloping land. Soil nutrients are soluble in water and tend to settle in the lower land areas. Therefore these lower soils tend to be more fertile than the higher areas. An ideal experimental site will be on flat land but this is not always possible – how many farmers' fields have you seen that are as flat as a football pitch?

If the land has previously been used for plot experiments, then this can lead to increased soil heterogeneity. Therefore areas of land that have previously been planted to different crops, different and varied fertility regimes, or subjected to varying cultural practices, should be avoided if possible. In cases where this has occurred, then the area should be planted with a uniform crop, with uniform management and fertilization for at least two years before it is reused for plot experiments. A second source of soil heterogeneity is related to unplanted alleys or roadways from previous experiments. If possible, unplanted alleys from previous research should be marked and avoided.

Grading (ground levelling) usually removes soil from elevated areas and redistributes it to the lower areas. This operation, which is designed to reduce slopes, results in uneven depths of topsoil and often exposes unfertile subsoil. These differences can prevail for many years and should be avoided unless soil heterogeneity trials determine that the grading effect is minimal.

Large trees and other structures can cause shade, which will affect plant performance, and also their roots spread further than their canopies and so will influence plant growth. Areas near buildings may be affected by soil movement and heterogeneity caused by the building operation. Plots adjacent to trees or wooded areas can also carry a greater risk of damage by birds or mammals.

The evaluation of soil heterogeneity requires growing **uniformity trials**. These involve growing a single cultivar (or a number of cultivars) in plots with very high levels of replication. Uniformity trials highlight soil fertility gradients and identify particularly productive or non-productive areas in fields. Uniformity trials can be used to produce contour maps of productivity. Statistical procedures such as serial correlation studies or least mean squares between rows, columns and diagonals can be applied to determine the significance of soil heterogeneity. It is also possible now to use various spectral imaging techniques to assess the crop's growth and performance.

Although uniformity trials have their place in field experimentation, they usually have little to offer a plant breeder. Uniformity trials indicate the response of specific genotypes to a given area in a given season. When these trials are repeated with different genotypes or in different years, then different results are often obtained (not surprisingly). In plant breeding evaluation trials, the number and diversity of genotypes under test are usually far greater than what can be considered in uniformity trials. Also it should be noted that often there is little choice of what land can or cannot be used for plot trials.

A plant breeding programme usually, at least in the advanced stages of selection, uses a number of different locations, sometimes spread over a large geographical region or 'target population of environments'. One main location may be identified where the majority of material is evaluated in the early and intermediate selection stages or where seed is increased. A number of different locations will be used (dispersed throughout the region where the new cultivars will be targeted) where advanced lines are tested for adaptability. Where many locations are used it is common to use farmers' fields for test plot evaluation. Some of the distinct differences between a farmer's field and the conditions prevailing at, say, an experimental research station, would include:

- Lack of experimental equipment or lack of small plot machinery. This can often be easily overcome by taking planting, spraying and harvest machinery from the research farm.
- Lack of experimental facilities such as precise irrigation control and pest or disease control, weather stations, and so on.
- Lack of post-harvest storage or assessment facilities. Therefore harvested produce needs to be transported to a central testing laboratory for post-harvest quality assessment.
- Large variation between farms and fields within farms. This is often not a major problem as the majority of trials on these farms are to select for such adaptability over a range of environments.
- The farm sites are usually further away from the base research laboratory and sometimes long trips are necessary to visit the plots. Therefore visits are usually limited and it can be difficult to identify potential problems as they arise and hence avoid their worst effects.

Despite all the potential difficulties with off-station or farm trials, it is possible to achieve very good results. Best results are usually obtained when the 'better' farmers are chosen for the tests and when these farmers are specifically interested in the results from the trials. Finally, when trials are to be carried out on farmers' land it is always advisable to keep the experiments simple and to have relatively large plot units, to make them more robust. It is always good advice to spend some time and provide farmers with at least a brief explanation about the trials underway, rather than just appear to plant, collect notes and harvest. This not only shows respect to the farmer, but also increases their interest and commitment to the trials, reducing the likelihood of trial losses or less than ideal agronomic management.

10.4.2 Plot size and replication

Historically, it has always been assumed that larger plots are more efficient and more representative than small plots in yield and other assessment trials. Nevertheless, this is not necessarily always the case. Research carried out at CIMMYT (International Maize and Wheat Improvement Center, or *Centro Internacional de Mejoramiento de Maíz y Trigo*) compared limited versus normal nitrogen soil supply using different plot sizes, and was able to conclude that the residual variance of progenies grown under nitrogen-limited conditions in small, non-bordered plots was usually less than when grown in larger, self-bordered plots, suggesting that soil heterogeneity was likely to be the main source of residual variance. Similarly there is no doubt that greater replication levels are always more desirable than fewer replicates. However, when the breeding field trials are being carried out under drought, low nitrogen or other abiotic stress conditions, which amplify the natural variation existing in agricultural soils, a larger number of replications might increase the likelihood of encountering unexpected soil variation patterns. In such circumstances, rather than using a larger number of replicates, it might be more effective to use appropriate statistical designs able to accommodate such variation patterns as well as a linear mixed models-based statistical analysis, as previously discussed. In other words, standard plot size and number of replications do not fit all requirements, and need to be adapted to the needs and resources available.

The difficulty of organizing efficient field trials is often related to some compromise in plot size and replication which will allow large numbers of test lines to be evaluated at low cost and on as small an area of land as may be available.

Land availability may not be the limiting factor in determining plot size or replication level. It would be pointless to organize more field plots than could be effectively managed by the staff available. Similarly, data need to be collected from effective field trials, and if too many unit plots are grown then it may not be possible to effectively evaluate all the plants or the produce from the trials. Finally, some crop species produce products that are bulky or perishable. It may be necessary to store the produce (or at least a sample of produce) from each plot, so the storage space available would then be a major determining factor.

In the early selection stages the amount of planting material available is often limited, and this puts practical constraints on the field trialling that is possible. For example, if only 2 g of seed are available for evaluations, and commercial seeding rates are 4 kg per acre, then only small plots with limited replication will be possible.

Increasing replication will be more efficient than increasing plot size in the majority of cases. Therefore if 200 plants were to be grown for evaluation purposes, then the most statistically efficient design would involve 200 replicates of randomized single plants. From a practical standpoint this may not, however, be the most effective or practical method or provide the most representative outcome. For example, there may not be the necessary machinery available that would allow for mechanized planting of completely randomized single plants. Therefore the dimensions of machinery available can be a determining factor when setting plot dimensions. If the only plot-seeder available plants six rows, then all plots are likely to be a factor of six rows wide. Similarly if a small combine harvester is available that has a cut of 1.5 metres, then plots are likely to match this harvesting capability. In addition, single plant evaluation may take greater land areas than would be available. Finally, single plants, if completely randomized, need to be spaced distinctly apart to differentiate one from another. The phenotypic performance of some crop species is markedly different when grown at wide spacing (wider than would be normal for commercial production) than if grown at narrow spacing.

Different plots in field trials invariably contain different genotypes. The performance of these genotypes can, in some cases, be affected by competition from the adjacent plots. For example, if a short genotype is grown next to a tall vigorous genotype, then the performance of the short type may be reduced compared with a single stand of the short stature plants. To a large extent these effects can be reduced by good experimental design and replication where the probability that adverse or advantageous competition occurring in all replicates is reduced with increasing replication. Sometimes nesting based on pedigree might alleviate such competition effects, since half or full sibs are likely to have a similar plant height, architecture or rooting patterns than sibs from unrelated pedigrees.

Some researchers suggest growing larger plots and harvesting or evaluating only the centre rows (i.e. that portion that is completely surrounded by plants of like type). It should be noted, however, that this would require greater amounts of planting material and larger land areas. It should also be noted that genotypes can suffer as much (or greater) competition by being grown by itself, and **ripple effects** can occur. To examine ripple effects, consider a five-row plot (rows A, B, C, D and E) where row A is grown adjacent to a different, tall and very competitive genotype. In this case then the A row may contain small stunted plants due to the competition from the tall genotype, and hence will result in lower yield. Row B, however, is likely to be affected by competition because although grown next to a like genotype, the like genotype (A row) is stunted and low-yielding. Therefore row B will be taller and more productive due to the lack of competition from row A. In a similar manner, row C will have to compete with the larger, more competitive B row plants and have reduced yield. The competition effects will be reduced, however, with increased distance from the tall different genotype and hence the term 'ripple effect'.

Breeders need to be aware that by harvesting only a portion of the total plot the error variance will be increased, as the error variance of the mean (average of all plants in the plot) is the error variance of a single plant divided by the number of plants.

It should be remembered that the value of field plot trials is to make comparisons and not to estimate definitive yield performance. Therefore field trials are used to compare the **relative performance** of different test lines in comparison to control entries. In this case increased or decreased yield as a result of competition will only become a factor if there is interaction between edge effects and genotypes.

10.4.3 Guard rows and discard rows

It is common practice to surround trials (and sometimes even individual plots) with **guard** or **discard rows**. These are areas planted to a specific cultivar or genotype, which is not part of the evaluation test. Guard rows are used for several reasons, including:

- If any mechanical damage occurs (e.g. a tractor spray unit accidentally runs over a plot), it is likely to happen to the edge plots. If these are to be discarded, then this damage is less likely to affect the performance of any of the test or control entries.
- Phenotypic performance can be greatly increased by avoiding differential edge effects. Therefore plots that are grown on the edge of a trial will not have any competition on one side, while all other test entries will be affected by competition from adjacent plots.
- In multifactor field trials, guard rows can be used to separate different treatment factors that may be difficult to apply to specific areas without having some effect on the immediately adjacent plot.
- Sometimes, guard rows avoid or reduce the theft of edible organs such as corn ears or fruits from experimental entries, as trespassers are tempted to steal edible plant organs from guard rows rather than from experimental breeding lines. Although this situation is not always considered in developed countries, in some developing countries it represents a significant issue that must be dealt with properly, as otherwise years of hard work and research investments might disappear overnight.

Guard rows are usually the same species as that under evaluation, but this is not always a necessity.

10.4.4 Machinery

Over past decades there has been an increase in the availability of small-scale machinery suitable for field plot trials. Most of the machines are designed as miniature versions of what is used in larger-scale agriculture. Tasks that can now be mechanically orientated include:

- Planting
- Weed, disease and pest control
- Harvesting

It is nearly always desirable to plant field trials mechanically as this is likely to result in more uniform plots than can be achieved by hand planting. This is almost always true for relatively small, seeded crops (e.g. barley, wheat and rapeseed). When the planting material is larger (e.g. potato tubers), hand planting can produce as good, or better, results compared with mechanical planting. The need for automatic planting will be dependent, therefore, on the size of seed to be planted, the density of seed sown, and the time that can be saved by automatic planting.

The most common small-seed plot planters are **cone planters** (Figure 10.6). This type of seeder can be used very successfully to plant either very small plots or much larger plots. A measured or counted quantity of seed is poured over a cone such that the seed is evenly distributed around the base of the cone that operates the seeder. During planting the cone revolves and seed passes through a hole to be subsequently dropped via disk or tube coulters into the soil at the required depth. It is often possible to plant several rows from the same unit seed lot. In this case, after the seeds drop from the cone they are evenly distributed to a number of tubes, which will each plant a single row. Cone planters are usually designed so that a range of plot lengths are possible. This is achieved by gearing the rate that the cone revolves. After one complete revolution then all the seed from one lot will have passed down the open hole. Planting can be done with continuous movement, with each plot being dropped onto the revolving cone at a designated **trip point**.

Cone planters are available where the seeds for each plot/row are loaded into a cassette or magazine. This is then mounted above a seeder unit with several revolving cones. With this system it is possible to plant several sets of rows simultaneously with each set being a different genotype. Cone seeders are particularly useful as they can be used with small seed lots, and all seed loaded is planted to completion. Therefore there is no need to maintain a seed reservoir, which would need to be emptied between different plots/genotypes. Cone planters are also self-cleaning.

With small seed, cone planters can result in an even distribution of planted seed, but it is

Figure 10.6 Planting yield assessment trials using a single cone planter.

sometimes desirable to have a more precise placement of seed. If this is necessary then **precision planters** can be used. With these machines it is possible to obtain spaced plants at relatively even density. Precision planters are in general of three types:

- Belt planters, where a reservoir of seed is maintained over a revolving belt. The belt has holes cut which are of precise size and shape so that only a single seed will fit through the hole. The density of planting is achieved by the number of holes in the belt and the rate at which the belt revolves (e.g. Stanhay seeders).
- Vacuum planters, where suction is applied to a revolving plate that has holes drilled to allow only a single seed to be sucked to the plate at each hole position. As the plant revolves, the vacuum is turned off at a specific place in the plate's rotation. At this point the single seed is dropped into the soil.
- Cup planters, where a series of rotating cups are dipped into a reservoir of seed. The size of each cup is such that only a single seed is scooped up as the cups revolve through the seed. At a specific point on the rotation the cups are tipped and the single seed is dropped into the soil.

The major limitation of precision planters, when used by plant breeders, is that they usually require a volume of seed in the reservoir in order to operate effectively. Therefore they have only limited use when small amounts of seed are available.

In some crops, transplanting is common, even on a commercial scale (e.g. fresh tomatoes). Small-scale transplanters are available that allow automatic transplanting of field plots. Seedlings are grown in 'seedling flats'. At transplanting time the seedlings are removed from the flats by hand and placed into the transplanter. Systems have also been developed where the seedling flat fits onto the transplanting machine and the whole operation is automated. In this latter case it is usually possible only to transplant large plots.

The areas between different plots in field trials are usually left unplanted. There is very little competition in these areas and weeds can be a major problem. Weed control in field plot trials can be carried out mechanically or chemically. Mechanical weed control can be by hand hoeing (a task often enjoyed by many summer student helpers!).

Automatic mechanical devices such as rota-tillers and harrow cultivators can achieve inter-row and inter-plot weeding. Often it requires a combination of chemical herbicide application, rota-tilling, harrowing and hand-hoeing to ensure that plots remain weed-free.

Evaluation of disease and pest resistance is an important factor in field testing. Test lines and controls will be grown in regions or areas where specific diseases or pests are common. In these trials, disease is often encouraged by including particularly susceptible genotypes as **spreaders** and by artificial inoculation of these spreader lines.

In other field studies it is not desirable to have disease or pest epidemics, and so these need to be controlled. Control is usually by chemical application, although some biological control of insects may be available. It should be remembered that many diseases are spread (and most are not helped) by having poor weed control.

A variety of harvesting machinery is available including small-plot combine-harvesters (e.g. Hege or Wintersteiger) that will cut, thrash and partially clean seed samples (Figure 10.7), and harvesters that will dig root crops such as potatoes. Often very small plots (or individual plants) need to be harvested separately. In some cases this can only be achieved by hand-harvest (e.g. pulling single plants or hand-digging individual produce). In the case of grain crops the small plots can be hand-harvested but the seed is removed from the selected plants by small-scale mechanical thrashing.

10.5 Use of computers in plant breeding

Many routines in a plant breeding programme follow a cyclic annual operation. Therefore the same tasks (or similar operations) are carried out on a seasonal basis. In general terms a simple breeding scheme may involve:

- Deciding which breeding lines are to be tested in which environments. Which characters are to be evaluated from each trial? What control or check genotypes will be included in each trial for comparisons?
- Designing experiments or evaluation trials. What types of experimental design will be used for each trial (unreplicated designs, randomized complete

Figure 10.7 Small-plot combine-harvesting canola trials.

block designs, lattice designs and split-plot designs)? Having decided on an appropriate design, then field plot plans need to be produced, planting material organized and arranged in order for planting.

- A number of clerical tasks will be required, such as producing plot labels or harvest labels (Figure 10.8), genotype lists and perhaps score sheets.
- After field (or other) trials have been planted, then data will be collected throughout the growing season, at harvest and post-harvest.
- As data are collected, each variate assessed needs to be analysed. More detailed analysis of over-site trials and to examine relationships between variates will be carried out.
- Statistical analysis is only one step in data interpretation. Further data examination techniques of scatter diagrams, histograms or bar charts can be used to obtain a better understanding of data.
- Selection will be applied based on information collected. Breeding lines will be divided into various categories (e.g. definitely select and advance to next stage, not quite sure so repeat in smaller trials, discard from the breeding scheme).
- Once the 'best' genotypes have been identified, the operation will begin at stage one again the following year.

This cyclic operation will continue over a number of years, starting in the early stages where perhaps many thousands of lines will be tested at limited sites and with few characters recorded, and moving to an intermediate and advanced stage, until after several selection rounds only one or two potential new cultivars have survived for varietal introduction.

Computers can be of great benefit to plant breeding in carrying out all of the tasks above, and perhaps even in others not mentioned. Often it is necessary to use different software packages for different aspects of the programme, although there are a few packages that have been specifically designed for managing plant breeding programmes or for field experimentation studies.

Computer software packages that are available are all roughly of the same form. There is a central data storage (database) that can be accessed by a number of routines. Each routine will perform specific operations. Various routines may add information to the database while others will take information from the database and carry out a specific task.

The following section will examine a number of the options or routines that are available and explain how they may be used to increase the efficiency of a plant breeding programme.

10.5.1 Data storage and retrieval

All plant breeding schemes will generate vast bodies of data. If these data are to be used effectively for

Figure 10.8   Assortment of planting and harvest labels used to organize breeding material.

selection of the most desirable genotypes, then reliable storage and retrieval of information is essential. Computers offer the option of storing data in such a manner that datasets can be tabulated for inspection in a number of different ways using **database management systems**.

In simple terms there are two types of databases used in plant breeding, called **breeding line databases** and **germplasm databases**. There are a number of differences between the database structures depending on the two types.

Plant breeding databases will store information on assessment trials. Therefore, a large proportion of the entries in these databases will be discarded after each selection stage. Early generations will have thousands of **records** (where one record is associated with information from a single genotype) with only a few data scores on each. Conversely, a genotype that survives to the advanced stages will have been assessed over several years, and in many of these years, assessment will have been carried out at a number of different locations. Therefore the amount of data storage space needed for each record will depend upon the stage that has been reached in the breeding scheme.

Germplasm databases, on the other hand, hold information on a wide range of different genotypes. However, unlike a breeding database, new accessions (or records) are added but very rarely are records deleted. Information stored in a germplasm database will have been collected over years and sites but not all accessions will have been assessed in a common environment. It is therefore necessary to **rate** accessions (e.g. on a 1 to 9 scale or A, B, C, etc.) so that comparisons can be made. It is very rare that actual yield data (e.g. t/ha) are stored on a germplasm database; it is more likely that a particular accession will be rated by a particular 'score' for yield.

Irrespective of the type of database, each record (test entry or accession) will be assessed for a number of characters or traits of interest. Variates can be of two forms, **numerical** (e.g. disease rating of 2, yield of 25.32 t/ha) or **character** (e.g. alpha-numeric character string like 'Yellow flowers'). In addition the different database types will hold other information not related to simple assessment, for example, an alpha-numeric string to identify the particular genotype (e.g. PI.23451 or 89.BW.11.2.34) and parentage of the line. Germplasm databases may store information not usually stored on a database, for example, species name, ploidy level, source of origin of seed, age of seed, amount of seed available, and so on.

When a particular genotype entry is introduced into a plant breeding programme it is usually identified by an alpha/numeric code. For example, cultivars will have specific names 'Jack's Wonder'. Genotypes, which have derived from other germplasm collections, will have an accession number. For example USA plant introduction lines all have PI numbers (e.g. PI.12342).

Different genetic lines derived from a breeding scheme will generally have similar identifying codes. Genotype codes can be assigned in numerical order (e.g. line 1, line 2, etc.). It is more useful to assign an identifying code that provides some information regarding the background of a specific genotype. For example, in a specific rapeseed/canola breeding group all crosses made are assigned a code identifier that includes the year of crossing, a two-letter code of the purpose behind the cross, and a numerical number. A cross identified by 93.WI.123 would indicate the 123rd cross made in 1993, with the purpose of developing a winter industrial (WI) type.

Specific individual genotypic selections from the cross would have different trivial numbers (e.g. 93.WI.123.23, 93.WI.123.69, etc.). If some form of pedigree selection scheme is used, then additional trivial numbers can be added to indicate the number of within-population selections made.

In setting up a suitable database the user must decide on a suitable **database structure**. This will determine the number of entries to be tested in each trial, the number of locations where evaluations will be carried out at each stage, and the number (and type) of data that are to be stored.

Irrespective of the type of database or form of data storage the primary aim is the same: to make information available for inspection in a clear and concise form.

10.5.2 Field plan design

Field trials and experiments are of major importance in a successful plant breeding programme. The ability to use computers for randomization has been realized for many years, and most programmes use some form of computer generation of field trials. These packages use entered information such as type of design, experiment title, number of entries, number of replicates, and so on, and produces a randomization along with a map representation of how the plots will appear in the field.

Once a computer has generated a field design it is possible to store all the trial details, number of entries, entry codes and actual randomization on a database system. This information can be retrieved later for analysis of data or producing plot labels.

10.5.3 Clerical operations

Despite advances made in database management, the ability to carry out complex selection strategies or analysis using computers, the simplest and most useful task a computer can do for a plant breeding programme is to perform as many of the routine clerical operations as possible. Several years ago, all plant breeding schemes produced all field maps, genotype lists, seed packet labels and score books by hand. Even in cases where highly methodical and dedicated staff are used, there are inevitable transcription errors, and more importantly it is time-consuming.

Computer systems can easily be used for:

- printing field plot plans (mentioned above);
- printing a variety of labels that can be used to organize experiments, store seed or planting material, or harvest experiments (Figure 10.8). Organization of large randomized experiments can be achieved with ease if computer labels are printed with the genotype identifier along with a field position (e.g. either a plot number or a two-dimensional array number);
- printing score books, which show simply a number of boxes in which data can be handwritten for a number of traits;
- genotype lists of lines, which have attributes in common (e.g. all bread-quality wheat, all industrial rapeseed, etc.);
- data summary tabulations of information stored in the database;
- keeping track of exactly where each test entry is (e.g. what stage of the breeding scheme, at which sites each entry is being evaluated, exactly where in the field each test line can be found).

10.5.4 Data collection

A breeding scheme is only as effective as the data collected on how the different genetic lines perform. The breeder must collect data on performance of different traits from appropriate assessment trials. Data collection and data management are areas that have received very little attention in a plant breeding context. However, the information-gathering stage is of great importance.

Data management is of three types:

- Collection of information into a form suitable for computer entry.
- Validation of data to ensure that errors can be corrected before data analysis.
- Sorting data into a form suitable for entry either into analytical software or for storage on a database.

Data collected from experiments can be hand-recorded onto score sheets and then entered (i.e. key to disk) at a later stage. This form of data collection and entry may appear inefficient compared with more direct systems (below). However, there are one or two advantages of hand-recording and later entry. There is always a hard copy of the information collected that can be referred to at a later date. Hand-recording of visually assessed data can often be achieved quickly compared with other means (although the data still need to be typed later). Therefore a combination of experienced assessor and experienced typist/data recorder may be as quick and efficient as directly logging data.

Information can be **logged** directly into a computer system. Data logging can be of two forms:

- Information is entered into a hand-held data logger. When combined with plot identification through bar codes, the use of data loggers can greatly speed up data collection in the field.
- Information is transferred direct from an analogue/digital machine (e.g. electronic balance or moisture meter).

In each case, the data are usually later transferred to the main computer system. Data validation (e.g. that the numbers are reasonable or within a certain range) can often be achieved during, or as a part of, the transfer operation. Alternatively, data may be validated or checked as entered into a hand-held unit.

Automatic transfer of data from analytical machinery is always an advantage as it reduces time and effort to achieve results. More important, however, is that this form of data collection usually avoids any additional transcription errors (e.g. writing down or hand-typing the wrong number).

Hand-held data loggers are rapidly becoming smaller, more sophisticated and cheaper. However, it can often take longer to enter data (particularly alpha-numeric character information) into a hand-held unit than to simply write the information. In addition, some expertise is required in the use of hand-held units, particularly accurate typing skills. Finally, if data are collected in a handheld data logger it is always best to have some form of hard-copy printout of the data as recorded. This would indeed apply to any data recording.

10.5.5 Data analysis

One primary consideration for the analysis of assessment trials is the ease and speed of processing. Often the most important traits (e.g. yield and quality) are not recorded until late in the season. A rapid throughput of analyses can therefore be critical to allow selection decisions to be made and new trails organized before planting time. This can best be achieved if all the genotype identification codes, experimental design details and randomization information have previously been stored in a database. In order to carry out an analysis of variance for a single variate assessed at one location, and produce an easily understandable but comprehensive output, it should only be necessary to enter parameters to identify which trial is to be analysed and the variate name.

As data are collected throughout the growing season, analysis of individual traits can be carried out soon after data collection. Inspection of de-randomized data and genotype averages can often serve as a good check that there are no major errors in the data. It is important that each variate be analysed to determine the variability within the genotypes for particular characters. Most database systems will automatically store means and statistics as analyses are performed.

The mode of data entry will, to a large extent, be determined by the method used to collect data (i.e. automatic logging, data logging, or pencil and paper). Irrespective of how the data are collected, eventually the data to be analysed will be available for entry into an analysis and storage scheme. The order in which numbers are entered can differ from one of no pattern (not a good idea), field plot order (either going across the trial or up the trial), or in standard order (e.g. genotype 1 replicate 1;

genotype 1 replicate 2; genotype 1 replicate 3; genotype 2 replicate 1; etc.). It is important that data be entered in the order expected by the software package.

Other features that will facilitate a rapid and efficient turnover of analysing individual traits and storing information will include:

- the ability to estimate missing values;
- the analysis of a subset of the total number of replicates (e.g. in a four-replicate trial, data for some traits may only collected from replicates three and four);
- to be able to transform (e.g. ARCSIN transformation for percentage data) or convert (e.g. convert dates to days after sowing, or to convert plot yield into t/ha) data before analysis;
- to derive variates from single or multiple datasets before analysis (e.g. if grain yield is recorded along with straw weight, total above-ground biomass can be derived by adding the two recorded characters).

If multiple environments are used (say at the advanced trial stage), then over-sites analysis (simple analysis of variance or joint regression analysis) can be performed using stored means from individual site analyses. If an assessment trial is grown at two (or more) locations, and yield per plot is recorded from multiple replicates at each site, the following procedure can be used to obtain an analysis of variance of yield over sites:

- Analyse data from each location separately and store the genotype means on a database, along with the error variance from the analysis.
- When each location has been analysed separately, then an analysis of variance with source terms: genotypes, locations, genotypes × locations and an error term can be produced quickly and easily. The error term is obtained by simply pooling the error terms from each of the individual analyses.
- When conducting across-locations or across-years analyses, it is sometimes overlooked that the default structure for the genetic variances and covariances in most statistical packages is that known as 'compound symmetry'. In many cases these assumptions do not hold true, and therefore analysis results would not be correct. Based on the particular nature of the multi-environment testing effort of breeding

programmes, breeders have to assess whether they should use variance/covariance structures more appropriate to their specific conditions. This is sometimes not considered because of fears that the statistical analysis will become more complex and/or that it will somehow slow down the speed of the breeding programme. Unfortunately, the consequence of this is that selection decisions departing from the correct ones are being made, which can decrease the rate of genetic gains being made and, more importantly, the ability of new cultivars to address the increasing needs of humankind.

To interpret data from assessment trials and provide indications of possible selection strategies, then joint regression analysis, over-site analysis, simple and multiple regressions and correlation analysis can all offer an insight into the variability of characters and also the relationship between traits. In addition, visual inspection of histograms and scatter diagrams can help in decision-making. Multivariate transformations (canonical analysis, principal components analysis, etc.) have been suggested as possible aids to plant breeders by reducing the dimensions of selection problems. If these transformations are readily and easily applied to breeding datasets, perhaps plant breeders will more readily use them.

Alongside complex analysis, it should be possible to carry out simple calculations, to include addition, subtraction, multiplication and division. Other calculations that may be helpful would include expressing data as a percentage of either the trial mean or the average performance of one or more control lines.

10.5.6 Selection

If many hundreds of lines are to be considered for selection, then computer simulation (by selecting a subset and comparing that subset to those lines rejected) can be a big help in either setting culling levels for different characters, or in setting weights in an index scheme.

The speed with which different selection strategies can be compared using computers offers the potential of investigating a number of different selection options within a narrow time schedule between final assessment of genotypes and preparations for planting the following stage trials.

10.5.7 Data transfer

The amount of data collected on individual genotypes in a plant breeding programme is directly
proportional to the stage of selection. By the most
advanced stages, data from surviving lines will have
been collected over several years and locations.
If plant breeding database systems are to be of a
useable size and if all information available is to be
stored together in a common database, then each
season either:

- records for discarded breeding lines must be
  deleted from the database and new data storage
  allocated to those selected lines; or
- selected lines must be transferred, along with
  any data collected, to a new database where
  additional storage is available.

Either of the above options can be used and both
are equally efficient. If the first option is chosen,
the old database can first be copied before the
unwanted records are deleted. This allows access to
data from discarded lines, which can often be used
in the future, for example to gain an indication of
particular defects of specific parents. It should be
stressed that any breeding programme must keep
updated, secure back-ups of its datasets, pedigrees
and related information at all times. If for whatever
reason the datasets of a breeding programme are
lost and a properly updated back-up is not available, many years of work and significant funding
might be lost as well, and in extreme cases the
affected breeding programme might be forced to
either start from zero or shut down. Fortunately, the
rapidly decreasing cost of computers and servers
makes easier and cheaper the secure storage of
the massive amount of data a breeding programme
accumulates over the years.

10.5.8 Statistical consultation

It is essential that agricultural experiments have
clear objectives and that they are well organized
and are designed based on sound statistical reason-
ing. Consultation with qualified statisticians should
be done before, and also throughout, the experimentation period. Plant breeding assessment trials
are no exceptions. There has been some concern
that statisticians will not be consulted if breeders
are capable of easily generating a number of different experimental designs and performing complex
analyses of data from these experiments. Care must
be taken to ensure that the appropriate design is
chosen to answer the questions required. Most
plant breeders' trials are, however, of a standard
form where a number of test genotypes are compared in performance with a number of standard
or control cultivars. Although the majority of plant
breeders are more than capable of using the appropriate experimental design and making the correct
interpretation of a standard analysis, it should be
noted that many analysis types (e.g. multivariate
analysis) are now readily available to non-qualified
workers but that interpretation of these results
often requires an experienced person. The point is,
therefore, that statisticians should not be ignored
and where possible they should be consulted and
encouraged to contribute ideas in data interpretation. The support provided by statisticians is often
much more valuable when they are engaged before
experiments are conducted and designed, rather
than once they have been harvested.

10.5.9 Ease of use

One feature of computers and computer software
that has not been discussed is the ease of operating the system. Many software packages are
user-friendly, which means that they can be used by
relatively inexperienced staff. This does not, however, imply that these database systems can be used
without computer training. There will be at least
minimal training required if a database scheme is
to be integrated into a breeding programme.

Most user-friendly computer packages give clear
and precise instructions in the form of prompted
messages, to which the user replies with one or
more operations or data entries. In many cases
these prompted instructions can partially eliminate
the need for 'user manuals'. It is, however, general
experience that a combination of prompted com-
mands along with a fully documented and concise
user's manual will normally be required.

10.6   Release of new cultivars

The ultimate goal of any plant breeding programme is to develop superior genotypes and to
release these into agriculture as new cultivars that
better serve the increasing needs of humankind.
The final stage of a breeding scheme therefore

involves the process of release, perhaps protection, and distribution of planting material to the seed industry and/or the farming community.

The first part of this process is when the breeder decides that a particular genotype has merit as a potential new cultivar. This decision will have been made by observing the performance of the potential new cultivar as it passes through all the stages of the breeding programme. This would entail a number of years and, in the more advanced stages, a range of different locations. It cannot be stressed too strongly that if there is any doubt regarding the worth of a potential cultivar, then these doubts must be addressed before deciding to release it for growing. The general agricultural community does not generally take kindly to being sold seed of a cultivar which proves to be of little, or no, use. In the seed market a good reputation is difficult to obtain but easy to lose!

In most cases the decision to 'release' or 'launch' a cultivar is not the exclusive decision of the breeder. Where a breeder is working for a commercial company, then the final decision to take the first steps towards commercialization is unlikely to be made only by the breeders. Others within the company, the board of directors, financial or marketing staff, will all contribute to the decision concerning the potential commercial impact that the cultivar may have, and more importantly, the potential profits that can be expected to the company if release is successful. If the new cultivar has been developed in a University department or other public research organization, then the final decision on release may involve heads of department and deans of the college or experimental station. Irrespective of whether public or private investment has financed the development of the line, then there is logic in the breeder having a major input into the final decision.

In this decision-making process, the requirements (often statutory) that are made of a new cultivar must be borne very clearly in mind. If the cultivar fails to meet the stipulated criteria, then it will not be possible to commercialize it and all the effort will have been wasted.

10.6.1 Information needed prior to cultivar release

Before a breeding line can be considered for release it must be shown to be **distinct** from other cultivars that already exist. Distinctness can be for morphological characters (e.g. flower colour) or a quality trait (e.g. low linoleic acid content in the seed oil). It is sometimes possible to say that a new cultivar is distinct for a quantitative trait such as high yield, but in this case the new cultivar must **always** express the high yield character if release is granted, and in practical terms this is not an easy way to proceed. More recently, breeders are using molecular techniques to distinguish new releases from already existing cultivars.

The new cultivar must also be **stable** and **uniform** (i.e. stable over several rounds of increase) so the genotype must always appear the same, irrespective of where it is grown. Therefore if a new cultivar is released which is described as having uniform white flowers, then all individual plants grown must have white flowers.

Careful attention to the final stages of seed increase and meticulous care in producing breeders' seed can be of great benefit in ensuring the uniformity and stability of the new variety.

10.6.2 Value in release

Prior to releasing a cultivar, breeders must demonstrate (from data collected from evaluation trials) that there is indeed merit in releasing the new cultivar. This will involve presenting data from several years testing and from a number of locations, but the exact requirements and procedures will vary from country to country.

In many countries, government authorities carry out official independent testing of all new cultivars before release is allowed. These trials are carried out over two or three years and at a number of locations throughout the target region. The aim of these trials (National List Trials) is to ensure that new cultivars are suitably adapted for the region. If breeding lines show sufficient **value for cultivation and use** (**VCU**) then they will be added to the National Variety List of the particular country. In the case of EU countries, when any new cultivar is placed on the National Variety List of any EU country, then it is automatically entered onto the EU Common Catalogue and can hence be increased and sold in other EU states.

In other countries, such as the US, there are no official National List Trials in which each new cultivar is evaluated. However, each US state has an appointed body of people who review performance data for all new cultivars and determine whether they merit release within the particular state. Breeders can submit data for release in more than a single state simultaneously.

10.6.3 Cultivar names

Any cultivar to be sold commercially must be given a unique name (or identifying code) prior to commercial release. Within any given crop species there should only be one with that particular name. So a wheat cultivar and a potato cultivar can both have the same name, say, 'Sunrise', but two barley cultivars cannot have a common name, say, 'Maltster'. Also, unless there is some unfortunate problem such as a cultivar mistakenly being allowed a duplicate name, it is difficult to change the name after release.

Hybrid cultivars are often given a code number rather than a recognizable name. In such cases the number code has a prefix that identifies the company responsible for its development.

In choosing names it is useful to select ones that are easy to remember, and convey, if possible, the right image (e.g. 'Star', 'Golden Supreme' or 'Bountiful'). Equally it is wise to avoid names that are obviously inappropriate, such as 'Usually Dies' or 'No Profit'.

Another point to bear in mind is, if a cultivar is to be marketed in a foreign country, or in an area where a second language is common, then it is important to check that the cultivar name does not have an unfortunate meaning in the other language, that it does have a desired image in that language, and that it be easily pronounced. For example, there is no 'w' in the Spanish alphabet, so it would be unwise to call the cultivar 'Wally's Wonder' if it is to be commercialized in a Spanish-speaking country, such as Spain or Mexico. The same would apply to the letter 'ñ' from Spanish, which is non-existent in the English language.

Think questions

(1) You are employed as a barley breeder for a commercial company. At present the company does not have any greenhouse facilities. You are trying to convince your peers that the breeding programme would benefit from having a greenhouse. List five uses you would have for a greenhouse in your breeding programme, if one were available.

Great job (we hope!), you have convinced the Board of Directors to proceed with purchasing a new greenhouse. List five features you would like to request to have in the new greenhouse facility.

(2) Briefly describe one problem of using a greenhouse in a plant breeding programme.

(3) What is the primary goal of conducting field evaluation trials in a plant breeding programme.

(4) List things to avoid when organizing plant breeding field plots.

(5) Why is experimental design important in plant breeding evaluation trials?

(6) List four problems that may be encountered in organizing and carrying out 'off-station' trials.

(7) You have been employed as Senior Breeder at the *Hass Bean Breeding Co.* for several years and have had to carry out your breeding operation with very little funding. Due to a recent takeover of the company by the *Human Bean Seed Association*, you have been given a vast sum of additional money. You have decided to spend this money upgrading the computer facilities used by your breeding group. To achieve this goal you have appointed a computer programmer to develop a computer-based plant breeding management system. List the five main features the computer package needs to perform that would help to increase the efficiency of breeding future new cultivars.

11
Current Developments in Plant Breeding

11.1 Intellectual property and ownership issues

Several USDA breeding groups, or similar public breeding efforts elsewhere, offer the cultivars they develop, free from royalties, to the farming community (although the seed itself still has to be bought in order to cover production costs) and therefore they do not hold any rights on the new varieties (in other words, the cultivar can be multiplied and sold by others). However, it is now very common to obtain some degree of proprietary protection of ownership of new cultivars that are released. All commercial companies require proprietary ownership of the cultivars they develop in order that they can control the supply of seed, who can grow the crop, and to obtain either profits from seed sales or royalties from seed sales which will sustain further research and breeding efforts.

When the cultivar produced is a synthetic or a hybrid, it is possible to have 'automatic' proprietary ownership by simply maintaining the parents that are used to generate the synthetic population or hybrid seed, and not allowing access to the parents. In the end, if a cultivar is not a hybrid then the most common method of protecting cultivars that are propagated by seeds is to apply for **Plant Variety Protection** (PVP), within the US or **Plant Variety Rights** (PVR) in some other countries. Also, in the US, when a clonal crop or other asexually propagated crop cultivar is released, then protection can be obtained by **Plant Patent** (PP).

In each case PVP, PVR or PP allows the developers of a new cultivar to control how much, where and who has the right to grow and sell the proprietary cultivar for a fixed period of time.

It is also wise to be aware of the information that is required to submit an application for the different protection types, before starting out on the trialling and selection procedures!

11.1.1 Patents

In fact there is a much more general consideration of patents as they affect plant breeding, which has recently become a major consideration in a number of ways. For example, it is possible to patent DNA sequences if they have a known biological role, as well as processes and methodologies, such as tissue culture protocols or transformation methods. There are basic criteria that apply to patent applications, but in essence you can apply for a patent if the 'invention' is not patented already, if you can demonstrate a significant novel step, and it is 'not obvious'. Clearly these have definitions (and interpretations of definitions) attached to them, but nevertheless these are the main principles. So increasingly, if researchers make any sort of breakthrough they will tend to consider patenting to protect their intellectual property (IP). In other words, they can potentially capture value

Plant Breeding, Second Edition. Jack Brown, Peter D.S. Caligari and Hugo A. Campos.
© 2014 John Wiley & Sons, Ltd. Published 2014 by John Wiley & Sons, Ltd.
Companion Website: www.wiley.com/go/brown/plantbreeding

from exploiting that technology either directly, by exploiting it and preventing others from doing so, or indirectly, by licensing others to allow them to use the technology, perhaps subject to royalty payments. This affects quite a number of protocols related to plant breeding – particularly related to biotechnology and molecular biology.

The patenting of gene sequences also has rather obvious effects on the freedom simply to use particular regions of DNA.

So you cannot, as a breeder or breeding organization, simply carry out all tissue culture procedures including many of the transformation protocols if you intend to produce a commercial product. This restriction also applies to many genes and gene constructs. If you simply carry out research there are few restrictions (especially if you can persuade the appropriate organization to give you the material), and methods are fairly freely used on the basis of research only. For commercial application, in the very broadest sense, you have to come to a written agreement with the patent holder.

There has been, and still is, continuing debate on the ethics of being able to patent DNA. Some argue that it is the basis of life, so cannot by definition be patented, while others claim that the work effort and technology required to identify and extract it means that they must have protection so they can recover their investment. Other controversies have been raised over who has rights when, for example, a gene is found in a species that has been collected as germplasm from another country.

Thus some years ago breeders worked on the agreed principle that 'pollen was free' but specific genotypes were basically protected (i.e. you could use any material as a parent but could not exploit the genotype itself directly). Now this has been overtaken by, for example, protection for a number of processes the breeder might consider using; a genotype has to be changed substantially before being recognized as no longer the property of the originator; and royalties are paid to include a particular gene in a cultivar for exploitation. There are, of course, no 'black and white' solutions, but it is an increasingly complex area within which to operate.

A more recent development than simply obtaining a patent for specific genes and gene sequences is patents on cultivars in their own right. In the case of cultivar patents, companies who develop the cultivar describe the process whereby the cultivar

is developed, including the specific parents, the breeding and selection schemes and the final seed increases. Cultivar patents are usually also protected with PP or PVP protection. So what is the benefit of a cultivar patent over, say, a PVP? The patent gives more protection to the applying company, particularly as it relates to using the cultivar in sexual crosses, or other cultivar development without permission of the patent-holder. Indeed, if anyone makes the same parent cross combination and selects a 'different' cultivar from the resulting progeny, they would equally be in infringement of the patent rights. So cultivar patents effectively limit using these cultivars as parents in crosses without obtaining specific permission. This inevitably leads to further restrictions on a limited germplasm exchange between different breeding groups.

Even without obtaining patent or other intellectual property rights, breeding groups retain far greater control of their specific breeding lines and cultivars through Material Transfer Agreements (MTAs), which are now commonly required, even by public institutions like Land Grant Universities in the US.

MTAs come in a wide range of styles and forms. Seed or other transferable plant products or plant tissues are usually referred to as "the Material" in MTAs, while the people who have ownership of the materials are referred to as "the Company", and those who will obtain or receive the material as "the Recipient".

The aim of MTAs is to protect the owner of the "Materials" that are transferred. As such, they all restrict how the "Recipient of Material" must handle the product. It is not uncommon to have MTAs with sweeping general statements such as: "the Material transferred or any progeny derived therefrom, including any biological material such as seed, leaves, stems and roots, as well as DNA sequences, genes or cells thereof, grown from such seed or its progeny will remain solely owned by the Company". Or "if the seed delivered to the Recipient includes hybrid seed, Material also includes any parental or inbred line that may be unintentionally included with the hybrid seed". Companies issuing MTAs may include statements such as "the Recipient of Material will be entitled to use the Material supplied solely for yield trial evaluations and any and all seed not planted and/or harvest products thereafter must be destroyed". MTAs often include more specific statements like: "Recipient will not: (1) give, transfer and/or distribute the Material to any third party; (2) use the Material with any third party, except as may be permitted by the Company; (3) give, transfer, distribute and/or use Material outside of the Recipient's laboratory or general testing facility; (4) propagate or increase seed of the Material; (5) breed with or make crosses or backcrosses with the Material; (6) select variant plants from the Material; (7) conduct biotechnology processes including, but not limited to, tissue culturing, mutagenesis, or transformation with the Material".

It is of course the right of specific breeding groups to retain ownership of breeding lines and cultivars that they develop. However, how far this 'ownership' should extend is perhaps more questionable. It is surely to the good of mankind that breeding companies have some exchange of germplasm. Without this, there is clearly a danger that cultivar or gene patents and MTAs will greatly limit exchange of germplasm and greatly restrict the choice of parent genotypes that breeders have access to.

11.2   The impact of biotechnology

Since 1996, the cumulative number of hectares grown with transgenic crops has reached over 1,500 million hectares on a worldwide basis, and every year thousands of farmers increasingly choose to use the results of this technology because of the advantages it carries. Indeed, it has been noticed that the speed of adoption of transgenic crops has the highest rate of technology acceptance throughout the history of agriculture. Several biological, agricultural and practical considerations associated with the use of genetically modified organisms (GMOs) have been addressed earlier. In spite of the multiple documented benefits of biotechnology in agriculture, some concerns have been raised in relation to their safety, risk assessment, public perception and more general social aspects. It will not be possible to cover all these here, but it is perhaps worth quoting what might represent an attempt at providing a balanced view of the situation. This is from a 2003 follow-up Discussion Paper of the Nuffield Council on Bioethics entitled 'The use of genetically modified crops in developing countries':

We conclude that the potential benefits of contemporary plant breeding, including those arising from the use of genetic modification of crops, have been empirically demonstrated in some instances, and have considerable potential in others, to improve agricultural practice and the livelihood of poor people in developing countries while reducing environmental degradation. There is an ethical obligation to explore these benefits responsibly, in order to improve food security, profitable agriculture and the protection of the environment in developing countries.

Advances in crop development by genetic engineering have been influenced in many countries by communication campaigns with different degrees of impact on public opinion, in particular in the European Union. In most countries, genetically engineered plants come under the control of regulatory government organizations. Working with recombinant DNA techniques for crop improvement requires a high degree of documentation and government approval. Permits need to be obtained before genetically engineered plants can be 'released' (grown in the field or in even in the greenhouse). Areas where field evaluations have occurred often need to be monitored for a period of 1 to 2 years after harvest, to make sure volunteer plants are taken care of.

However, as mentioned above, there have been a number of concerns raised and public perception issues over the use of genetically modified organisms. Some of these were justified, although others were based on ill-informed and irrational scares, while yet others reflect public concerns that surround any new technology – particularly one that uses recombinant DNA techniques. Addressing these concerns and the issues that surround them is important but cannot be given full justice here. But what must be stressed is that any breeder needs to take into account both the general prevailing scientific knowledge as well as the general perception of the issues involved. As breeding technologies evolve, breeders are particularly well placed to help clarify in their own minds, and hence inform others, about these advances, their potential and their safety. They need to keep up-to-date with what is being made possible by scientific discovery, what it might enable them to do, what risks it might present, what risk assessments need to be carried out, what benefits it might confer, and what needs to be checked before they are confident to use particular genes or types of genes. They must clearly be governed by the scientific facts, but they can also clearly identify risks and benefits from their particular intimate knowledge of the crop and its applications.

As an example of one of these issues, most of our crop species have close relatives that exist as weeds in the same region as the crops are cultivated. One of the environmental concerns that have been expressed is that the genetically engineered traits might be transferred to these weedy species, making them difficult to control. This phenomenon, called pollen-mediated gene flow, is a natural process by which sexually compatible species exchange genes. Transgenic traits will be transferred to related species as with any other gene, and this is probably inevitable. This is why one of the aspects of the risk assessment of GM crops is to assess the existence of wild relatives in the cultivation areas for the new crop, to evaluate the risk for trait introgression and its consequences. In fact, the question that arises is really how much the plants in wild populations will be affected by the presence of such a gene, and, if beneficial, hence spread. Clearly when invading a farming environment in which a particular herbicide is used, any genotype with resistance to that herbicide will have a distinct advantage and spread. It is clear that this situation is not exclusive to GM traits, as any herbicide-resistant non-GM variety will raise the same issues.

Herbicide resistance in weeds is a common biological phenomenon known to agronomists. However, there are several management practices that may mitigate its spread. Combining and rotating herbicidal active molecules and applying appropriate agronomic practices are essential to this end. In the absence of selective pressure (where no herbicide is used), the persistence of the transferred gene will depend upon whether it affects any other traits (positively or negatively in terms of natural selection) and particularly whether it carries a 'penalty' to its carrier's fitness. Similar scenarios will exist for other genes that are transferred into crop plants. The cost–benefit analysis in the broadest sense is needed, and is part of the environmental risk assessment process that is applied to every GM event under development before their use by farmers is allowed. The benefits brought about by herbicide-tolerant crops include enabling the widespread adoption of no-tillage farming practices, which largely reduces soil losses due to erosion, allowing a build-up of organic matter and reducing the compaction of agricultural land driven by traditional soil management. Additionally, such soil cultivation methods reduce the CO_2 footprint of agriculture as it increases the pool of carbon sequestered by soils which is not released into atmosphere.

Perhaps a greater concern in pollen-mediated gene transfer relates to traits which at present do not exist in natural weed populations. These may include higher productivity or reproducibility (i.e. yield) with limited nitrogen application, salt tolerance, tolerance to other abiotic stresses, and so on. If these traits do transfer to related weedy species, then the 'hybrid weeds' that develop may indeed have a genetic advantage and be more difficult to manage. Should we therefore stop such developments? Most likely no, we should not. However, we do need to realize that these are new technologies, and as such may need specific or different management strategies.

Consider the first automobiles: when first introduced to our cities they were considered so dangerous that someone had to walk ahead of the new "horseless carriage" with a large red flag to warn people that something new was coming. Now if any audience of people is asked "are automobiles safe?", the answer must be no, they are not. Many people are hurt or killed annually in automobile accidents. So should we ban or restrict the use of cars? Of course not. Before the internal combustion engine, many people were hurt, killed and maimed by the horses and carts that cars have now replaced. So automobiles have not been banned in any country, but almost every country that allows automobiles have specific rules to which users should adhere. "Don't drive too fast; don't overtake on a bend, don't drink and drive" … the list is long. So, perhaps as new technology is introduced to plant breeding and through breeding efforts into agriculture and general use, new management strategies need to be considered. This is not because GM technology in itself is inheritably high-risk; it is just a different risk.

When transgenic cultivars are managed according to good agronomic practices, theoretical risks such as the development of weeds carrying tolerance to herbicides can be greatly reduced. This explains why these concerns have not materialized into the development of, or formal report of, "super weeds", despite the large-scale use of herbicide-tolerant plants developed through biotechnology. In contrast, however, the number of glyphosate-tolerant weeds in regions that produce Roundup Ready® crops can exceed the number from other regions where these crops are not propagated, even though glyphosate may be excessively used in these areas as a burn-down.

Developing transgenic crops is a long and costly process, and biotechnology-based seed companies are no different from traditional breeding companies in that they need to recover their costs. Currently, there are also an increasing number of academic and publicly funded efforts, not only in developed but also in developing countries, focused on developing commercial transgenic crops, which will have to face similar challenges in acceptance and deregulation of GM cultivars. It remains to be determined whether the advantages of transgenic lines will be sufficient to offset the costs incurred and whether investments are recouped. An important part of such costs addresses the regulatory studies needed to comply with the risk assessment process. This situation has raised several concerns as it could delay access to the benefits of transgenic crops developed by academic institutions addressing the needs of small farmers or crops with relatively small acreage.

Many of the early transgenic plants in species such as potato have resulted from genetic engineering of cultivars for which there is no plant variety protection (i.e. they have no legal owner). If transgenic crops are to have a large impact on agriculture, then genetic engineering companies need to develop cultivars transformed for specific traits that cannot readily be produced by more traditional means, and to work in collaboration with traditional breeding programmes/companies in order to keep development of the myriad of other characters increasing in performance, in order to compete successfully in commercial contexts.

Plant transformation methods complement more traditional plant breeding work by increasing the diversity of genes and germplasm available for incorporation into crops, and by (perhaps) shortening the time required for the development of new cultivars. Genetic engineering of plants also offers exciting opportunities for the agrochemical, food processing, specialty chemical and pharmaceutical industries in developing new products within crop species and offering new manufacturing processes. However, it is highly unlikely that these techniques will replace the traditional techniques that have been used in the past. Recombinant DNA techniques will, rather, add to the 'array of possibilities' available to plant breeders in future cultivar development. Successful biotechnology efforts are those combining gene discovery, genetic engineering, the routine use of molecular markers and effective breeding programmes. A well-run breeding programme is uniquely placed to play the critical role of assembling a cultivar desired by farmers which requires tolerance to biotic and abiotic stress, adaptation to the farming environment, and generally higher yields: all the features sought by farmers in the cultivars they decide to purchase.

No individual technology can realistically offer a solution to all agricultural issues, and transgenic crops do not represent an exception. Nevertheless, they represent yet one more component of the toolkit required by agriculture to address pressing issues such as supplying a rapidly increasing population, the need to reduce the environmental footprint, the limited availability of new farming regions, and the growing need for accessible food, fibre and fuel.

11.3 The regulation of genetically modified plants

The development of regulatory frameworks was initiated by scientists and has been key in the development and deployment of GM crops. These play a critical role in trying to assess all of the effects of any transgenic plant and the possible risks involved. They provide a mechanism for independently assessing each type of transgenic plant, and only permit releases which, in their view, are safe both in terms of the wider environmental aspects as well as impact on human health and nutrition. Transgenic plants are therefore subjected to detailed analysis before their commercialization is approved by any of the local regulatory agencies. For this purpose, some countries have developed specific committees or panels for GM scrutiny within existing regulatory authorities (such as EFSA, the European Food Safety Authority, in the European Union) or have established specific inter-institutional bodies (like CTNBio, Comissão Técnica Nacional de Biossegurança, or Brazilian National Technical Commission on Biosafety in Brazil), whereas yet others, like the US, split the regulatory assessment across several agencies such as the FDA (Food and Drug Administration), EPA (Environmental Protection Agency) and USDA (United States Department of Agriculture), depending on the nature of the transgenic plant. Regardless of the structure in place, regulatory authorities have the mandate to assess the merit of the evidence in requests to release to the environment transgenic events submitted by developers of transgenic plants, be they of private or public nature. Requests received by regulatory authorities may be related to field research activities, or to commercial release, but whatever type of release is envisaged, it can only occur after receiving full clearance from regulatory authorities.

11.4   Plant breeding as a career

Despite many misconceptions and concerns of the general public about agriculture, and plant breeding specifically, there is but one simple fact to state to any person who seeks to enter plant breeding as a career in the 21st century. The fact is that the sole reason for agriculture is simple: without it there would not be civilization, or certainly not how we

have come to know it. The world's population is still rising: will this continue? Will plagues, wars or famine wipe out vast quantities of our global population? We can all hope that these are answered with a resounding no. We eat better, eat more, and pay less now for our daily calories than ever before in civilized history. Plant breeders are not responsible for agricultural success, but there is no doubt that plant breeding has added tremendously to our ability to succeed in food production. The tools and technologies of traditional plant breeding approaches involving the screening of thousands of different breeding lines, yield trials, quality evaluation and disease screening will still be required and will continue. Plant breeders in the future will be no different than in any other critical industry; they will simply have better and more tools available to succeed in their task. Plant breeding will therefore continue, in fields, greenhouses, and in laboratories, so it is critical that universities and other educational institutions see the need to train and develop new plant breeders to continue our cultivar development. Those who become the next generation of plant breeders will still have that overwhelming satisfaction that comes from looking over a vast stretch of the countryside and being able to state that "all of this has arisen from a single plant that I had the good fortune to select and decide would become a successful new cultivar".

Further Reading

Acquaah, G. 2012. *Principles of Plant Genetics and Breeding* (2nd edn). Wiley-Blackwell: Oxford.

Allard, R.W. 1999. *Principles of Plant Breeding* (2nd edn). John Wiley & Sons, Ltd.: Chichester.

Bevan, M.W. and Uauy, R. 2013. Genomics reveals new landscapes for crop improvement. *Genome Biology* **14**: 2016–17.

Borém, A. and Fritsche-Neto, R. (eds) 2014. *Biotechnology and Plant Breeding: Applications and Approaches for Developing Improved Cultivars*. Academic Press: New York.

Bos, I. and Caligari, P.D.S. 2008. *Selection Methods in Plant Breeding* (2nd edn). Springer Science.

Darbeshwar, R. 2012. *Biometrical Genetics: Analyses of Quantitative Variation* (3rd edn). Alpha Science Intl. Ltd.

Falconer, D.S. and Trudy, T.F.C. 1996. *Introduction to Quantitative Genetics* (4th edn). Garland Science. Benjamin Cummings.

Kearsey, M. and Pooni, H. 2004. *Genetical Analysis of Quantitative Traits* (2nd edn). Routledge.

Mather, K. and Jinks, J.L. 1982. *Biometrical Genetics* (3rd edn). Chapman and Hall: London.

Michael, T. and Jackson, S. 2013. The first 50 plant genomes. *Plant Genome* **6**: 1–7.

Murphy, D. 2007. *Plant Breeding and Biotechnology: Societal Context and the Future of Agriculture*. Cambridge University Press: Cambridge.

Nicholl, D.S.T. 2008. *An Introduction to Genetics Engineering*. Cambridge University Press: Cambridge.

Slater, A., Scott, N.W. and Fowler, M.R. 2008. *Plant Biotechnology: The Genetic Manipulation of Plants* (2nd edn). Oxford University Press: Oxford.

Smartt, J. and Simmonds, N.W. 1995. *Evolution of Crop Plants*. Longman Group UK Ltd.: London.

Suggested Answers to Think Questions

Chapter 1: Introduction

(1) **Different crop species originated in different regions of the world. List the centres of origin of the following ten crop species: onion (*Allium* spp), alfalfa (*Medicago sativa*), rice (*Oryza sativa*), potato (*Solanum tuberosum*), soybean (*Glycine max*), sorghum (*Sorghum bicolor*), cotton (*Gossypium* spp.), sunflower (*Helianthus* spp.), wheat (*Triticum* spp.), and apple (*Malus* spp.).**

Answers:
Onion (*Allium* spp), Iran, Afghanistan, Pakistan;
Alfalfa (*Medicago sativa*), Iran, northern Pakistan;
Rice (*Oryza sativa*), Thailand, southern China;
Potato (*Solanum tuberosum*), Peru;
Soybean (*Glycine max*), northern China;
Sorghum (*Sorghum bicolor*), Equatorial Africa;
Cotton (*Gossypium* spp.), Central America, Brazil;
Sunflower (*Helianthus* spp.), western USA;
Wheat (*Triticum* spp.), Syria, Jordan, Israel, Iraq;
Apple (*Malus* spp.), Asia Minor, Central Asia.

(2) **A combination of natural selection, and selection directed by plant breeders (early and modern), has influenced the crops we now grow. List five characteristics that mankind has selected which would not have been selected by a process of natural evolution.**

Answers:
Less spiny, reduced toxins, larger seeds or fruits, reduced seed dormancy, reduced seed dispersal mechanism, etc.

(3) **"The yield of many crops species has risen dramatically over the last 50 years. This has been the direct result of plant breeding during this period and hence the trend is likely to continue over the next 50 years." Briefly discuss this statement.**

Answer:
Plant breeding has had a significant impact by improving productivity, and experiments have suggested that 50% of the increase in a number of crops can be shown to be attributed to breeding. However,

Plant Breeding, Second Edition. Jack Brown, Peter D.S. Caligari and Hugo A. Campos.
© 2014 John Wiley & Sons, Ltd. Published 2014 by John Wiley & Sons, Ltd.
Companion Website: www.wiley.com/go/brown/plantbreeding

many other changes have occurred in parallel in agriculture that have contributed to the rapid rise in productivity of our agricultural crops. These include: inorganic fertilizers, pesticides and herbicides, improved machinery and husbandry, irrigation management, and harvesting techniques. As to whether this increase can be continued is difficult to answer; however, given the recent advances in recombinant DNA techniques, and other molecular marker selection tools, most breeders would suggest that they can still make significant progress in genetically improving all our crops, particularly some minor crop species that are still to be exploited.

(4) **"The place of origin of crops, their history and evolution are events from the past and therefore have no relevance to modern plant breeding." True or False? Discuss your answer.**

Answer:

The place of origin of our modern crops can have tremendous impact on breeding success. Where they evolved originally gives a number of insights into their characteristics and the type of environmental conditions that they were originally selected for. Examining the evolution of our crops provides valuable cytological and genetic information to breeders. In addition, and perhaps as importantly, the geographical origin of our crops gives insight into possible ancestors which have added valuable genes (i.e. disease resistance) into our modern cultivars and could be a source of further variation.

(5) **Many believe that civilization (of humans) started with the beginnings of agriculture. Basically there are two forms of agriculture: (1) rearing animals for meat, milk, etc., and (2) raising crops for human or animal feed. No one knows which form of agriculture evolved first (or maybe both types started together). Explain why (in your opinion) one form came before the other or both forms evolved at the same time.**

Answer:

It is thought that agriculture began at a similar time throughout the world. Some would argue that humankind first domesticated plants, produced surplus food, and later domesticated livestock since they had the capacity to feed them. But, perhaps more likely, they may have domesticated livestock but moved somewhat nomadically, and then moved towards a more settled lifestyle and domesticating crops to allow them to feed animals. There is some evidence that early farmers were smaller and sicklier compared with those who continued to hunt and gather. When agriculture began, the world had just come out of a mini-ice age and there would probably be a shortage of easy prey animals. It is easier for somewhat weaker individuals to catch young animals and feed these to maturity. This would also be the case for some forage crops (i.e. alfalfa) that surely were first propagated as livestock feed.

(6) **A combination of natural selection, and selection directed by plant breeders (early and modern) have influenced the crops we now grow. Have modern plant breeders improved the genetic fitness of our agricultural crop species, or have they simply selected plant types that are more suited to modern agricultural systems?**

Answer:

This is something that could be argued either way. There is little doubt that plant breeders have dramatically improved quality and disease/pest resistance of our modern crops, which could surely be considered as improving their genetic fitness of agriculture. On the other hand, the introduction of genetic dwarfism into cereal cultivars, which occurred at the same time that the introduction of inorganic fertilisers allowed higher nitrogen application rates without crop lodging, could be considered as modification of crop morphology to better fit changes in production systems.

(7) **A tremendous amount of plant diversity exists within Australia. Why, therefore, are there few Australian aboriginal farmers, and almost none of today's world crop species had their centre of origin in the Australian continent?**

Answer:
One argument for the lack of agriculture in Australia would be that there was not the need for it. Australia was affected much less by ice ages that must have caused extinction of many unadapted plant and animal species. Food could have been readily available year round, and there was also less need to domesticate crops of livestock to stockpile food for lengthy winter months. Similarly, there was less need for warm clothing in the form of animal pelts.

(8) **You are drifting your way through life when there is a crash and a tremendous flash of light, and from nowhere an alien spaceship lands. The alien, speaking perfect English, says "*I bring to you a gift: the Universe's most productive and perfect crop species*". Thereafter the alien gives you a bag of 200 seeds, returns to the spacecraft and whoosh, it is gone. Describe what you would do with the seeds and what information you would collect that would allow you to develop new cultivars from these seeds.**

Answer:
The first thing that would need to be determined, if this truly is a productive and perfect crop, is whether it is actually edible and which plant parts are the food source. Then one would have to determine how and where the plants should be grown. This would include the reproductive lifecycle. Is this plant a summer annual, winter annual, or biennial? Is it self-pollinating or out-pollinating? What are the plants' requirements for water, nutrients and other compounds that are needed to maximize productivity? From a breeding standpoint it needs to be determined whether the 200 seeds are genetically variable, as without genetic variability it will be impossible to make genetic improvements. Additionally, the most appropriate breeding approach would have to be identified and deployed. It would also be valuable to know whether this species is related to others on Earth that may have some degree of sexual compatibility. Then, of course, protect the intellectual rights of cultivars developed as appropriate, and seek the backing of a large seed organization to fund production and distribution of the seed.

Chapter 2: Modes of Reproduction and Types of Cultivar

(1) **Complete the following table by assigning a YES or NO to each of the 16 cells.**

Answer:

	Pure-line cultivar	Open-pollinated cultivar	F_1 Hybrid cultivar	Clonal cultivar
Are these cultivars composed of only one single genotype?	Yes	No	Yes	Yes
Is heterosis a major yield factor in resulting cultivars?	No	Yes	Yes	Yes
Are resulting cultivars propagated by means of botanical seeds?	Yes	Yes	Yes	No
Can the seed, or plant parts, of a cultivar be used for its own propagation?	Yes	Yes	No	Yes

(2) **Inbreeding and out-breeding species tend to have different characteristics. Explain factors that would determine whether a given species should be classified as inbreeding or out-breeding.**

Answer:

- Inbreeding species: tolerant to inbreeding; few deleterious recessive alleles; '*closed*' flowers; little heterosis.
- Out-breeding species: intolerant to inbreeding; many deleterious recessive alleles; flower morphology that promotes cross-pollination; high heterosis.

(3) A number of different cultivar types are available in agriculture. Outline the major features of the following cultivar types:

(a) hybrid cultivars;

(b) pure-line cultivars;

(c) clonal cultivars;

(d) multiline cultivars;

(e) open-pollinated cultivars; and

(f) synthetic cultivars.

Answer:

(a) Hybrid cultivar: out-breeding mating system; annual or biennial; homogeneous and highly heterozygous; selection and inbred parents; select for combining ability.

(b) Pure-line cultivar: inbreeding mating system; annual seed propagation; homozygous and homogeneous; selection and move towards homozygosity; includes multi-lines.

(c) Clonal cultivar: out-breeding mating system; perennial or quasi-annual propagation; highly homogeneous and heterozygous; selection of genetically fixed genotypes.

(d) Multiline cultivar: mixtures or blends of a number of different cultivars or breeding lines; may result from developing near-isogenic lines and using these to initiate the mix.

(e) Open-pollinated cultivar: out-breeding mating system; annual, biennial, or perennial; heterozygous and heterogeneous; selection to change gene frequency (population improvement).

(f) Synthetic cultivars: out-breeding mating system; must show heterosis; similar to a hybrid cultivar with many parents (maybe 50–70); most associated with forage cultivars.

(4) List two inbreeding crop species and two out-breeding crop species that have been exploited as:

(a) pure-line cultivars;

(b) open-pollinated cultivars;

(c) hybrid cultivars;

(d) clonal cultivars;

(e) multiline cultivars; and

(f) synthetic cultivars.

Answer:

(a) pure-line cultivar: wheat; barley; canola[1]; etc.

(b) open-pollinated cultivar: corn; alfalfa; canola[1]; etc.

(c) hybrid cultivar: corn; tomato; canola[1]; etc.

(d) clonal cultivars: potato; strawberry; apple; etc.

(e) multiline cultivar: wheat; barley; pea; etc.

(f) synthetic cultivars: corn; alfalfa; canola[1] ; etc.

(5) Describe the major features of the following types of apomixis: diplospory; semigamy; advantageous embroyony; pseudogamy; parthenogenesis; apospory

---

[1]One crop species can be subject to very diverse breeding approaches, and therefore may give rise to different types of cultivars.

Answer:
Diplospory has the origin of embryo and endosperm as the 2n megaspore mother cell. Semigamy is where a pollen grain enters the embryo sac, penetrates egg, no fusion, sperm and egg cells develop independently. Advantageous embryony is when an embryo sac is formed during seed development (citrus). Androgenesis is where seed develops from sperm nucleus of the pollen grain in the embryo sac. Parthenogenesis is development of haploid from egg cell without fertilization. Pseudogamy requires pollination, and stimulates embryo development. For apospory, somatic cells of ovary divide mitotically to form 2n embryo sac, and is the most common form (e.g. Kentucky bluegrass).

(6) List five different plant parts that can be used for asexual reproduction.

Answer:

- A bulb is a large bud with a stem at its lower end (e.g. an onion).
- A corm is very much like a bulb in size and form but has a different internal structure (e.g. crocus).
- A rhizome is a horizontal stem that grows at or below the soil surface (e.g. iris).
- A stolon is a stem that grows horizontally on the soil surface (e.g. strawberry).
- A tuber is a swollen stem that enlarges beneath the soil surface. The eyes of a tuber are the sites for buds from which stems and roots can develop (e.g. potato).

Chapter 3: Breeding Objectives

(1) Explain the difference between vertical and horizontal disease resistance and outline the advantages or disadvantages of each form of resistance as it relates to cultivar development.

Answer:
Resistance that is controlled by many genes shows a continually variable degree of resistance and is referred to as horizontal resistance. Vertical resistance is associated with the ability of single genes to control specific races of a disease or pest. Horizontal resistance is more durable as the pest must overcome all the resistance genes. Conversely, vertical resistance is less durable, particularly to diseases with a rapid spread, like airborne fungi. However, single gene resistance to many pests has remained functional and effective over many years (i.e. potato cyst nematode). The advantage of vertical resistance is that individual alleles can be readily identified and transferred from one genotype to another. Indeed, developing horizontal resistance can be slow as it is difficult to accumulate multiple resistance genes into a single genotype.

(2) Outline the major features involved in selecting for end-use quality, and indicate any particular problems that breeding for improved quality might cause.

Answer:
The major difficulty in breeding for improved quality characteristics is being able to test many (often several thousand) lines quickly, cheaply, and in an effective manner which simulates actual large-scale quality (for example, consider breadmaking quality characters). In addition, in many cases some stages of breeding programmes cannot supply the amount of plant material needed for quality assessments. Quality in many crops is a subjective judgement whereby individual users have specific preferences for what constitutes 'good quality'. Indeed, for many processed foods, processors want raw products to have no taste as it is easier for them to produce a uniform product by adding flavour, or indeed colour. Finally, it is often difficult to obtain a definition of what constitutes good quality from industry, as they can be very covert about their processing systems. Although unrelated to breeding, an additional issue of breeding for end-use quality traits is that the market is not always willing to pay a premium for enhanced levels of quality.

(3) When breeding new cultivars it is often necessary to try to predict events that may occur in the future. Briefly outline four factors that may influence cultivar breeding and hence need to be considered in setting the breeding objectives of a cultivar development programme.

Answer:
Politics; consumer preference; different and new pests and diseases; economics.

(4) In plant breeding, two main disease resistance mechanisms exist, *inhibition of infection* and *inhibition of growth after infection*. Explain each of these mechanisms.

Answer:
Inhibition of infection refers to plant resistance to pest or disease initiation whereby the pest or disease cannot enter the plant and hence gain food or initiate infection. These mechanisms would include hypersensitivity, whereby the plant cells surrounding the site of infection die and form a barrier that means the pathogen cannot infect the rest of the plant; or physical barriers that prevent the pest or disease entering plant tissues – examples would include plant trichomes or leaf wax barriers.

Plant pest or disease resistance due to inhibition of growth after infestation/infection is resistance due to lack of spread after infection, and relates to the ability of a plant to slow down the development of the pest/disease or to interfere with the pest/disease reproduction. These include: antibiosis, where the plant produces a substance that reduces survival, growth, development, or reproduction of the pest or disease; and antixenosis, which refers to plant resistance that reduces pest preference or acceptability of the plant.

(5) Choose any agricultural/horticultural crop (e.g. wheat, barley, potato, apple, hops, etc.) and outline a set of breeding objectives to be used to develop new cultivars. Indicate potential markets for the new cultivars.

Answer:
Include increased yield, increased quality, reduced grower costs (i.e. genetic disease and pest resistance, ease of harvest and storage, etc.).

(6) Explain, using examples as necessary, the meaning of the terms *plant tolerance* and *plant escape* in relation to pest and disease resistance and plant breeding.

Answer:
Plant tolerance is a form of plant '*resistance*' whereby a tolerant genotype will suffer less crop injury or yield loss than a susceptible plant when both are equally infested with a pest or infected with a disease. Tolerant plants have alleles that enable them to withstand the effect of pests and diseases without expressing significant yield loss.

Plant escape in relationship to pests and diseases refers to a change in the genotypes growth cycle or growth pattern whereby the plant and the pest are not synchronous in development, and hence the plant fails to be infected by the disease or infested by the pest. Whenever feasible, it is best to select for genetic resistance to a given pest or disease rather than for the ability to escape or avoid it, since in many cases escaping a disease, for instance through early flowering or a faster crop cycle, can represent a penalty in terms of yield or other important traits. In addition, changes in synchrony of pests and plants could mean that the pests will modify their lifecycle to better fit (say) earlier flowering crops.

(7) Four loci affecting powdery mildew disease have been identified, at which single, dominant, disease-resistance alleles (coded, A, B, C and D) each offer complete immunity to a specific race of powdery mildew (with alleles a, b, c and d each showing susceptibility). Virulence genes have also

been identified in the pest (A', a', B', b', C', c', and D', d'). Given the information below, indicate whether each genotype would be resistant or susceptible to the disease.

Answer:

Plant genotype	Mildew genotype	Plant response
AAbbCCdd	A'a'B'B'c'C'd'd'	Resistant
AaBbCcDd	A'a'B'b'C'c'D'd'	Resistant
aaBBccDD	A'A'b'b'C'c'd'd'	Susceptible
Aabbccdd	A'A'B'B'C'C'D'D'	Resistant

(8) **You have been offered a job as Senior Plant Breeder with the *'Dryeye Onion Company'* in southern Idaho, United States. Your first task in this new position is to set breeding objectives for onion cultivar development over the next 12 years. Outline the main points to be considered when setting your breeding objectives; indicate what questions you would like answered to enhance the breeding objectives you will set. What might be the main sources of information that you would use to arrive at your response?**

Answer:
When setting breeding objectives for an onion breeding programme it is necessary to determine the target region and end-use of the product (i.e. processing or fresh market). It will be necessary to get to know the farming and processing industry of onions and how they are produced. Other factors will include the problems within existing cultivars (i.e. which pests or diseases exist), and then determining whether genetic variation for resistance to these are available. Breeding objectives will almost always have questions relating to people, politics and money. Gathering previous experience from onion breeding programmes located elsewhere might be relevant as well. Who is the end-user and what do they need or want from a cultivar? What morphological factors are important, for example, should the onions be large and few, or small and numerous? At what time(s) in the year are they required? Germplasm will be a key factor. What is available? How can you broaden the genetic base? What 'novel' techniques are on hand (i.e. tissue culture, molecular markers, plant transformation)? Finally, the budget cannot be ignored, as this will determine staff available, facilities available, and what can be achieved with limited funding. Production information is available from the industry, by talking to seedsmen, growers, packers, processors, and so on. Germplasm can often be found in repositories (i.e. USDA, etc.). Then of course there will still be a degree of crystal-ball gazing!

Chapter 4: Breeding Schemes

(1) **Disease resistance to powdery mildew has been identified in a primitive, uncultivated species related to cultivated barley (*Hordeum vulgare*). Research has shown that the resistance is controlled by a single dominant allele (notation = *RR*). There is a resistance screen test available which shows 100% reliability. Outline a breeding scheme that could be used to introgress this character into an existing cultivar ('Golden Promise') which is susceptible to powdery mildew (notation = *rr*). The aim is for the resulting resistant lines to comprise at least 96% of the Golden Promise genotype.**
Consider now that the disease resistance is controlled by a single recessive allele – how would this affect the breeding method you have described above?

Answer:
Assume the uncultivated (wild type) resistant line is parent W, and the adapted (recurrent) parent is A (i.e in this case Golden Promise).

- Cross $W \times A = F_1 = 50\%$ genetic contribution of each parent.
- $F_1 \times A = BC_1 = 75\%$ genetic contribution of A (Golden Promise) and 25% genetic contribution of W (wild type);

- Screen BC_1 population and select plants with good resistance to mildew (i.e. *Rr* types).
- use the resistant plants in cross of $BC_1 \times A = BC_2 = 87.5\%$ genetic contribution of A and 12.5% genetic contribution of W;
  - Screen BC_2 population and select plants with good resistance to mildew (i.e. *Rr* types).
- use the resistant plants in cross of $BC_2 \times A = BC_3 = 93.75\%$ genetic contribution of A and 6.25% genetic contribution of W;
  - Screen BC_3 population and select plants with good resistance to mildew (i.e. *Rr* types).
- use the resistant plants in cross of $BC_3 \times A = BC_4 = 96.87\%$ genetic contribution of A and 3.12% genetic contribution of W.
- Now self-pollinate BC_4 plants to produce BC_4 F_2 and identify lines with homozygous mildew resistance (*RR*). These plants should have 96.87% of the Golden Promise with the resistance gene. However, this may be somewhat over estimated due to linkage drag pooling a high percentage of the wild type when selecting for the resistance gene.

If mildew resistance is recessive (i.e. only expressed as homozygous *rr*) then it will be necessary to self-pollinate the populations between each round of back-crossing and then to screen the progeny from these self-pollinations for mildew resistance (i.e. *rr* types). So it will involve more work!

(2) Crop cultivars can be divided into pure-line cultivars, open-pollinated populations, hybrid cultivars or clonal cultivars. Briefly, describe the major problems that can be encountered or the attributes of the crop types that can be utilized, when breeding: a pure-line cultivar; an open-pollinated population; a hybrid cultivar; and a clonal cultivar.

Answer:

- Pure-line cultivars: difficulties that arise when selecting amongst segregating populations where dominance effects can be large; also selecting on a single-plant basis.
- Open-pollinated populations: difficulty is to achieve effective mass selection to increase desirable gene frequencies.
- Hybrid cultivar difficulties are cost and increased effort in selecting first suitable inbred parents, and thereafter choosing parents that have good combining ability when used in hybrid development.
- Clonal cultivars: difficulties arise from limited and slow rates of propagation; risk of stocks being infected with diseases and bulk of planting and harvested produce.

(3) You have been appointed as alfalfa (*Medicago sativa* L.) breeder to develop crops for the Alfred Alfoncia Seed Company Ltd. Outline a suitable breeding scheme you will use to develop superior synthetic cultivars for your new employer. Show each stage of the scheme on a year-by-year basis.

After several rounds of clonal selection, you have identified four potential highly productive parent lines for developing a new synthetic (lines are coded as A, B, C and D). Hybrid seed was produced from all two-parent cross combinations possible and these were evaluated in yield trials along with the parent lines. From the yield (kg) results (below), indicate the three parent synthetic most likely to be highest yielding and state the expected yield.

> **Parent A = 18; Parent B = 22;**
> **Parent C = 21; Parent D = 25**
> **A × B = 25 A × C = 26 A × D = 24**
> **B × C = 30 B × D = 21 C × D = 24**

Answer:
Many different breeding schemes could be designed and, although different, they may all be successful. However, a successful scheme would most likely have three distinct stages, including:

(1) Develop a clonal nursery of genetically different alfalfa genotypes and evaluate each potential genotype phenotypically while increasing the nursery lines vegetatively, so that each is genetically fixed.

(2) Evaluate potential selected parents from the clonal nursery genetically in a randomized mating design where each parent/clone is replicated multiple times and allowed to randomly cross-pollinate (usually with specific pollinators, i.e. leaf cutter bees). Then harvest seed from each potential parent and evaluate production potential.

(3) Determine which of the selected parents will be included in the 'new' synthetic cultivar, and increase these parents vegetatively for commercial seed production.

The best synthetic line with three parents would be to combine A, B and C. Expected yield of synthetic would be $F_1 - (F_1 - P)/n$, where F_1 would be $(25 + 30 + 26)/3 = 27$, $P = (18 + 22 + 21)/3 = 20.33$ so $F_1 - (F_1 - P)/n, = 27 - (27 - 20.33)/3 = 24.78$ kg.

(4) Outline three characteristics of a crop species that would merit consideration before developing hybrid cultivars from that species.

Answer:
Must show economically viable heterosis, must be difficult to produce (i.e. show inbreeding depression) high-yielding inbred lines, and must have an economically viable mechanism to produce hybrid seed.

(5) Outline the advantages and disadvantages of a bulk breeding scheme and a pedigree breeding scheme to develop pure-line cultivars.
   Many breeding schemes for pure-lines are a combination of bulk and pedigree schemes. Design a suitable bulk/pedigree breeding scheme (it is not necessary to make notes on when specific characters are selected) that could be used to develop superior pure-line cultivars. Outline any advantages or disadvantages of your breeding scheme.

Answer:
Disadvantages of bulk breeding schemes include: time between initial cross and first yield trials; and natural selection during the bulked towards homozygosity stage is not always advantageous. Advantages of bulk schemes include: conscious selection is not attempted until near-homozygosity; avoids difficulty of selecting amongst segregating lines where dominance effects can be large; least inexpensive. It should be noted that bulk breeding schemes have become more popular with the use of doubled haploids and SSD. Disadvantages of pedigree breeding schemes include: that they are more laborious with considerable record-keeping; expensive, requires more land and experienced staff needed with 'good eye'; selection inefficient on single plants; no actual early generation yield testing. However, advantages of pedigree breeding schemes are that if single plant selection is efficient, inferior genotypes are discarded without expensive yield testing. Many different pedigree/bulk schemes exist and most can be highly successful.

(6) A hybrid breeding programme has identified six superior inbred lines (A, B, C, D, E and F) which were intercrossed in all combinations and the F_1 progenies evaluated for productivity in a field yield trial. The results from each single cross in the trial were:

A × B = 61	A × C = 65	A × D = 51	A × E = 56
A × F = 62	B × C = 63	B × D = 65	B × E = 59
B × F = 60	C × D = 59	C × E = 58	C × F = 60
D × E = 57	D × F = 68	E × F = 54	

Which parent combination will provide the most productive single cross hybrid and what would the expected yield be? Which parent combination will provide the most productive three-way cross hybrid and what would the expected yield be? Which parent combination will provide the most productive double cross hybrid and what would the expected yield be?

Explain why the most productive double cross combination (from the prediction equation) may not result in the most commercially suited hybrid cultivar.

Answer:

From the dataset the 'best' two-parent hybrid would be the hybrid $D \times F$, with expected yield of 68.0 t/ha. The 'best' three-way hybrid would be $(B \times F) \times D$, with expected yield of 66.5 t/ha (i.e. [(B $\times$ D) + (D $\times$ F)]\2 = [65 + 68]/2 = 66.5). The 'best' double-cross hybrid would be $(B \times F) \times (A \times D)$, with expected yield of 64.0 t/ha (i.e. [(A $\times$ B) + (B $\times$ D) + (A $\times$ F) + D $\times$ F)]/4 = [61 + 65 + 62 + 68]/4 = 64.0). The 'best' parent hybrid combination may not make the best commercial hybrid if parent seed production is particularly difficult.

(7) **In order to justify development of hybrid cultivars it is necessary to have some method of producing inexpensive hybrid seed. Outline three systems that have been used to produce hybrid seed and indicate the advantages or disadvantages of each system.**

Answers can include:

- *Mechanical*: Hermaphroditic species would require female emasculation and pollination; diclinous species are easier, but cost and labour inputs are often prohibitive.
- *Nuclear male sterile (NMS)*: Found in many species; usually recessive (msms), MsMs and Msms are fertile; problems in introducing NMS and 100% restoration.
- *Self-incompatibility (SI)*: SI usually recessive (s_1s_1), S_1s_1 are self-compatible. SI system often not functional before flower opening; obtaining 100% SI is rare; offers cheap seed production.
- *Cytoplasmic male sterility (CMS)*: Most common for hybrid development; wide natural occurrence; foreign cytoplasm can effect performance; mostly dominant; lack of pollen in male parent can affect insect pollinators.

(8) **'Russet Burbank' has been the predominant potato cultivar in the Pacific Northwest for many years. You have been asked to design a breeding programme that will produce cultivars that can replace some of the Russet Burbank acreage. Outline the breeding design you would use, indicating: (a) five breeding objectives; (b) the breeding scheme (use diagrams if necessary); and (c) indicating the number of clones evaluated and selected at each stage.**

Answer:

The possibilities are wide here, and so it is not possible to give any specific thoughts as alternatives are numerous and many different answers could be correct.

(9) **In order to produce a hybrid cultivar using cytoplasmic male sterility you require three types of genotypes. Describe the differences between the three types required, and describe how commercial production will be conducted.**

Answer:

Start by considering how to increase seed quantities of the female parent which is cytologically male sterile (A-cms-rfrf). This is done by producing an isogenic line (the maintainer parent) which differs only in that it is male fertile (A-n-rfrf). Both A-cms and A-n have no fertility restoration genes. After crossing together the CMS parent and the maintainer parent we have AA'-cms-rfrf which is male sterile, with no restoration genes (rfrf). This becomes the female cross in the hybrid scheme and is crossed to B-n, which is pollen fertile and is homozygous for the dominant restoration gene (RfRf). This cross results in hybrid seed of AA'B-cms-rfRf which is pollen fertile as the restoration allele overrides the CMS.

(10) **You are to embark on a backcrossing study designed towards convergence breeding. You have at your disposal two genotypes coded C and M. Genotype C is a cytoplasmic male sterile line, but one that does**

not have very good hybrid combining ability; genotype M has great hybrid combining ability and is male fertile (without a male fertile restorer gene). Design a scheme (listing each operation necessary) that will result in a genotype which is 95% the nuclear genotype of the M line but is cytoplasmically male sterile.

Answer:
$C \times M = F_1 = 50\%$ genetic contribution of each parent. $F_1 \times M = BC_1 = 75\%$ genetic contribution of M and 25% genetic contribution of C; $BC_1 \times M = BC_2 = 87.5\%$ genetic contribution of M and 12.5% genetic contribution of C; $BC_2 \times M = BC_3 = 93.75\%$ genetic contribution of M and 6.25% genetic contribution of C; $BC_3 \times M = BC_4 = 96.87\%$ genetic contribution of M and 3.12% genetic contribution of C, while the cytoplasm has consistently derived through the maternal side from C.

(11) You are walking along a deserted beach in southern California (are there any of these left?) and you inadvertently kick a brass oil lamp. On picking up the lamp you gently rub the side and – *poof!* – a 20 foot tall genie appears! '*Oh master!* ' says the genie, '*I can grant you two wishes*' (well, times are hard and the old three wishes rule does not apply any more). You think for a few minutes, then say '*I would like you to conjure me up the most wonderful plant that exists within the Universe, so that I can grow it on my farm in Idaho and make millions of dollars in profits*'. *Poof!* This plant appears before you – a plant species that has never been seen on Earth before.

As a plant breeder, list five questions you would ask the genie about the biology of this plant that would help you to design a cultivar development programme to increase the value of it as a crop.

Having sorted that out, you begin to think of your other wish. '*I would like you to tell me the formula for a chemical apomict*' (a chemical that, when applied to a crop, will result in 100% apomictic seed from the crop). '*Well*' says the genie. '*I can do this, but you must specify which crop species the chemical will work on*' (it can only work on one chosen species). What crop species would you choose to have apomictic seeds, and why?

Answer:
Questions to ask: Is the species: in-breeding or out-breeding; vegetatively reproduced; expressing inbreeding depression or heterosis; spring or winter annual or biennial; day-length sensitive?

Chemical apomict: Many crops would benefit from being apomictic. Easiest answer may be maize where heterosis is very high and hybrids are routinely grown. Apomixis would allow maize heterosis to be fixed and avoid the need for expensive seed production systems. Potatoes would be a second crop where apomixes would have tremendous value as vegetative potato production has several negative effects, for example seed tubers are bulky, can easily be infected with virus or bacterial disease and must be grown under strict conditions. In addition, if true botanical potato seeds were to replace potato tubers for planting, this would result in more harvested tubers to eat and also it would be possible to plant one hectare of potatoes with a few kg of seed rather than a few tonnes of tubers.

Chapter 5: Genetics and Plant Breeding

(1) A cross is made between two homozygous barley plants. One parent is dominant for a tall stature gene (*TT*) and has a single dominant resistance to yellow rust (*YY*), but highly susceptible to powdery mildew (*mm*). So the genotype of the first parent can be represented by *TTYYmm*. The other parent has short stature (i.e. recessive dwarfing gene, *tt*) with single (completely dominant) resistance to powdery mildew (*MM*), but is susceptible to yellow rust (*yy*). So the genotype of the second parent can be represented by *ttyyMM*. Consider if a sample of F_1 plants from this cross were self-pollinated and used to produce a population of 12,800 F_2 plants. At the harvest of these F_2 plants the short (dwarf) plants were selected and only seed from these plants was planted out in head-rows at F_3. How many of the F_3 head-rows would be resistant to both yellow rust and powdery mildew? How many will be resistant to powdery mildew but susceptible to yellow rust? And, how many will be susceptible to both diseases?

Answer:

Assume first that we will select only for the dwarf at F_2 (i.e. *tt*). This being the case, we will retain $1/4$ of the initial population based on this height selection. Thus from the 12,800 F_2 plants we will (theoretically) select 3,200 and so have that many F_3 head-rows planted. Now that we have selected for a single recessive allele for height, we can forget about that trait, as it will be 'fixed'. We need then only concern ourselves with the two segregating loci being rust and mildew resistance. The F_3 head rows will segregate phenotypically as:

$$25 \ Y\_M\_tt: 15 \ Y\_mmtt : 15 \ yyM\_tt: 9 \ yymmtt$$

Therefore:

Resistant to both diseases	$= 25/64 \times 3{,}200 = 1{,}250$
Resistant to rust, susceptible to mildew	$= 15/64 \times 3{,}200 = 750$
Resistant to mildew, susceptible to rust	$= 15/64 \times 3{,}200 = 750$
Susceptible to both diseases	$= 9/64 \times 3{,}200 = 450$

(2) In a breeding experiment similar to that in question 1, a cross is made between two homozygous barley plants. One parent is tall and with short leaf margin hairs (*TTss*), and the other is short with long leaf margin hairs (*ttSS*). Single genes control both plant height and leaf margin hair length, tall plants (*T_*) being completely dominant to short (*tt*), and long leaf margin hairs (*S_*) dominant to short (*ss*). A sample of F_1 plants were self-pollinated and 1,600 F_2 plants evaluated for plant height and leaf margin hair length. After evaluating the phenotypes it was found that there were:

893 tall plants with long leaf margin hairs;
313 tall plants with short margin hairs;
0 short plants with long leaf margin hairs; and
394 short plants with short margin hairs.

Give a genetic explanation of what underlies the observed frequency of phenotypes and test your theory using a suitable statistical test.

Answer:

Perhaps the best possible explanation would be recessive epistasis where *tt* is epistatic to *SS* and *Ss*. A statistical test of this hypothesis would be a χ^2 test, assuming an F_2 segregation, 9:3:0:4.

	Phenotypes		
	Tall/long margins	Tall/short margins	Short/short margins
Observed (O)	893	313	394
Expected (E)	900	300	400
$(O-E)^2/E$	0.05	0.65	0.09
	$\chi^2_{2df} = 0.76$ = not significant		

As the χ^2 value is less that the corresponding table value we accept the hypotheses that the observed and expected are the same and accept our theory that the observed ratio is explained be recessive epistasis where *tt* is epistatic to *SS* and *Ss*.

(3) A spring wheat breeding programme aims to develop cultivars that are resistant to foot-rot (controlled at a single locus with a completely dominant allele (*FF*) conferring resistance), and that are also resistant to yellow stripe rust (controlled at a single locus with a single dominant (*YY*) allele conferring resistance). A cross is made between two parents where one parent is resistant to foot-rot and susceptible to yellow stripe rust (*FFyy*) while the other is resistant to yellow stripe rust but susceptible to

foot-rot (*ffYY*). The heterozygous F_1 is crossed to a double susceptible homozygous line with genotype *ffyy* and 2,000 BC$_1$ progeny were evaluated for disease resistance. The following numbers of phenotypes were observed:

Resistant to both foot-rot and yellow stripe rust	= 90
Resistant to foot-rot but susceptible to yellow stripe rust	= 893
Susceptible to foot-rot but resistant to yellow stripe rust	= 907
Susceptible to both foot-rot and yellow stripe rust	= 110

Determine the percentage recombination between the foot-rot and yellow stripe rust loci.
Given the frequency of phenotypes with the above percentage recombination, how many F_2 plants would need to be evaluated to be 99% certain of obtaining at least one plant that is resistant to both foot-rot and yellow stripe rust? In contrast, how many would need to be evaluated to be 99% certain of obtaining at least one plant that is resistant to both foot rot and yellow stripe rust, given independent assortment of the two loci involved?

Answer:

$$\% \text{ recombination} = (90 + 110)/2,000 = 0.1 \text{ or } 10\%$$

Gametes	$FY - 0.05$			$fy - 0.05$
		$Fy - 0.45$	$fY - 0.45$	
$FY - 0.05$	*FFYY*	*FFYy*	*FfYY*	*FfYy*
	25	225	225	25
$Fy - 0.45$	*FFYy*	*FFyy*	*FfYy*	*Ffyy*
	225	2,025	2,025	225
$fY - 0.45$	*FfYY*	*FfYy*	*ffYY*	*ffYy*
	225	2,025	2,025	225
$fy - 0.05$	*FfYy*	*Ffyy*	*ffYy*	*ffyy*
	25	225	225	25

Pooling like phenotypes, we have:

$$F\_Y\_ = 5,025$$
$$ffY\_ = 2,475$$
$$F\_yy = 2,475$$
$$ffyy = 25$$

Probability of resistance to both diseases = $5,025/10,000 = 0.5025$. To obtain 99% certainty of obtaining at least one of this type = $\ln(1 - 0.99)/\ln(1 - 0.5025) = 6.6$, so we need 7 plants to be over 99% certain that one plant is resistant to both foot-rot and yellow stripe rust.

If we assume that foot-rot and yellow stripe rust gene are on different chromosomes (i.e. independently assort) then the probability of both diseases = $9/16 = 0.5625$. In this case to have 99% certainty of obtaining at least one plant of this type = $\ln(1 - 0.99)/\ln(1 - 0.5625) = 5.6 =$ or 6 or more plants, so the number needed is fewer, but not by many!

(4) Potato cyst nematode resistance is controlled by a single locus with a completely dominant allele (R). A cross is made between two auto-tetraploid potato cultivars where one parent is resistant to potato cyst nematode and the other is completely susceptible. Progeny from the cross are examined and it is found that 83.3% of the progeny are resistant to potato cyst nematode. What can be determined about the genetic status of the resistant parent in the cross?

A breeding programme aims to produce potato parental lines that are either quadruplex or triplex for the potato cyst nematode resistant allele *R*. A cross is made between two parents, which are known to be duplex for the resistant allele *R*. What would be the expected genotype and phenotype of progeny from this duplex × duplex (i.e. *RRrr* × *RRrr*) cross?

Answer:

It can be assumed that the 'resistant' parent carried two of the dominant resistance alleles (i.e. was duplex). If it had carried only one copy, then a 50:50 ratio would be expected, and if it carried three or four copies then all the progeny would be resistant.

Gametes	1:duplex	4:simplex	1:nulliplex
1:duplex	1:quadriplex	4:triplex	1:duplex
4:simplex	4:triplex	16:duplex	4:simplex
1:nulliplex	1:duplex	4:simplex	1:nulliplex

Pooling all like types together we get: a genotype ratio of = 1 quadriplex : 8 triplex : 18 duplex : 8 simplex : 1 nulliplex; and a phenotype ratio of = 35 R_- : 1 *rrrr* (or nulliplex).

(5) **Why are plant breeders interested in conducting scaling tests for quantitatively inherited characters of importance in the breeding scheme?**
A properly designed experiment was carried out in canola where parent lines (P_1 and P_2), are grown alongside F_1 and F_2 progeny plants from the cross $P_1 \times P_2$. The average yield of plants from each of the four families, the variance of each family and the number of plants evaluated from each family were:

Family	Mean yield	Variance of yield	Number of plants
P_1	1,901	74	21
P_2	1,502	68	21
F_1	1,429	69	21
F_2	1,888	102	31

Use an appropriate test to determine whether the additive – dominance model of inheritance is adequate to explain the genetic variation in yield in canola.
What can you conclude from the results obtained above? What could have caused this to occur?

Answer:

Scaling tests show the adequacy of the simple additive/dominance model. If such a model is adequate to explain progeny performance, then there is a good chance that breeders can predict performance (i.e. using heritabilities). If this is not the case, then they cannot predict so straightforwardly.

Use a C-scaling test where: $4F_2 - 2F_1 - P_1 - P_2 = C$ and $\sigma^2_c = 16\sigma^2_{F2} + 4\sigma^2_{F1} + \sigma^2_{P1} + \sigma^2_{P2}$

$C = 1,291, \sigma_c = \sqrt{2050} = 45.3, t_{90\ df} = 28.5$. So C is significantly different from zero and hence the additive dominance model is not adequate to explain the progeny means observed.

This could be a result of epistasis, chromosomal abnormalities or cytoplasmic effects.

(6) **Assuming an additive – dominance mode of inheritance for a polygenic trait, set out the parameters to describe P_1, P_2, and F_1 in terms of *m*, [*a*] and [*d*] and then similarly for F_2, B_1 and B_2.**
From a properly designed field trial that included B_1, B_2 and F_1 families, the following yield estimates were obtained.

$$B_1 = 42.0 \text{ kg/acre}; \quad B_2 = 26.0 \text{ kg/acre}; \quad F_1 = 38.5 \text{ kg/acre}$$

From these family means, estimate the expected value of P_1, P_2 and F_2, based on the additive – dominance model of inheritance

Answer:

$P_1 = m + a$;
$P_2 = m - a$;
$F_1 = m + d$;
$F_2 = m + \frac{1}{2}d$;
$B_1 = m + \frac{1}{2}a + \frac{1}{2}d$;
$B_2 = m - \frac{1}{2}a + \frac{1}{2}d$.

Now:

$B_1 - B_2 = a$, so $a = 42.0 - 26.0 = 16.0$, and
$B_1 + B_2 = 2m + d$, but
$F_1 = m + d$, so
$B_1 + B_2 - F_1 = m = 29.5$

Finally:

$F_1 = m + d$, so
$d = F_1 - m = 9$

So expected values:

$P_1 = 45.5$
$P_2 = 13.5$
$F_2 = 34.0$

(7) A cross is made between two homozygous barley plants. One parent has a two-row ear (controlled at a single locus by a recessive allele (*hh*) (as opposed to the dominant six-row (*HH*)) and is dominant for a tall stature at the 'T' locus (*TT*), is resistance to yellow rust (*YY* – completely dominant), but highly susceptible to powdery mildew (i.e. *TThhYYmm*). The other parent has a six-row ear (*HH*), short (dwarf, *tt*), resistant to powdery mildew, but is susceptible to yellow rust (i.e. *ttHHyyMM*). All trait loci are located on different chromosomes. A sample of F_1 plants from this cross were self-pollinated, and a population of 92,160 F_2 plants grown. At harvest the short (dwarf) plants with two-row ears are selected and only seed from these plants are grown out in head-rows at F_3. How many of the F_3 head-rows would be resistant to both yellow rust and powdery mildew? How many will be resistant to powdery mildew but susceptible to yellow rust? And how many will be susceptible to both diseases?

Answer:

At the F_2 harvest only the dwarf (*tt*) plants with 2-row ears (*hh*) are selected. As both these traits must be homozygous recessive (i.e. *tthh*) their genotype will be genetically fixed in further rounds of self-pollination. 1/16 of the F_2 population will be dwarf with 2-row ears, so approximately 5,760 plants will be selected.

At F_3, the population will segregate in the normal way, despite the fact that 15/16 of the population have been discarded as the traits segregate independently (i.e. all loci are on different chromosomes). So phenotypically we would expect:

Resistant to yellow rust and mildew (i.e.*tthhY_M_*)	= 26/64 = 2,250
Susceptible to yellow rust and resistant to mildew (i.e.*tthhyyM_*)	= 15/64 = 1,350
Susceptible to yellow rust and mildew (i.e.*tthhyymm*)	= 6/64 = 540

(8) Two homozygous squash plants were hybridized and an F_1 family produced. One parent had long, green fruit (*LLGG*), and the other had round, yellow fruit (*llgg*). 1,600 F_2 progeny were examined from selfing the F_1s and the following numbers of phenotypes observed:

LLGG	L-gg	llG-	llgg
891	312	0	397

Complete an appropriate analysis to explain and interpret this segregation pattern.

Answer:

The observed frequency of phenotypes could be explained by recessive epistasis whereby ll is epistatic to G_-. To test this hypothesis a chi-squared test would be carried out where we compare the observed ratio 891:312:0:397 to a 9:3:0:4 ratio, or given that our total observed is 1,600, we compare the observed to an expected ratio of 900:300:0:400 ratio. The chi-squared value is the sum of the (observed − expected)^2/expected = $(891 − 900)^2/900 + (312 − 300)^2/300 + (397 − 400)^2/400 = 0.59$, and as it is smaller than zero we conclude that the observed data do not deviate significantly from expected, given a situation involving recessive epistasis whereby ll is epistatic to G_-.

Chapter 6: Predictions

(1) **Given values for the variance of the mean of the F_2, and variance of the F_1, estimate h_b^2 and explain what this tells us about the genetic determination of the trait.**

$$V_{\overline{F}_2} = 436.72$$
$$V_{\overline{F}_1} = 111.72$$

Given below are the variances of the mean from two parents (P_1 and P_2), the F_1, F_2, and both backcross families (B_1 and B_2). Estimate h_n^2 and explain what the value means in genetic terms.

$$V_{\overline{P}_1} = 14.1 \quad V_{\overline{P}_2} = 12.2$$
$$V_{\overline{F}_1} = 13.3 \quad V_{\overline{F}_2} = 40.2$$
$$V_{\overline{B}_1} = 35.2 \quad V_{\overline{B}_2} = 34.6$$

Answer:

$\sigma_e^2 = \sigma_{F1}^2 = 111.72$;
$h_b^2 = \sigma_g^2/\sigma_p^2 = \sigma_g^2/\sigma_g^2 + \sigma_e^2 = (436.72 − 111.72)/436.72 = 0.744$,
or 74.4% of the variation is genetic in nature.

$h_n^2 = \frac{1}{2}A/$ [total phenotypic variation];
σ_e^2 = error variance = $[\sigma_{p1}^2 + \sigma_{p2}^2 + 2\sigma_{F1}^2]/4 = [14.1 + 12.2 + (2 \times 13.3)]/4 = 13.2$;
$D = 4[\sigma_{B1}^2 + \sigma_{B2}^2 − \sigma_{F2}^2 − \sigma_e^2] = 4[35.2 + 34.6 − 40.2 − 13.2] = 65.6$;
$A = 2[\sigma_{F2}^2 − \frac{1}{4} D − \sigma_e^2] = 2[40.2 − 16.4 − 13.2] = 21.2$;
$h_n^2 = \frac{1}{2} A/[\frac{1}{2}A + \frac{1}{4}D + \sigma_e^2] = [10.6/(10.6 + 16.4 + 13.2)]$ or $[10.6/(\sigma_{F2}^2)]$
  $= 10.6/40.2 = 0.264$ or 26.4% of total variation is additive genetic variance.

(2) **List the six assumptions necessary for a straightforward interpretation of a Hayman and Jinks' analysis of diallels.**
Below are shown values of array means, within array variances (V_r) and covariances between array values and non-recurrent parents (W_r) from a Hayman and Jinks' analysis of a 7×7 complete diallel in dry pea.

Parent name	V_r	W_r	Array mean
'Souper'	34.1	19.3	456
'Dleiyon'	99.9	79.3	305
'Yielder'	21.0	11.2	502
'Shatter'	99.4	68.4	314
'Creamy'	49.6	39.4	372
'SweetP'	59.1	48.8	361
'Limer'	61.8	49.2	393

The trait of interest is pea grain yield. Regression of W_r against V_r resulted in the equation: $W_r = 0.837 \times V_r - 4.817$, with standard error of the regression slope equal to $se_b = 0.0878$. An analysis of variance of $W_r + V_r$ showed significant differences between arrays, while a similar analysis of $W_r - V_r$ showed no significant differences between arrays. What can be deduced regarding the inheritance of pea yield from the information provided? If you were a plant breeder interested in developing high-yielding dry pea cultivars, on which two parental lines would you concentrate your breeding efforts? Briefly explain why.

Answer:
Diploid segregation; homozygous parents; no difference between reciprocal crosses; no epistasis; no multiple alleles; genes are distributed independently between the two parents.
V_i/W_i regression: t test $[0.837 - 1.000]/0.0878 = 1.86$ = ns, therefore an additive/dominance model of inheritance would appear appropriate to explain yield variation in pea.
The 'best' two parents to choose might be 'Souper' and 'Yielder', as yield in plant breeding is often of highest priority, and these have markedly higher yield than the others. Note that parents with low V_i and low W_i will tend to have a high degree of recessive alleles for that trait (here yield). So to produce inbred cultivars it is often easier to select for recessive traits. Here we would like a high average yield associated with low V_i and W_i values. Both 'Souper' and 'Yielder' show these desirable characteristics.

(3) Below is shown an analysis of variance of plant yield from a 6×6 half diallel (including parents). The analysis of variance is from a Griffing's analysis (Model 2). GCA = general combining ability, SCA = specific combining ability, Error = error obtained by replication, df = degrees of freedom, and SS = sum of square.

Source	df	SS
GCA	5	4,988
SCA	15	6,789
Error	21	5,412

Discuss the results from the analysis, given that the six parents were: (a) specifically chosen, and (b) chosen completely at random.

Answer:
Mean squares for GCA, SCA and error would be 998, 453, and 258, respectively.
(a) Parents specifically chosen: means a fixed model and we would test the GCA and SCA in an F-test against the error mean square, giving F-value of 3.87 and 1.76, for GCA and SCA, respectively, the former being significant at the 5% level while the latter is not significant. As GCA is significant and SCA not, there should be good prospects to predict progeny performance given the GCA of each parent.

(b) Parents chosen completely at random: this means we can assume a random model and then the GCA F-value is calculated by dividing the GCA mean square with the SCA mean square or $998/453 = 2.20$, which is not significant, and so in this case the GCA is not significant and hence it will not be reasonable or possible to accurately predict progeny performance from the parental general combining abilities.

(4) **It is desired to determine the narrow-sense heritability for flowering date in spring canola (*B. napus*). Both parents and their offspring from ten cross combinations were grown in a properly designed field experiment. At harvest, yield was recorded for each entry, and using these data the average pheno-type of two parents (i.e. $[P_1 + P_2] / 2$) was considered to be x, the independent variable, while the performance of their offspring was considered as y, the dependent variable. A regression analysis is to be carried out by regression of the offspring (y) onto the average parent (x). The following data are derived: $SP(x, y) = 345.32$; $SS(x) = 491.41$; $SS(y) = 321.45$. Estimate the slope of the regression line (b), test whether this slope is greater than zero, and estimate the narrow-sense heritability from the regression equation.**

How would the relationship between the regression and the narrow-sense heritability differ if the regression were carried out between only the offspring and the male parent?

Answers:
Heritability can be estimated by regression of progeny performance onto the mid-parent values. Now the regression slope (b) = $[SP(x, y)/SS(x)] = 345.32/491.41 = 0.70$, so $h_n^2 = 0.70$. To perform a suitable t-test it is necessary to estimate the standard error of the regression slope (b), from $se(b) = \sqrt{\{\{[SS(y) - [b \times SP(x, y)]]\} (n - 2) \times SS(x)\}} = \sqrt{\{\{[321.45 - [0.70 \times 345.32]\}/\{8 \times 491.41\}\}} = \sqrt{\{79.73/3,931.3\}} = 0.14$. Now to test whether we have a significant heritability, it must be different from zero, so t-value = $|b\text{-}0.0|/se(b) = |0.7/0.14| = 5.00$, which is larger than the required target at 8 df (2.31), and as such we can say that flowering date in these canola has a narrow-sense heritability significantly higher than zero.

(5) **Four types of diallel can be analysed using a Griffing's analysis. Describe these types. Families from a 5×5 half diallel (including selfs) were planted in a two-replicate yield trial at a single location. The parents used in the diallel design were chosen to be the highest yielding lines grown in the Pacific-Northwest region. Data for yield were analysed using a Griffing's analysis of variance. Family means, averaged over two replicates, degrees of freedom and sum of squares (SS) from that analysis are shown below. Interpret the results from the Griffing's analysis. What differences would there be in your analytical methods if the parents used had been chosen at random?**

	Parent 1	Parent 2	Parent 3	Parent 4	Parent 5
Parent 1	62.0				
Parent 2	71.0	69.5			
Parent 3	55.5	52.5	50.5		
Parent 4	72.5	80.5	56.5	76.5	
Parent 5	70.5	66.5	36.5	71.0	64.5

Source	df	SS
GCA	4	6,694.058
SCA	10	825.676
Replicates	1	8.533
Error	14	317.467
Total	29	7,845.733

From the same diallel data (above), within-array variances (V_r) and between-array and non-recurrent parent covariances (W_r) were calculated. The values of V_r and W_r for each parent along with the mean of V_r, mean of W_r, sum of squares of $V_r (\Sigma V_r^2)$, sum of squares of $W_r (\Sigma W_r^2)$ and sum of products $\Sigma V_r W_r$ are shown below.

	V_r	W_r
1	98.0	102.0
2	66.3	53.2
3	161.8	207.4
4	50.3	65.2
5	71.4	83.1

Mean of $V_r = 89.56$; mean of $W_r = 102.19$; $\Sigma V_r^2 = 7,702.01$; $\Sigma W_r^2 = 15,192.05$; $\Sigma V_r W_r = 10,535.29$. From these data, test whether the additive – dominance model is adequate to describe variation between the progenies. What can be determined about the importance of additive versus dominance genetic variation in this study? From all the results (Griffing, and Hayman and Jinks, above), which two parents would you use in your breeding programme and why?

Answer:
Full diallel, including selfs, Griffing method 1; half diallel, including selfs, Griffing method 2; full diallel, excluding selfs, Griffing method 3; and half diallel, excluding selfs, Griffing method 4.
Mean squares from the Griffing analysis would be 1,673.5, 82.6, and 22.7, for GCA, SCA and error respectively. If a fixed effect model is used, then both the GCA and SCA effects are tested for significance against the error mean square, giving F-value GCA = 1,673.5/22.7 = 73.7 (highly significant); F-value SCA = 82.5/22.7 = 3.63 (significant at 5% level). However, if a random model is used, then the GCA F-value is calculated by dividing the GCA mean square by the SCA mean square or 1,673.5/82.5 = 20.3 (also highly significant). In this case the GCA effect is so large that it is irrelevant which model is used, since the conclusion is the same – GCA is highly significant, but SCA is significant at the 5% level. Although statistically GCA and SCA are significant, it should be noted that the GCA term accounts for [[sum of square for GCA/sum of square for total] × 100]or [6,694.058/7,845.733] × 100 = 85.3% of the total variation observed. In comparison the SCA effect accounted for only [825.676/7,845.733] × 100 = 10.5% of the total variation. From a practical plant breeding viewpoint, this is good for selection.

(6) Given the variance from two parents (P_1 and P_2), the F_1, F_2, B_1 and B_2 families, estimate the narrow-sense heritability (h_n^2) and explain what the value means in genetic terms.

$$\sigma^2_{P_1} = 9.5; \quad \sigma^2_{P_2} = 7.4$$
$$\sigma^2_{F_1} = 8.6; \quad \sigma^2_{F_2} = 17.7$$
$$\sigma^2_{B_1} = 14.3; \quad \sigma^2_{B_2} = 15.2$$

Answer:
$h^2_n = \frac{1}{2}A/[\text{total phenotypic variation}]$;
σ_e^2 = variation within families without genetic variability = error variance
$= [\sigma_{p1}^2 + \sigma_{P2}^2 + 2\sigma_{F1}^2]/4 = [9.5 + 7.4 + (2 \times 8.6)]/4 = 8.5$;

$D = 4[\sigma_{B1}^2 + \sigma_{B2}^2 - \sigma_{F2}^2 - \sigma_e^2] = 4[14.3 + 15.2 - 17.7 - 8.5]$
$= 4 \times 3.3 = 13.2; A = 2[\sigma_{F2}^2 - \frac{1}{4}D - \sigma_e^2] = 2[17.7 - 3.3 - 8.5] = 2 \times 5.9 = 11.8;$
$h_n^2 = \frac{1}{2}A/[\frac{1}{2}A + \frac{1}{4}D + \sigma_e^2] = [5.9/(5.9 + 3.3 + 8.5]$ or $[5.9/(\sigma_{F2}^2)] = 5.9/17.7$
$= 0.333$ or 33.3% of total variation is additive genetic variance.

(7) **A new oil crop (*Brassica gasolinous*) has been discovered that may have potential as a biologically renewable fuel oil substitute. This diploid species is tolerant to inbreeding and is self-compatible. A preliminary genetic experiment was designed to examine the inheritance of seed yield (Yield) and percentage oil content (%Oil). This experiment involved a 4 × 4 half diallel (including selfs). The four homozygous parental lines are represented by the codes AAA, BBB, CCC and DDD. The half diallel array values (averaged over two replicates), array means, general combining ability (GCA) values, mean squares from the analyses of variance (Griffing style), V_r and W_r values (Hayman and Jinks' analysis), variance of array means (V_{xr}) and parental variances (V_p), and the one-way analyses of variance for $V_r + W_r$ and $V_r - W_r$ are shown below for each character.**

Yield					%Oil			
AAA	40.5				20.5			
BBB	38.5	29.5			20.5	25.0		
CCC	37.0	28.0	19.5		23.0	26.0	30.5	
DDD	32.5	20.5	18.5	10.0	24.5	27.5	31.0	36.0
	AAA	BBB	CCC	DDD	AAA	BBB	CCC	DDD
Array means	37.1	29.1	25.8	20.4	22.1	24.7	27.6	29.8

Source	df	Yield	%Oil
GCA	3	796.5	180.6
SCA	6	90.5	30.1
Replicate blocks	1	0.4	0.1
Replicate error	9	51.5	5.7

V_r and W_r values

	Yield		%Oil	
	V_r	W_r	V_r	W_r
AAA	46.2	171.3	15.6	50.7
BBB	218.2	376.7	36.3	75.0
CCC	297.7	436.7	58.3	98.0
DDD	344.3	472.3	97.3	132.3

V_{xr} and V_p values

	Yield	%Oil
V_{xr}	196.9	44.4
V_p	687.6	180.7

		Yield		%Oil	
Source	df	$V_r + W_r$	$V_r - W_r$	$V_r + W_r$	$V_r - W_r$
Between	3	8760	28.0	628	7.68
Within	4	133	17.0	115	4.77

Without using regression, estimate the narrow-sense heritability (h_n^2) for seed yield, and explain this value in terms of genetic variance. Explain the analyses for Yield and outline any conclusions that can be drawn from these data. Explain the analysis for %Oil (percentage of seed weight that is oil) and outline any conclusions that can be drawn from these data. Which one of these four genotypes would you choose as a parent in your breeding programme? Explain your choice. Describe any difficulties suggested from these analyses in a breeding programme designed for selecting lines with high yield and high percentage of oil.

Answer:
In these parents there is a suggestion (visual inspection of data) that they are negatively correlated (i.e. parents with high yield tend to have low oil content and *vice versa*). This might mean that the characters themselves are either pleiotropic (in which case we can never break this relationship), or are strongly linked, which means we will probably need many cycles of crossing to break the linkage, before selection is imposed, making it more difficult to select a new cultivar with high yield combined with high oil content.

Finally, a plant breeder perhaps would choose parent AAA as yield has highest priority, and although oil content is important, no one (to-date) pays a premium for high oil content. Also, within the breeding programme it would be most likely that the breeder would cross AAA with a high oil type.

(8) F_1, F_2, B_1 and B_2 families were evaluated for plant yield (kg/plot) from a cross between two homozygous spring wheat parents. The following variances from each of the four families were found:

$$\sigma^2_{F_1} = 123.7; \quad \sigma^2_{F_2} = 496.2$$
$$\sigma^2_{B_1} = 357.2; \quad \sigma^2_{B_2} = 324.7$$

Calculate the broad-sense ($h_b{}^2$) and narrow-sense ($h_n{}^2$) heritabilities for plant yield. Given the heritability estimates you have obtained, would you recommend selection for yield at the F_3 stage in this wheat breeding programme, and why?

Answer:
$\sigma_e{}^2 = \sigma_{F1}{}^2 = 123.7; h_b{}^2 = \sigma_g{}^2/\sigma_p{}^2 = \sigma_g{}^2/\sigma_g{}^2 + \sigma_e{}^2 = (496.2 - 123.7)/496.2 = 0.751$, or 75.1% variation is genetic in nature.

$h^2{}_n = \frac{1}{2}A/[$total phenotypic variation]; $\sigma_e{}^2 =$ error variance (above) $= \sigma_{F1}{}^2 = 123.7; D = 4[\sigma_{B1}{}^2 + \sigma_{B2}{}^2 - \sigma_{F2}{}^2 - \sigma_e{}^2] = 4[357.2 + 324.7 - 496.2 - 123.7] = 4 \times 62.0 = 248; A = 2[\sigma_{F2}{}^2 - \frac{1}{4}D - \sigma_e{}^2] = 2[496.2 - 62.0 - 123.7] = 2 \times 310.5 = 621$.

$h^2{}_n = \frac{1}{2}A/[\frac{1}{2}A + \frac{1}{4}D + \sigma_e{}^2] = [310.5/(310.5 + 62.0 + 123.7)]$ or $[310.5/(\sigma_{F2}{}^2)] = 310.5/496.2 = 0.626$ or 62.6% of the total variation is additive genetic variance.

(9) Griffing has described four types of diallel crossing designs. Briefly outline the features of each of Methods 1, 2, 3 and 4. Why would you choose Method 3 over Method 1? Why would you choose Method 2 over Method 1?
A full diallel, including selfs, was carried out involving five chickpea parents (assumed to be chosen as fixed parents), and all families resulting were evaluated at the F_1 stage for seed yield. The following analysis of variance for general combining ability (GCA), specific combining ability (SCA) and reciprocal effects (Griffing's analysis) was obtained:

Source	df	MS
GCA	5	30,769
SCA	10	10,934
Reciprocal	10	9,638
Error	49	5,136

Complete the analysis of variance and draw conclusions from the analysis. If it was actually the case that the parents were chosen at random, how would this change the results and your conclusions?

Plant height was also recorded on the same diallel families and an additive – dominance model found to be adequate to explain the genetic variation in plant height. Array variances (V_rs) and non-recurrent parent covariances (W_rs) were calculated and are shown alongside the general combining ability (GCA) of each of the five parents, below:

Parent	V_r	W_r	GCA
1	491.4	436.8	− 0.76
2	610.3	664.2	+ 12.92
3	302.4	234.8	− 14.32
4	310.2	226.9	− 15.77
5	832.7	769.4	+ 17.93

Without further calculations, what can be deduced about the inheritance of plant height in chickpea?

Answer:
If a fixed effect model is used, then both the GCA and SCA effects are tested for significance against the error mean square, giving F-value GCA = 5.99, with 5 and 49 degrees of freedom (significant at the 99% level); F-value SCA = 2.12 (not significant). However, if a random effects model is used, then the GCA F-value is calculated by dividing the GCA mean square by the SCA mean square or 30,769/10,934 = 2.81, with 5 and 10 degrees of freedom (not significant). So, in conclusion, with a fixed effect model we have significant GCA and non-significant SCA, but with the random effect model, both GCA and SCA are not significant. From a breeding standpoint, let's hope the fixed model is appropriate, since a significant GCA is usually necessary to obtain a good response from selection!

(10) A 4×4 half diallel design (with selfs) was carried out in cherry and the following seed yields of each possible F_1 family were observed:

	Small reds			
Small reds	12	Big yields		
Big yields	27	36	Jim's delight	
Jim's delight	21	35	27	Jacks' best
Jack's best	28	27	26	21

From the above data, determine the narrow-sense heritability for yield in cherry.

Answer:
A simple inspection of the data table would initially suggest the presence of a high level of additive genetic effects, since the arrays tend to have an internal consistency. In order to estimate heritability it is necessary to perform a regression analysis of the progeny performance (dependent variable, y) onto the mid-parent value (independent variable, x). Taking the parents in pairs, this would be for x and y, respectively: $(12 + 36)/2 = 24.0$; $(12 + 27)/2 = 19.5$; $(12 + 21)/2 = 16.5$; $(36 + 27)/2 = 31.5$; $(36 + 21)/2,27$; and $(27 + 21)/2 = 24.0)$ or x & y written as 24 & 27; 19.5 & 21; 16.5 & 28; 31.5 & 35; 28.5 & 27; and 24 & 26. From this the regression slope can be calculated and turns out to be 0.52; therefore as we used the mid-parent performance, then $h^2_n = 0.52$, or 52% of the variation observed is additive in nature.

(11) A crossing design involving two homozygous pea cultivars was carried out, and both parents were grown in a properly designed field experiment with the F_2, B_1 and B_2 families. Given the following standard deviations for parents (P_1 and P_2), the F_2, and both backcross progeny (B_1 and B_2), determine the

broad-sense heritability and narrow-sense heritability for seed size in these dry peas. What can be concluded from the values of heritability you obtain?

Family	Standard deviation	Variance
P_1	3.521	12.39
P_2	3.317	11.00
F_2	6.008	36.10
B_1	5.450	29.70
B_2	5.157	26.59

Answer:

$\sigma^2_e = (\sigma^2_{P1} + \sigma^2_{P2})/2 = 11.69$;

so $h^2_b = (\sigma^2_{F2} - \sigma^2_e)/\sigma^2_{F2} = (36.10 - 11.69)/36.10 = 0.68$, so 68% of total variation is genetic variation.

$D = 4(\sigma^2_{B1} + \sigma^2_{B2} - \sigma^2_{F2} - \sigma^2_e) = 34.00$;

$A = 2(\sigma^2_{F2} - 1/4D - \sigma^2_e) = 31.82$;

so $h^2_n = 1/2A/(1/2A + 1/4D + \sigma^2_e) = 0.44$, so 44% of total variation is additive genetic variation.

(12) It is common in plant breeding programmes to carry out diallel analyses to determine the genetics of specific parents and their progeny. With particular reference to diallel interpretation, explain the terms **fixed** and **random** parents, and how choosing one or other would influence the analyses of a Griffing ANOVA.

Griffing's analysis of diallels is a statistical analysis based on the following model:

$$Y_{ijk} = \mu + g_i + g_j + s_{ij} + e_{ijk}$$

Describe the meaning of the terms in the g_j and s_{ij} in the Griffing model, above.

Below are the mean seed yields (kg/plot), averaged over four replicates, of the parents and offspring from a 3×3 half diallel, with selfs.

	Sunrise	Premier	Clearwater	Hyola.401
Sunrise	16.4			
Premier	14.3	11.1		
Clearwater	18.3	15.2	17.6	
Hyola.401	13.2	10.7	16.2	15.3

Using these data, determine the g_i value of each parent and estimate the s_{ij} value in the cross Hyola.401 × Premier.

Answer:

In theory, fixed or random relates to how the parents were chosen. A random set of parents should be a random sample of all possible parents, while a fixed sample of parents is specifically chosen. However, the choice is also related to the inference that can be made from the data analyses. If parents are fixed, then the conclusions from the analyses relate only to those specific parents. If the sample is truly random, then inference can be made to a wider base of parents. In breeding, the latter is usually preferred; however, as breeders we rarely (if ever) choose parents at random. Often, then, we bend the rules by, for example, "choosing a set of parents from the cultivars which have high adaptability to the growing conditions". It is also worth noting that parents in diallel designs *must* have differences between them, otherwise there is little likelihood of significant variation amongst the progenies.

The terms g_i and g_j are general (or average) combining abilities of the *i*th and *j*th parent respectively. The term s_{ij} is the specific combining ability (not explicable by general combining ability). The term e_{ijk} is an error term obtained by replication.

Grand mean of progeny excluding selfs = 87.9/6 = 14.65, so $GCA_{\text{-Sunrise}}$ = array mean − Grand mean (excluding selfs) = 15.27 − 14.65 = +0.62. So $GCA_{\text{-Premier}} = -1.25; GCA_{\text{-Clearwater}} = +1.92$; and $GCA_{\text{-Hyola.401}} = -1.28$. The $SCA_{\text{-Hyola.401} \times \text{Premier}}$ = Table value (10.07) − Grand mean − $GCA_{\text{Hyola.401}}$ − $GCA_{\text{Primier,}} = 10.07 - 14.65 - (-1.25) - (-1.28) = -2.05$.

(13) **A newly appointed barley breeder was interested in the inheritance and heritability of yield in spring barley. He chose five cultivars from a range that were grown throughout his target region and inter-crossed them in all possible cross combinations (excluding reciprocals) including selfs, and he completed a Hayman and Jinks' analysis. Some of the results are shown below.**

Cultivar	Mean yield	V_i	W_i
Premier	62	1.8	−8.1
Sunrise	44	64.1	51.5
Reaper	39	43.8	38.0
Lifeline	66	7.1	9.4
Star	39	71.0	51.1
Mean		37.56	28.38

The regression equation of W_i onto V_i is: $W_i = 0.8134 \times V_i - 2.15$, and the standard error of the slope is 0.103.
The variance of parents (σ_p^2) is 42.37, and the variance or array means (σ_{xr}^2) is 16.77
Complete the Hayman and Jinks' analysis of variance, determine whether the additive – dominance model is appropriate, calculate the narrow-sense heritability, and describe what can be determined about the inheritance of yield in spring barley.

Answer:
V_i/W_i regression: *t*-test for difference from slope of 1.0 is [0.8134 − 1.0]/0.103 = 1.81 = ns, therefore an additive – dominance model of inheritance would appear appropriate.

Now, A = $4/7[\sigma_p^2 + mW_i + \sigma_{xr}^2] = 4/7[42.37 + 28.38 + 16.77] = 50.01; D = 4mV_i - A = 4 \times 37.56 - 50.01 = 100.23$. As we know that the error (E) = 9.53 and so $h^2 = \frac{1}{2}A/[\frac{1}{2}A + \frac{1}{4}D + E] = 25.0/59.59 = 41.9$, therefore 41.9% of the variation is additive in nature.

Inspection of the V_i and W_i values in combination with the average parent values high V_i and W_i values is associated with high average progeny yield. Therefore, high yield is related to an accumulation of dominance effects.

Chapter 7: Selection

(1) **Selection in a plant breeding programme can be divided into three different stages: early generation selection, intermediate generation selection and advanced generation selection. Briefly, state the major differences between these three stages.**

Answer:
Early generation selection usually involves thousands of breeding lines with low quantities of planting material being available. This means growing small plots, often single plants at one site, which often produces ineffective selection. Frequently the lines to be selected are highly heterozygous genotypes. Selection is usually carried out by visual inspection of phenotype of the plants or plant produce.

Intermediate generation selection involves hundreds of breeding lines in a small number of locations (often only one) and occurs when more planting material (i.e. seeds) is available. More efficient evaluation of quantitatively inherited characters (i.e. yield, quality) can therefore be carried out. In inbreeding species, the breeding lines will be more homozygous. At this stage, there is still sufficient genotypic variation to allow for some response to selection. The number of sites is generally somewhat limited.

Few (tens) of lines are remaining and large quantities of planting material are available. Testing is conducted throughout the target region, with maybe 10–12 sites/locations. Agronomic trials (e.g. variable nitrogen levels, pesticide treatments, etc.) may be conducted. Breeding lines are confirmed and quality testing assured. On-farm testing begins.

(2) **A 10×10 half diallel crossing design was used to examine the potential of each of ten parents in a wheat breeding programme. F_1 families along with each of the parents were grown in a properly designed field trial. Yield (kg per plot) was recorded and from the data, array means, within-array variances (V_rs) and between-array covariances (W_rs) were estimated. These statistics are shown below. From this information, determine which three of the ten potential parents would be best suited for use in a cultivar development scheme designed to increase wheat yield. Explain your choices.**

Parent	Mean	V_r	W_r
1	32.3	234.0	215.2
2	15.2	45.2	19.2
3	21.3	150.4	298.1
4	24.5	17.3	19.2
5	29.3	100.1	90.1
6	17.4	210.9	250.3
7	16.3	199.0	99.1
8	19.1	26.9	15.6
9	17.1	292.8	211.2
10	22.3	379.5	403.1

Answer:
It should be obvious that a plant breeder wishes to select high-yielding wheat cultivars and so would have a preference for progenies derived from high-yielding lines. From this, lines 1 and 5, and, maybe, lines 4 and 10 would be contenders as they produce, on average, high-yielding progeny. Now parents with low V_r and W_r values would have yield determined by a high frequency of dominant alleles, whereas parents with high V_r and W_r values would have yield determined by a high frequency of recessive alleles. Given that wheat is an inbreeding species and that in inbreeding species it is easier to genetically "fix" recessive alleles rather than dominant alleles, then given similar yield potential it would be best to select parents with high average performance and with high V_r and W_r values. In this case it would suggest selecting parents 1 and 10. Given that Parent 5 has high average performance and intermediate V_r and W_r values, this could be the third choice over, say, parent 9 with higher V_r and W_r values but markedly lower average progeny performance.

(3) **In a bean breeding programme, it is desired to produce new cultivars that are short (dwarf) in stature and have oval shaped beans (rather than round). Both bean shape and dwarfism are controlled by single loci with two alleles at each. The allele for oval beans is dominant to the alternative allele for round beans. The dwarfing allele is recessive to the non-dwarf (tall) allele.**

   A cross is made between two homozygous parents where parent 1 is dwarf with round beans (*ttrr*) while parent 2 is tall with oval beans (*TTRR*). A number of F_1 plants are selfed to produce F_2 seeds from which 1,600 F_2 plants are grown. Assuming independent assortment of genes, outline a selection

scheme that will result in harvesting F₄ seeds that are homozygous for oval beans and dwarf stature. Indicate the number of plants selected at each selection stage.

Answer:

Since we start with 1,600 F_2 plants from the cross between two homozygous parents, one *ttrr* and the other *TTRR*. The 1,600 F_2 plants will be expected to segregate phenotypically into 900 tall/oval (*T_R_*); 300 tall/round (*T_rr*); 300 dwarf/oval (*ttR_*): 100 dwarf/round (*ttrr*). If we then select all the phenotypically dwarf/oval F_2 (*ttR_*) plants, we would expect to have seed from 300 selected plants. These plants are now fixed genetically for dwarf status as the dwarf allele is recessive and requires homozygosity at the loci for the dwarf allele (*t*) to be expressed. The F_3 families derived from the F_2 selections will therefore be expected to consist of homozygous dwarves, but two types of families for seed shape (1/3 (100) being homozygous *RR* and 2/3 (200) being *Rr*) will all be phenotypically indistinguishable. The latter 200 families (which we would wish to discard) will be expected, however, to segregate in the F_4 with the genotypic ratios of $^1/_4RR$ (50) : $^1/_2Rr$ (100) : $^1/_4rr$ (50). We now need a means of identifying the 100 homozygous (*RR*) oval seed families and discarding the heterozygous (Rr) families. Thus it is necessary to progeny-test F_3 plants derived from each of the selected 300 single F_2 plants. To enable this, when we harvest we must collect the seed separately from each selected F_2 plant. This seed we then grow in short rows of, say, 40 F_3 plants – in other words in a head-row. At harvest all the seed from each of the 300 F_3 head-rows is harvested separately and the ones where all the seed is uniformly oval in shape are selected, while those with mixed oval and round seed are discarded. This therefore results in identifying the expected 100 genotypes with homozygous dwarf stature and oval seeds.

Alternatively, if more resources were available, then seed from each of the F_2 plants could be harvested separately and used to plant 1,600 F_3 head-rows. At harvest, select all the dwarf status head-rows (24/64 or 3/8 or 600 head-rows) and only harvest seed from these rows. From amongst these 600 selected rows it would be expected that 225 (9/24) would be dwarf status and all oval seeds (F_3 plant was ttRR); 150 (or 6/24) would be dwarf status and have mixed oval and round seed (F_3 plant was ttRr); and the remaining 225 (9/24) would have dwarf status and all round seeds (F_3 plant was ttrr). This second method would use more resources but have the advantage of identifying more of the desired genotypes (i.e. 225 compared with 100) but would involve growing and examining 1600 head-rows compared with only 300.

A further possibility, if time and money are no object, would be that the breeders could self-pollinate the population over, say, 6 generations (6 years if this was an annual species and the growth was in the field, or less if single seed descent techniques have been developed) to obtain F_7 progenies (close enough to being F_∞ for practical purposes) (or produce doubled haploid lines), plant these out, and approximately $^1/_4$ would be expected to be the desired genotype of being dwarf status and oval seeds.

(4) **You are a potato breeder working in a publicly funded organization. Due to the breakup of the former Soviet Union, you have inherited 500 potato lines from the Siberian Potato Research Centre. Briefly outline (using diagrams if necessary) how you would screen these genotypes for their potential as new parents in your breeding programme.**

Answer:

There are of course many different screening schemes that could be used to complete this task. Firstly you could simply make a large number of cross combinations (say in a $^1/_2$ diallel design, resulting in a total of 124,750 possible cross combinations, i.e. $[500^2 - 500]/2$) and then spend the rest of a lifetime examining the progeny. Not perhaps the best use of one's time and effort.

Conversely, one might consider the 500 possible lines/parents as breeders usually do, and first eliminate any of the potato lines which you "suspect" might be terrible parents because phenotypically they have a severe defect or are completely unadapted to the proposed growing conditions (although you might save these lines in a germplasm collection to screen them for other useful characteristics at a later stage). Simultaneously the 500 potato lines could be screened for disease and pest resistance or quality characters to identify particular lines which may have 'specific' traits of interest. So let's begin by including all 500 potato lines in a single replicated yield evaluation trial. Harvest the tubers from this trial and do quality assessment. Simultaneously screen the 500 possible lines in a disease nursery or other pest resistance trials. After this initial screen the 500 lines can be reduced to a more manageable number (say the 'best' 50 lines or 10%). Now these need to be tested in more detail to identify the best phenotypically, again perhaps in more detailed field tests and disease and quality testing in order to reduce this further, down to say 10 lines. It would now be necessary to screen these potential parents genotypically by examining the performance of their progeny in cross combinations, either by intercrossing the 10 selected parents, or more likely by crossing these 10 Russian potato lines to the best parents from the existing breeding programme and thereafter examining the progeny, while at the same time screening them as possible new cultivars in their own right through the standard breeding and selection scheme.

(5) **In a winter rapeseed breeding programme, 3,000 near-homozygous breeding lines were evaluated for yield and oil content from a properly designed field trial. The correlation coefficient between yield and oil content was found to be $r = 0.41$. Using independent culling you want to select the highest-yielding 10%, and select for oil content retaining the best 15%. How many genotypes would you expect to be: (1) selected for high yield and high oil content; and (2) selected for high yield but discarded for high oil content?**

Answer:
This question can be answered using the inverse tetrachoric correlation table (see Table 7.2 in book). The correlation between yield and oil content to be $r = 0.41$, selection for yield is at the 10% level, and selection for oil content is at the 15% level. From Table 7.2 we have a corresponding number value of 0.36 (see in box 0.15 y-axis and 0.10 x-axis and fourth value in that box). So, from 1,000 lines, given this correlation, we would expect to select 36 lines with high yield (in the top 10%) and high oil content (in the top 15%). But this number (36) is based upon selection amongst 1,000, and we are here examining 3,000, so the actual number that would be expected to be selected is 36×3 or 108. So (part 1) we would expect 108 lines to be selected for high yield and high oil content out of the 3,000 lines evaluated. Now (part 2), 10% of the lines (i.e. 300 lines) were selected for high yield, so by subtraction $300 - 108$ must have been selected for high yield and discarded as lower than the top 15% oil content.

(6) **Independent culling can be a very effective means to select for two or more characters simultaneously and can be easily applied to breeding data. Under what circumstances would independent culling not be very effective? If index selection is used, list two methods that could be used to weight the traits.**

Answer:
Independent culling is not very effective when selecting for two or more traits that are strongly negatively correlated (i.e. say where high yield is associated with poor quality and *vice versa*).
The most difficult aspect of index selection is choosing suitable weights for each of the traits of interest. Index selection weights can be assigned either as:

• Economic values, where the potential economic impact of each trait is estimated and the datum recorded of each variable expression is weighted by that value. The problem with economic weights is that they change from year to year.
• Statistical values, where the weight values are derived according to some statistical procedure. The most commonly used routines have involved *multivariate transformations* such as principal component analysis, canonical analysis or discriminant function analysis. The problem with statistical weights is that they will be different from dataset to dataset. In some cases the weights do indeed show some biological meaning, but in other cases there appears to be no coherent association.

(7) 50 progeny from ten potato crosses were evaluated for yield in a properly designed field trial. From the results the following cross means and genetic variances (σ_g^2) were obtained:

	Mean	σ_g^2
Cross 1	25.60	27.34
Cross 2	19.33	19.40
Cross 3	27.71	13.31
Cross 4	12.06	10.39
Cross 5	13.11	15.63
Cross 6	26.56	14.21
Cross 7	27.45	25.69
Cross 8	19.21	15.21
Cross 9	23.21	39.13
Cross 10	19.32	17.31

Also grown in the same trial were ten commercial cultivars. The average performance of the commercial cultivars was 20.14 and the standard deviation was 3.26. Using univariate cross prediction procedures, determine which three crosses should be used in breeding for cultivars that would have high yields. Rank your choices as first, second and third and include the probabilities used for your decision, under the following criteria:

- Greater than average controls plus one control standard deviation.
- Greater than the average control plus twice the control standard deviation.

Using the data presented above, how many clonal lines would need to be raised from Cross 6 to be 90% certain of having one line that would have a yield potential exceeding 3 standard deviations from the control mean?

Answer:

To answer this problem we must first convert these means and genetic variances to the standard normal distribution using: (mean − target)/σ_g. So we have the first target (T1) set at the control plus one standard deviation (20.14 + 3.26) of 23.40, and at the second (T2) it is the control mean plus twice the standard deviation (20.14 + 6.52) at 26.66. Then we need to calculate the probability that each of the progeny from the 10 crosses will exceed the set target values as shown in the table below:

Cross	Mean	σ_g^2	Prob (> T1)	Prob (> T2)
Cross 1	25.60	27.34	0.66	0.42
Cross 2	19.33	19.40	0.18	0.05
Cross 3	27.71	13.31	0.88	0.61
Cross 4	12.06	10.39	0.00	0.00
Cross 5	13.11	15.63	0.00	0.00
Cross 6	26.56	14.21	0.80	0.48
Cross 7	27.45	25.69	0.79	0.56
Cross 8	19.21	15.21	0.14	0.03
Cross 9	23.21	39.13	0.49	0.29
Cross 10	19.32	17.31	0.16	0.04

Using these probabilities in relation to T1 (our prediction of the control mean plus one standard deviation), we would select: Cross 3 with an expected frequency of 0.88 (88%) equalling or exceeding the target value, Cross 6, with an expected 80% equal or greater than the target value, and Cross 7, with 79% equal or greater than target value. With the increased target value (T2) of control mean plus twice the standard deviation, then we would select: Cross 3 (61% expected to equal or exceed the target value), Cross 7 (56% equalling or exceeding the target value) and Cross 6 (48% equalling or exceeding the target value). So, in this case, although the absolute ranking of the top three crosses based on different targets differed somewhat, in fact the result, in terms of selecting the best three crosses, was the same irrespective of which target value was set.

The control plus three times the standard deviation gives a target value of 29.92, so the (mean − target)/σ_g for Cross 6 is $(26.56 − 29.92)/3.77 = −0.89$, so the probability of exceeding the target value is 0.1867 or 18.67%. Using the equation $\ln(1 − p)/\ln(1 − x)$, (see page 81, chapter 5) we have $\ln(1 − 0.95)/\ln(1 − 0.1867) = (−2.9957)/(−0.2006) = 14.5$, therefore it would be necessary to evaluate 15 lines from Cross 6 to be 95% certain of identifying at least one selection that exceeded the target value for this trait. Not a large number in breeding terms!

(8) **A half diallel crossing design is carried out involving 30 homozygous parents. F_3 progeny from each of the 435 possible cross combinations were raised and grown in a two replicate yield trial (planting F_3 and harvesting F_4 seed). Also grown within the trial were all 30 parental lines. Based on the yield results from the trial it was found that the regression of progeny performance (y) onto mid-parent value (x) was:**

$$y = 0.832x + 0.002$$

Also from this trial, the phenotypic variance of the F_3 families was found to be $\sigma_p{}^2 = 65.216$, and the average yield of all 435 F_3 lines was 301.5 kg. What would be the expected gain from selection if the F_3 families were selected at the 10% level (i.e. discard 90% of families), and what would be the expected yield of these 43 selected lines at F_4, according to yield performance? Would you expect the same response to selection if the 43 selected F_4 families were further selected for yield the following year? (Explain your answer).

Analysis:
From the regression of the offspring onto the mid-parent we have that the heritability (h^2), given by the slope of the regression line, = 0.832. Also the phenotype variance is given as $\sigma_p{}^2 = 65.216$, from which we can take the square root to give the standard deviation, $\sigma_p = 8.076$. Selecting the 'best' 10% would result in a selection intensity (i) of 1.755. From these three estimates we can calculate the expected response from selection (R), where $R = i\,\sigma\,h^2 = 1.755 × 8.076 × 0.832 = 11.79$, so the best 10% selected for yield would expect to have a yield gain of 11.79 kg better than the average yield of the initial unselected population. Thus the expected yield of the selected lines would be 301.5 (the average of the initial population) + 11.79 = 313.29.

Would we expect a similar gain from selection if we chose the highest yielding F_4 lines, say the best 10%, again? No, we would not. Even if the heritability was the same (which it would not be) and the selection proportion (10%) the same, it would be very unlikely that the standard deviation of the selected 43 lines would be as big as the initial 435 lines, so we would expect a reduced selection response.

(9) **In chickpea breeding it is assumed that the correlation between yield performance of F_4 families in one year and F_5 families in the following year is $r = 0.57$. 4,000 F_4 breeding lines of chickpea were evaluated for seed yield in a properly designed field trial. All 4,000 lines were re-evaluated at the F_5 stage the following year. If the highest yielding 10% were selected based on F_4 performance and the highest yielding 15% were selected based on their F_5 performance, how many of the original 4,000 lines would you expect to be selected based on both F_4 and F_5 performance? Explain what Type I and Type II errors mean in the context of selection. Estimate the Type I and Type II errors expected by selecting F_4 families for yield at the 10% level and selecting F_5 families at the 15% level.**

Answer:

The correlation in chickpea for yield between F_4 and F_5 is $r = 0.57$. With 4,000 F_4 lines and selecting for yield at F_4 at the 10% (0.10) level and at F_5 at the 15% level (0.15), how many lines would we expect to be selected for high yield at both the F_4 and F_5 selections? We must use the inverse tetrachoric correlation table (Table 7.2). With selection at 0.10 and 0.15 and a correlation of almost 0.6, we would expect 051 (51) repeat selections for every 1,000 lines selected. As we started with 4,000 F_4s, we would expect to select $51 \times 4 = 204$ lines at both the F_4 and F_5 stages.

Now, a Type I error is when you discard a line in the first generation when you should have selected it in the second (i.e. discarded at F_4 but selected at F_5). A Type II error is where you selected a line at F_4, but discarded it in the later selection generation, here F_5. As 10% of the F_4 were selected, then we selected 400 F_4 lines of which $400 - 204 = 196$ were discarded at the F_5 stage, so the Type II error here is $196/400 = 0.49$, or 49%. As we selected 15% at the F_5 stage, that means we selected 600 F_5 lines. Of these, $600 - 204 = 396$ were discarded at the F_4 stage, so the Type I error $= 0.66$, or 66%. As we selected a higher proportion of F_5 lines than F_4 lines, it is hardly surprising that the Type I error is larger than the Type II, and of course both errors would be equal if we were to select the same proportion in both generations, or selection stages.

(10) Parental selection can generally be divided into two different types. What are these types and, briefly, indicate the differences between them.

Answer:

Phenotypic parental selection is where the parents are evaluated on their own performance and the ones with the best expression (i.e. say highest yield) are used as parents. Genotypic parental selection is where the progeny of specific parents are evaluated to determine the worth of each parent when used in cross combinations. In genotypic parental selection the parents with the better performing progeny are used in crosses.

(11) Describe the main features of North Carolina I, North Carolina II and diallel crossing designs. Explain the terms in the model for the analysis of a diallel according to the method described by Griffing:

$$Y_{ijk} = \mu + g_i + g_j + s_{ij} + e_{ijk}$$

and indicate the importance of these terms and the use of Griffing analysis in selecting superior parental lines.

Answer:

A diallel cross design is described as all possible crosses amongst a group of parent lines. Diallel designs can be complete, all possible combinations including parents selfs, or indeed without parent selfs, or partial, being without reciprocal crosses, and again either with or without parent selfs.

North Carolina I crossing designs are where a number of parent genotypes are crossed to one or more tester lines. In such a design it is not necessary to hybridize every parent to a common set of tester lines. From the design an analysis of variance can estimate general combining ability of parents, which is tested for significance against the testers within parents' mean square.

North Carolina II designs are where a number of parent genotypes are crossed to one or more tester and every parent is crossed to the same tester lines. In this case an analysis of variance can partition the total variation into differences between general combining ability of the parents, combining ability of the testers and an interaction term (parent × tester) which indicates specific combining ability.

$$Y_{ijk} = \mu + g_i + g_j + s_{ij} + e_{ijk}$$

where μ is the overall mean of all entries in the diallel design, g_i is the general combining ability of the ith parent, g_j is the general combining ability of the jth tester and s_{ij} is the specific combining ability between the ith parent and the jth tester, and e_{ijk} is the error term from the model.

(12) **You have been appointed as Assistant Professor/Plant Breeder in the Crops Division of McDonalds University in Frysville, MD, United States. It appears that the breeding programme has been trying to select improved genotypes with decreased sugar content in the tubers. However, in the 10 years previous to your appointment, there appears to have been no genetic improvement resulting from breeding. Outline three reasons that could individually, or in combination, have caused this lack of response.**

Answer:
The lack of genetic improvement could be explained for a number of reasons, including:

(1) there is no genetic difference between the breeding lines being selected, in other words there is zero heritability;

(2) there was zero, or negative, repeatability between successive rounds of selection; or

(3) the breeder simply failed to recognize suitable selections. The latter may seem unlikely but it has been known to happen!

(13) **In a series of properly designed experiments, a team of plant breeders produced estimates of narrow-sense heritabilities (h_n^2) for plant height and plant yield from two different segregating families of spring barley (95.BAR.31 and 95.BAR.69). Both families were grown on two farms (Moscow and Boise, in Idaho, United States) in three successive years (1993, 1994 and 1995). The h_n^2 values from each year and site are summarized below.**

Character	Year	95. BAR. 31		95. BAR. 69	
		Moscow	Boise	Moscow	Boise
Plant	1993	0.20	0.22	0.83	0.86
height	1994	0.01	0.21	0.31	0.74
	1995	0.21	0.30	0.52	0.79
Plant	1993	0.11	0.25	0.52	0.57
yield	1994	0.02	0.15	0.12	0.46
	1995	0.21	0.24	0.21	0.48

Which family is likely to give better responses in a breeding programme, given equivalent selection intensities, and why? In such a breeding programme, which of the two characters (plant height or plant yield) is likely to give the better response to equivalent selection intensities, and why? Consistently higher average values of h_n^2 were obtained at the farm near Boise. Which site, if either, is likely to provide the more accurate estimate of h_n^2, and why? The heritability estimates (particularly those for plant height for 95.BAR.69 in Moscow) varied greatly over the 3-year period. What could be the cause?

Answer:
Over all sites and years 95.BAR.69 had an average heritability for yield and plant height of 0.393 and 0.675, respectively, while 95.BAR.31 had average heritabilities for yield and plant height as 0.163 and 0.192. A better response for selection will be achieved from 95.BAR.69 with its higher heritabilities, but in terms of the potential to actually select suitable lines, the mean performances of the progenies would also need to be considered. Yield heritabilities were generally lower than those for plant height and hence would likely give a lower response to selection.

Boise consistently shows higher heritabilities compared to Moscow. Perhaps there is less variability in soils or climate at that site, and so produces less micro-environmental variance. As heritability is genetic variance/total variance, a lower "error" would result in higher heritability.

(14) **4,000 F_3 lentil breeding lines were evaluated for yield and the highest yielding 500 genotypes were selected. The average yield of the 500 selections was 1,429 kg/ha, while the average yield of the discard genotypes was 1,204 kg/ha. A random sample from the discards and the 500 selected lines were grown in a properly designed F_4 trial where the average yield of the random genotypes was 1,199 kg/ha and the yield of the selected 500 lines was 1,362 kg/ha. Determine the narrow-sense heritability for yield at the F_3 of these lentil progenies.**

Answer:
This question relates to determining the heritability for yield between two generations F_3 and F_4 of lentil. The average yield of 4,000 F_3 lines was ($[500 \times 1,429$ kg ha$^{-1}] + [3,500 \times 1,204$ kg ha$^{-1}])/4,000 =$ 1,232 kg ha^{-1}. The difference between selected and non-selected at F_3 was 197 kg ha^{-1} = S, or the selection differential. At the F_4 stage the difference between the base population, the random sample (1,199 kg ha^{-1}) and the selected population (1,362 kg ha^{-1}) was 163 kg ha^{-1} = the response to selection (R). Therefore $h_n{}^2$ = R/S = 163/197 = 0.827, or 82.7% of the total variation for yield in lentil at F_3 was genetic in origin. This is good!

(15) **In a barley breeding programme for "high diastase power", it is known that the correlation between yield and diastatic power is $r = 0.60$. In this breeding programme 3,000 doubled haploids are evaluated for yield and diastatic power in a properly designed field trial. If you wish to retain approximately the 'best' 150 lines, what proportion would you discard based on yield and what proportion would you discard based on diastase power to achieve this selected number?**

Answer
The correlation between yield and diastatic power in barley breeding is $r = 0.60$, and we wish to select 150 lines from 3,000 based on selecting the highest yield and diastatic power. Here we need to use the inverse tetrachoric correlation table (Table 7.2). As we wish to retain 150 lines selected for both criteria and as the table is based on 1,000 and not 3,000 as we have, then we need to seek a cell in the table whereby a correlation of 0.60 (6th number in the cell) is approximately equal to (150/3)50. This is best approximated in the cell relating to selecting one character at 10% and the other at 15%. So we could achieve this retention value of 150 for 3,000 by selecting yield at the 10% level and diastatic power at the 15% level. Similarly we could select diastatic power at the 10% level and yield at the 15% level. Either way what would result is approximately 150 double selections, but of course of different composition!

(16) **In 1998, 10,000 F_3 head-rows were grown in an unreplicated, but randomized, design from *Bobby Z's Wheat Breeding Program*. At harvest, seed from all rows were thrashed and weighed. The average yield of all 10,000 lines was found to be 164.24 kg. After weighing, 200 rows were taken at random and retained irrespective of their yield performance. From the remaining lines, the 'best' 1,000 rows for yield were selected for further evaluation in the breeding scheme. The average yield of these selected lines was 193.74 kg. In 1999, the bulk F_4 seed from the 200 random lines and the 1,000 selected lines were grown in a randomized complete block design. At harvest, each plot was harvested separately and the yield recorded. From this 1999 trial it was found that the average yield of the selected lines was 168.11 kg, while the average yield of the 200 randomly chosen lines was 133.41 kg. Using the above information, determine the narrow-sense heritability for yield in this material. Explain how this estimated value of heritability might influence your selection strategy for future F_3 head-row selection stages.**

Answer:
This question relates to determining the heritability for yield between two generations F_3 and F_4 of wheat. Average yield of 10,000 F_3 lines was 164.24 kg plot^{-1} and that for the selected best 1,000 F_3 lines was 193.74 kg plot^{-1}. The difference between selected and non-selected at F_3 was 193.74 − 164.24 kg plot^{-1} = 29.5 kg plot^{-1} = S, or the selection differential. At the F_4 stage the difference between the base population, the random sample (133.41 kg plot^{-1}) and the selected population (168.11 kg plot^{-1}) was 34.7 kg plot^{-1} = the response from selection (R). Therefore the h_n^2 = R/S = 29.5/34.7 = 0.85 or 85% of the total variation for yield at F_3 in wheat was genetic in origin.

(17) **In a Douglas fir breeding programme you have only sufficient resources to screen segregants from two crosses each year. However, in 1995, you have been provided data of family mean (MEAN), phenotypic variance (VAR) and narrow-sense heritability (h^2) from each of six crosses. From these statistics, determine which two crosses you will concentrate your efforts on in 1995 – assuming that you will subsequently select at only the 10% level. Explain, briefly, your choices.**

Cross code	MEAN	VAR	h^2
DF.33.111	22.3	16.7	0.45
DF.66.123	24.6	14.3	0.51
DF.97.37	28.1	6.3	0.47
DF.97.332	26.1	15.3	0.11
DF.99.1	22.5	10.2	0.84
DF.99.131	18.9	26.1	0.75

Answer:
The response from selection is given by $R = i\,\sigma\,h^2$, and the expected yield of each family is the initial population mean + the response from selection. The response to selection (R) at the 10% level and the expected family mean after selection is given for each family below. Also shown is the ranking for each family based on yield.

Cross code	Mean	Variance	h^2	R	Expected yield	Rank
DF.33.111	22.3	16.7	0.45	3.2	25.5	6
DF.66.123	24.6	14.3	0.51	3.4	28.0	2
DF.97.37	28.1	6.3	0.47	2.1	30.2	1
DF.97.332	26.1	15.3	0.11	0.8	26.9	4
DF.99.1	22.5	10.2	0.84	4.7	27.2	3
DF.99.131	18.9	26.1	0.75	6.7	25.6	5

From this, as a breeder, I would choose to select families DF.97.37, ranked as highest, and DF.66.123, ranked as second highest, based on their expected yield (in other words, a combination of their expected response to selection in combination with the family mean yield).

(18) **25 progeny from ten winter wheat crosses were evaluated for yield in a properly designed F_3 cross-prediction study under field conditions. From the results, the following means and additive genetic standard deviations (σ_A) for seed yields were obtained for each cross:**

	Mean	σ_A
92.WW.46	236.60	127.34
92.WW.53	199.33	191.40
92.WW.54	241.82	91.43
92.WW.61	142.00	119.10
92.WW.71	133.11	125.46
92.WW.74	236.73	102.14
92.WW.93	233.55	281.77
92.WW.108	201.22	106.63
92.WW.111	229.37	299.39
92.WW.116	169.11	119.32

Also grown in the same trial were five commercial cultivars. The average performance of the commercial cultivars was 211.10 kg and the standard deviation was 41.83 kg. Using univariate cross prediction procedures, determine which three crosses should be used in breeding for cultivars that would need to produce yields:

- Greater than the average of the controls plus one control standard deviation.
- Greater than the average control plus twice the control standard deviation.

Rank your choices as first, second and third and include the probabilities used for your decision.

Answer:

To answer this problem we must first convert these means and genetic standard deviations to the standard normal distribution using: (mean − target)/σ_A where we have the target (T1) set at the control plus one standard deviation, as 211.10 + 41.83 = 252.93, and when it is set for T2 at the control mean plus twice the standard deviation (211.10 + 83.66 = 294.76). The probability that each of the progeny from the 10 crosses will exceed the set target values is shown in the table below:

Progeny code	Mean	σ_A	Prob > T1	Prob > T2
92.WW.46	236.60	127.34	0.448	0.323
92.WW.53	119.33	191.40	0.242	0.179
92.WW.54	241.82	91.43	0.452	0.281
92.WW.61	142.00	119.10	0.176	0.106
92.WW.71	133.11	125.46	0.169	0.099
92.WW.74	236.73	102.14	0.436	0.284
92.WW.93	233.55	281.77	0.472	0.413
92.WW.108	201.22	106.63	0.312	0.189
92.WW.111	229.37	299.39	0.468	0.413
92.WW.116	169.11	119.32	0.242	0.147

Based on these probabilities for equal or greater than T1 (mean + standard deviation), a breeder would probably choose the progenies 92.WW.93, ranked highest, 92.WW.111, ranked second highest, and 92.WW.54, ranked third highest. If a breeder were basing selection on the probability of equal to or greater than the mean plus twice the standard deviation (T2), then the progenies selected would be 92.WW.93 and 92.WW.111, ranked joint first with 41.3% of the progeny expected to exceed the desirable target value, along with progeny 92.WW.46 ranked third with a 32.3% expectation.

(19) **500 dry pea F_6 breeding lines were evaluated for yield potential at two locations (Hillside and Nethertown). After harvest and weighing, the 15% highest yielding lines were selected at each location.**

When results from this selection were examined it was found that 39 lines were selected in the top 15% at each site. From this information, estimate the narrow-sense heritability.
In this same study, the following phenotypic variances and site means for yield (over all 500 F_6 lines) were:

	Hillside	Nethertown
Phenotypic variance	96 kg^2	124 kg^2
Site mean	27 kg	29 kg

Given these data and the heritability estimated above, determine the expected response to selection at 10% level.

Answer:
Given that when the top 15% were selected, based on highest yield at each location, we selected 39 from the 500 lines with high yield at both locations. We can use the inverse tetrachoric correlation table (Table 7.2) to determine the expected correlation between yield at the two sites and hence the heritability. Selecting 39/500 is the same as 78/1,000. So we look up in Table 7.2, in the 0.15 and 0.15 cell, the number closest to 078, and find that this would be obtained with an expected correlation $r = 0.7$ (i.e. 7th largest number in that cell). From this we can estimate the heritability as $h^2 = r^2 = 0.49$.

We can now predict the expected response from selection at each of the locations, assuming equal heritabilities, if we select at a level of retaining the top 10% (i.e. $i = 1.755$) highest yielding lines. Now, $R = i \sigma h^2 = 1.755 \times \sqrt{96} \times 0.49 = 8.42$ kg at the Hillside site, and $= 1.755 \times \sqrt{124} \times 0.49 = 9.57$ kg at the Nethertown site. Similarly the expected yield of the 'selected' lines would be $27.0 + 8.42 = 35.42$ kg at Hillside and $29.0 + 9.57 = 38.57$ kg at Nethertown.

(20) Describe the difference between general combining ability and specific combining ability.
The following are average plant heights (over four replicates) of a 5×5 half diallel (including selfs) between sweet cherry cultivars.

Golden Glory	112				
Early Crimson	72	53			
Sweet Delight	102	64	99		
Dwarf Evens	56	41	65	49	
Giant Red	130	100	109	107	115
	Golden Glory	Early Crimson	Sweet Delight	Dwarf Evens	Giant Red

Estimate the general combining ability of each parent and calculate the specific combining ability of each cross. Which parents would be 'best' in a breeding programme to develop cultivars with short heights?

Answer:
The general combining ability (GCA) measures the average performance of parental lines in cross combination. GCA is therefore related to (but not directly equal to) the proportion of variation that is genetically additive in nature. The specific combining ability (SCA) is the remaining part of the

observed phenotypic variation that is not explained by the general combining ability of both parents that constituted the progeny. By definition, SCA is related to the portion of genetic variability that is not additive.

The general combining ability of each parent in the half diallel can be estimated as: g_i = the mean of array in which the ith parent is a parent – the overall mean (μ). The overall mean is $(72 + 102 + 56 + 130 + 64 + 41 + 100 + 65 + 109 + 107)/10 = 84.6$. Consider the general combining ability of Golden Glory (g_{GG}), where the array mean is $(72 + 102 + 56 + 130)/4 = 90.0$, and so $g_{GG} = 90.0 - 84.6 = +5.4$. This means that, in general, when Golden Glory is used as a parent, the progeny averages 5.4 cm taller than expected. The general combining effects of the other parents are shown below. Note that they sum to zero. Negative general combining ability affects shown parents which, on average, produce shorter progeny.

Also shown in the table below are the specific combining ability (SCA) effects. These are calculated from: $Y_{ijk} - \mu - g_i - g_j$. So for the progeny from the cross Early Crimson × Golden Glory the SCA effect $= 72 - 84.6 - 5.4 - (-15.3) = -2.7$. This can be repeated for each cell in the table. Note again that the sum of the SCA effects is zero. Note also that the magnitude of the SCA effects indicates a departure from a simple additive model of inheritance. For example, if all the SCA effects were very small or zero, it would indicate that generally the effects were additive in nature.

	Golden Glory	Early Crimson	Sweet Delight	Dwarf Evens	Giant Red
Golden Glory	–				
Early Crimson	−2.7	–			
Sweet Delight	11.6	−5.7	–		
Dwarf Evens	−16.7	−11.0	−2.7	–	
Giant Red	13.1	3.8	−2.9	12.8	–
GCA =	5.4	−15.3	0.4	−17.3	26.9

(21) What is the difference between a cultivar with general environmental adaptability and one with specific environmental adaptability?

Eight yellow mustard F_8 breeding lines were grown at 20 locations throughout the Pacific Northwest region. Significant genotype × environment interactions were detected for yield by analysis of variance, and when a joint regression analysis was carried out. The following are line means and environmental sensitivity $(1 + \beta)$ values from the analysis.

Line	Overall mean	Environmental sensitivities
90.EW.34.5	1,234	0.55
90.JB.456	1,890	1.05
90.JB.562	2,345	1.34
91.HG.12	1,897	0.76
91.HH.145	1,976	0.52
92.22.12	2,567	1.42
92.AE.1	2,156	0.83
92.HK.134	2,152	1.72

Select the two '*best*' breeding lines that show general environmental adaptability. Select the two '*best*' breeding lines that show specific environmental adaptability.

Answer:
A cultivar with general environmental adaptability is one that does equally well (or indeed poorly) irrespective of the environment in which it is grown. Therefore such cultivars show very little variability in performance when grown in different environments. A cultivar with specific environmental sensitivity will be very variable when grown in different environments. Usually, of course, specific environmentally sensitive cultivars perform poorly when grown under adverse (or poor) environments, but perform exceptionally when grown in the absence of adversity (good environments). Breeders need to specify which environments a cultivar is targeted for when releasing new cultivars.

Breeding lines with low environmental sensitivity values will tend to be stable over environments, while those with high environmental stability values will be more variable over different environments. The '*best*' two lines for general environmental stability would be 90.EW.34.5 and 91.HH.145. However, 90.EW.34.5 has very low yield and as such it would not be selected. Instead a breeder would chose 92.AE.1 with higher (0.83) sensitivity, but much higher yield potential. The '*best*' two lines with specific environmental sensitivity would be 92.HK.134 (1.72) and 92.22.12 (1.42), and both have a high yield potential. A breeder may decide, however, to choose 90.JB.562 over 92.HK.134, as 90.JB.562 has second highest yield potential, and also a high environmental sensitivity. Selection is always somewhat of a compromise, often somewhat subjective and often based on the feel of the specific breeder for the lines being selected.

Chapter 8: Broadening the Genetic Basis

(1) **Under what circumstances would a plant breeder choose to use mutation breeding?**

Answer:
When a gene (or allele of a gene) cannot be found for a particular character within the species being bred or within related species that would be used for intergeneric hybridizations.

(2) **List two methods that have been used by plant breeders to induce increased frequency of mutagens in plants.**

Answer:
Chemical (i.e. EMS) and radiation (i.e. X-rays).

(3) **List three problems that a plant breeder may encounter when using mutagenesis in a breeding programme.**

Answer:
Mutagenesis is indiscriminate and generates large numbers of undesirable mutations. Desirable mutations are rare. Isolation of desirable mutations is difficult in many crop species.

(4) **What is possible with plant transformation (recombinant DNA) techniques that is not possible with more traditional breeding techniques.**

Answer:
Could include: Possible to transfer single genes from other species (including non-plant species) into a plant and have such transgenes express function successfully. Bypass natural barriers that limit sexual gene transfer. Allow breeders to utilize genes from completely unrelated species. Create new variability beyond that currently available in germplasm.

(5) **You have been asked to use recombinant DNA techniques in your alfalfa breeding programme to increase profitability. Describe all the processes you will have to complete to achieve this goal.**

Answer:
Find an agronomically desirable gene and develop a DNA clone of the gene. Establish a mechanism to transfer the gene into the target plant and related tissue culture techniques. Develop a suitable DNA construct that includes: a promoter region recognized by the host; the gene of interest; a selectable marker gene (commonly used genes provide tolerance to herbicides like glufosinate or glyphosate, or genes that enable transformed cells to metabolize compounds like mannose, which otherwise are not usable by non-transformed cells); and a gene encoding the correct end of transcription of the gene of interest. Select cells that have been transformed. Regenerate fertile whole plants from single transformed cells. Fully characterize in molecular terms these transformed plants. Obtain appropriate regulatory approvals to field-test these plants under biosafety guidelines. Identify a promising event. Gather all the information that appropriate regulatory authorities request and submit an application seeking regulatory approval to commercialize such an event. Once the event has been approved for commercialization, deploy appropriate stewardship/good agronomic practices guidelines.

(6) List three vectors that have been used to transform plants with genes from other organisms.

Answer:
Agrobacterium tumefaciens – agent of crown galls in several fruit species; physical or mechanical: protoplast electroporation, particle gun, direct injection; carbon whiskers.

(7) Briefly describe three problems that plant breeders may encounter in developing transgenic cultivars.

Answer:
Could include: Public acceptance: food labelling; contamination and segregation of crops with different end use or regulatory status. Environmental risks: gene transfer to related weedy species. Costs: excessive costs of regulatory approvals.

(8) Name three types of mutation that can occur in plants, and briefly describe the features of each type. Outline two difficulties that might be problematic when using mutagenesis in a plant breeding programme.

Answer:
Genome mutation: changes in chromosome number due to addition or loss of whole chromosomes; structural changes in chromosome; translocations (interchange between non-homologous chromosomes); inversions, changes in the location of loci but not their number; deletions and duplications.
Gene mutation (point mutation): change in a single gene. Often the result of a single base-pair change at the DNA level.
Extranuclear mutation: mutation in one of the cytoplasmic organelles (i.e. plasmids, mitochondria).

(9) Interspecific and intergeneric hybridization can sometimes be useful techniques in plant breeding by introgression of characters and genes from different species. In an interspecific cross between *Brassica napus* and *B. rapa*, there was no evidence that any *B. napus* egg cells had been fertilized. List three factors that could have caused this non-fertilization.

Answer:
Could include: Inability of pollen grains to germinate on the receptive stigma. Failure of pollen tubes to develop successfully down the style and be attracted to the ovary. Male gametes do not fuse with the egg cell. Abortive endosperm causes embryo to abort. Failure of the endosperm to develop sufficiently to support the embryo development.

Chapter 9: Contemporary Approaches in Plant Breeding

(1) List four uses for molecular markers in plant breeding.

Answer:
Could include: Marker assisted selection, for difficult-to-evaluate characters; quantitative trait loci (QTL); DNA fingerprinting to identify genotypes (or cultivars); securing proprietary ownership; selecting parents with known genetic distance; cytological information (mainly in interspecific hybrids); genomic selection.

(2) Describe any advantages of using molecular markers in plant breeding.

Answer:
Easy, quick and inexpensive to score; neutral in terms of phenotype; no deleterious effects on fitness; high levels of polymorphism; stable expression over environments; assessable early in plant development; non-destructive assessment; and codominant in expression.

(3) Describe one difficulty that might be encountered in utilizing QTL in plant breeding selection.

Answer:
A QTL tends to be specific to the cross it was identified in and also to specific environments. So a QTL for yield derived from a cross between A × B may not be a useful marker for selecting progeny from a different cross (say C × D). Also a QTL derived from yield data collected in 2012 may not be a good selectable marker in terms of yield when progeny are grown in 2013.

(4) A cross is made between two parents in a hybrid wheat breeding programme, where P_1 is the female parent and P_2 is the male parent. When you run RFLP analysis you see electrophoretic patterns like those below.
P_1 is the lane pattern for the female parent, P_2 the male parent, and other lanes (a to l) were observed from a sample of the seed of the F_2 progeny.

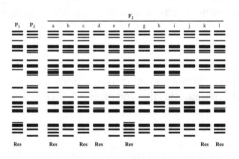

Explain what could have caused the pattern shown in lane f. If it is know that Parent 1 has a single gene conferring resistance to yellow stripe rust and that the stripe rust resistance of each genotype/lane is given below each lane (i.e. Res = resistant to stripe rust) indicate on the diagram a possible molecular marker for the stripe rust resistance gene.

Answer:
Lane f would have the same band pattern as an F_1 would have, with all the bands of both parents.
The arrow indicates a molecular marker that appears to be linked to yellow stripe rust resistance. Lines could be screened and selected based on the presence of the molecular marker to identify lines with yellow stripe rust resistance.

(5) Describe three applications involving tissue culture or *in vitro* plant propagation in a plant breeding scheme.

Answer:
Could include: Producing homozygous lines through doubled haploidy, i.e. through anther, microspore, ovary or ovule culture; maintaining and increasing disease-free clonal materials (i.e. potatoes or strawberries); regenerating whole plants from single cells in plant transformation; massive scale propagation; international transfer (many fewer phytosanitary complications).

Chapter 10: Practical Considerations

(1) You are employed as a barley breeder for a commercial company. At present the company does not have any greenhouse facilities. You are trying to convince your peers that the breeding programme would benefit from having a greenhouse. List five uses you would have for a greenhouse in your breeding programme, if one were available.

Great job (we hope!), you have convinced the Board of Directors to proceed with purchasing a new greenhouse. List five features you would like to request to have in the new greenhouse facility.

Answer:
Uses: Artificial hybridization; seed increase of breeding lines, including increases towards homozygosity; evaluation of characters that are difficult to assess in the field; and transfer from *in vitro* to *in vivo*.
Features: Adequate lighting, cooling and shading; irrigation; insect proofing; suitable ventilation; rolling benches to conserve space; all controlled by a suitable computer-based climate control system.

(2) Briefly describe one problem of using a greenhouse in a plant breeding programme.

Answer:
Limitation in availability of space; agriculture is conducted under field conditions and the greenhouse may not truly mimic field conditions; and high running and maintenance costs.

(3) What is the primary goal of conducting field evaluation trials in a plant breeding programme.

Answer:
Predict the performance of genotypes as if in large-scale agriculture; try to mimic commercial-scale agricultural situations.

(4) List things to avoid when organizing plant breeding field plots.

Answer:
Fertility gradients; previous test plots; lack of crop rotation, trees, buildings, access areas; and know the history of the land (e.g. old river beds, tracks, etc.).

(5) Why is experimental design important in plant breeding evaluation trials?

Answer:
To remove bias from phenotypic performance; and to allow better estimation of the error variance.

(6) List four problems that may be encountered in organizing and carrying out 'off-station' trials.

Answer:
Could include: Lack of on-site equipment; lack of experimental facilities; lack of post-harvest storage and seed-sorting facilities; large variation from farm to farm and farmer to farmer; and distance from research base.

(7) You have been employed as Senior Breeder at the *Hass Bean Breeding Co.* for several years and have had to carry out your breeding operation with very little funding. Due to a recent takeover of the company by the *Human Bean Seed Association*, you have been given a vast sum of additional money. You have decided to spend this money upgrading the computer facilities used by your breeding group. To achieve this goal you have appointed a computer programmer to develop a computer-based plant breeding management system. List the five main features the computer package needs to perform that would help to increase the efficiency of breeding future new cultivars.

Answer:
Produce a variety of randomized experimental designs and produce printable plot plans; complete clerical tasks like lists, score books, etc.; aid in data collection/validation and pedigree management; complete statistical analyses; data summaries and interpretation; and aid in selection.

Index

Plant Breeding, Second Edition. Jack Brown, Peter D.S. Caligari and Hugo A. Campos.
© 2014 John Wiley & Sons, Ltd. Published 2014 by John Wiley & Sons, Ltd.
Companion Website: www.wiley.com/go/brown/plantbreeding